AF358660

Bacteriophage Therapy: Recent Developments and Applications of a Renaissance Weapon

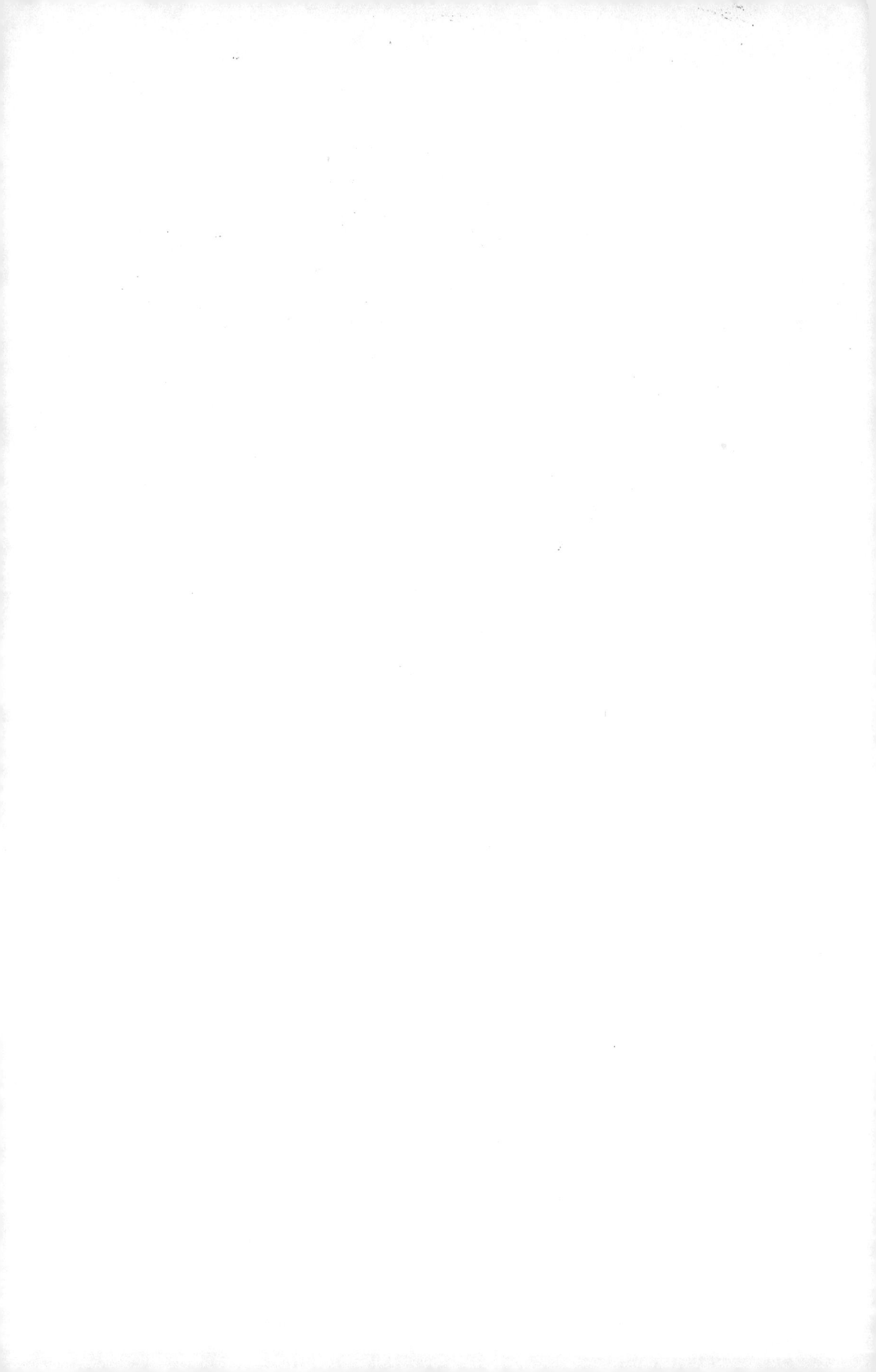

Bacteriophage Therapy: Recent Developments and Applications of a Renaissance Weapon

Editors

Aneta Skaradzińska
Anna Nowaczek

Basel • Beijing • Wuhan • Barcelona • Belgrade • Novi Sad • Cluj • Manchester

Editors

Aneta Skaradzińska
Department of Biotechnology
and Food Microbiology
Wrocław University of
Environmental and Life Sciences
Wrocław
Poland

Anna Nowaczek
Department of Veterinary
Prevention and Avian Diseases
University of Life Sciences
in Lublin
Lublin
Poland

Editorial Office
MDPI
St. Alban-Anlage 66
4052 Basel, Switzerland

This is a reprint of articles from the Special Issue published online in the open access journal *Antibiotics* (ISSN 2079-6382) (available at: www.mdpi.com/journal/antibiotics/special_issues/bacter_phage_therapy).

For citation purposes, cite each article independently as indicated on the article page online and as indicated below:

Lastname, A.A.; Lastname, B.B. Article Title. *Journal Name* **Year**, *Volume Number*, Page Range.

ISBN 978-3-7258-0300-2 (Hbk)
ISBN 978-3-7258-0299-9 (PDF)
doi.org/10.3390/books978-3-7258-0299-9

Contents

Review

Automating Predictive Phage Therapy Pharmacology

Stephen T. Abedon

Department of Microbiology, The Ohio State University, Mansfield, OH 44906, USA; abedon.1@osu.edu

Abstract: Viruses that infect as well as often kill bacteria are called bacteriophages, or phages. Because of their ability to act bactericidally, phages increasingly are being employed clinically as antibacterial agents, an infection-fighting strategy that has been in practice now for over one hundred years. As with antibacterial agents generally, the development as well as practice of this phage therapy can be aided via the application of various quantitative frameworks. Therefore, reviewed here are considerations of phage multiplicity of infection, bacterial likelihood of becoming adsorbed as a function of phage titers, bacterial susceptibility to phages also as a function of phage titers, and the use of Poisson distributions to predict phage impacts on bacteria. Considered in addition is the use of simulations that can take into account both phage and bacterial replication. These various approaches can be automated, i.e., by employing a number of online-available apps provided by the author, the use of which this review emphasizes. In short, the practice of phage therapy can be aided by various mathematical approaches whose implementation can be eased via online automation.

Keywords: active treatment; bacteriophage therapy; biocontrol; biological control; JavaScript; MOI; passive treatment; pharmacodynamics

1. Introduction

Phage therapy is the application of bacterial viruses, more commonly known as bacteriophages or phages, especially toward the control or eradication of bacterial infections such as in animals, including in humans [1–9]. This is a subset of the use of phages more generally to control or eradicate nuisance bacteria found in broader environments [10], resulting in so-called phage-mediated biocontrol or biological control of bacteria. More broadly still is the use of viruses as biocontrol agents against organisms other than just bacteria [11]. Key to the successful use of antibacterial, antimicrobial, or biological control agents generally is the attainment of sufficient densities or concentrations of those agents in situ. But what concentrations are sufficient?

Here, I provide means toward answering that question for phages, which to some degree is situation-specific, and particularly so to the extent that one is attempting to minimize or at least reduce the amount of biocontrol agent applied. This involves discussion of a number of mathematical approaches toward gaining an appreciation of the impact of specific phage titers on targeted bacteria. In addition to providing equations that can be readily applied to different phage-treatment scenarios—and which generally are relatively simple, that is, fairly basic in their composition—I provide links to online JavaScript-based calculators which provide numerical solutions (Table 1). Most of these models can be considered to be of phage therapy pharmacodynamics [12], that is, of the anticipated degree of negative impact of a given in situ phage titer on a population of targeted bacteria.

Table 1. Summary of web pages referred to and their URLs (Uniform Resource Locators).

Topic	Section	URL
Multiplicity of Infection	Section 2.1	moi.phage.org
Phage Adsorptions	Section 2.2	adsorptions.phage-therapy.org

Citation: Abedon, S.T. Automating Predictive Phage Therapy Pharmacology. *Antibiotics* **2023**, *12*, 1423. https://doi.org/10.3390/antibiotics12091423

Academic Editor: Grzegorz Wegrzyn

Received: 30 July 2023
Revised: 2 September 2023
Accepted: 3 September 2023
Published: 8 September 2023

Table 1. *Cont.*

Topic	Section	URL
Bacterial Half-Life	Section 2.3.1	b-half-life.phage.org
Decimal Reduction Time	Section 2.3.2	decimal.phage-therapy.org
Phage Half-Life	Section 2.3.4	p-half-life.phage.org
Inundative Phage Quantities	Section 2.4	inundative.phage-therapy.org
Poisson Frequencies	Section 2.5	Poisson.phage.org
Killing Titers	Section 2.5.2	killingtiter.phage-therapy.org
Active Phage Therapy	Section 2.6	active.phage-therapy.org

2. Predictive Phage Therapy Pharmacology

In this section, I discuss a number of simple mathematical models that collectively can be predictive of the potential for a given phage titer to negatively impact a targeted bacterial population (Sections 2.1–2.5), along with scenarios toward attaining those titers in situ (Section 2.6). Many of these models I have previously discussed, e.g., [12–14]. Here, however, the primary aim is one of describing the basis of online calculators which I have developed that implement the various underlying calculations (Table 1).

Abbreviations of terms used in these calculations are summarized in Table 2, and introduced as well throughout the text. It is important to recognize, however, that for clinical or in vivo phage therapy, many of their values can be poorly described in practice, though exceptional can be determinations or at least estimations of initial, in situ phage titers. As a consequence of modeling-input values not necessarily being definite, it can be difficult to match model outputs to therapeutic outcomes. Nevertheless, it can be useful literally to play with models, entering variable and parameters values using the online calculators, as listed in Table 1, to gain a better "feel" for the pharmacodynamics of systems being worked with, that is, especially in terms of the potential for a given in situ phage titer to impact a targeted bacterial population. Alternatively, the presented models may be qualitatively and even quantitatively predictive of in vitro phage therapy experimentation.

Discussed specifically in this section are concepts associated with determining or estimating phage multiplicities of infection (MOIs; Section 2.1), the likelihood of a bacterium being phage adsorbed for a given phage titer (Section 2.2), rates of bacterial declines in number also as functions of phage titers (Section 2.3), how to estimate what phage titers may be required to reduce bacterial numbers to predetermined sufficient levels (Section 2.4), and the use of Poisson distributions in considering the impacts of phage titers on bacterial survival (Section 2.5). This is followed by consideration of in situ phage population growth (Section 2.6).

2.1. Multiplicity of Infection

Often seen in the phage therapy literature is the concept of multiplicity of infection (MOI). MOIs are relevant due to the statistical nature of phage adsorptions, i.e., such that phage adsorptions [15] are Poissonally distributed across susceptible bacteria [16] (Section 2.5). Though this Poissonal tendency can be quite useful toward appreciating phage therapy pharmacodynamics, the use of MOIs in the phage therapy literature can, in my opinion [12,17], often be problematic. In this section, I consider two different ways of defining phage multiplicities of infection—MOI_{input} vs. MOI_{actual} (Figure 1)—and a way of predicting the latter. An appreciation of these concepts can be useful toward the development of subsequent calculations of phage titer impacts on bacteria.

Table 2. Relevant Parameters and Variables.

Abbreviation	Description	Comments
A_c	Bacterial probability of being adsorbed	Likelihood of an individual bacterial cell being adsorbed per unit time, e.g., 1 min; the "c" stands for "cell"
A_t	Adsorptions over time	Number of phage adsorptions that occur over some interval of time, t
B	Burst size	Number of virions produced per phage infection; might range from 10 to well in excess of 100
e	Base of the natural logarithm	=2.718… (a non-repeating decimal)
I_P, I_N	Decay rate	Rates of loss of free phages (I_P) or bacteria (I_N) that occur for reasons that are independent of phage adsorption
IPD_{min}	Inundative phage density	Minimum phage titer required to reduce a bacterial population from some starting number to some ending number over some specified interval of time, not assuming 100% phage adsorption
IPN_{min}	Inundative phage number	Minimum phage titer to achieve the same as IPD_{min} except here assuming 100% phage adsorption
k	Adsorption rate constant	Probability that one virion will adsorb one bacterium as suspended in a unit volume of fluid (e.g., 1 mL) over the course of some unit time (e.g., 1 min), hence, e.g., $mL^{-1} min^{-1}$ units, though often expressed instead as $mL\ min^{-1}$
L	Latent period	Measure of the length of infection by a phage a bacterium
ln	Natural logarithm	For example, $\ln(2) = 0.69 = -\ln(0.5) = -\ln(1/2)$; $\ln(e) = 1$
MOI_{actual}, n	Actual multiplicity of infection	Number of adsorbed phages divided by the number of adsorbable bacteria; equivalent to n as used in Poisson calculations
MOI_{input} or $MOI_{addition}$	Input multiplicity of infection	Number of phages added to targeted bacteria divided by the number of those bacteria
M	Malthusian parameter	A measure of bacterial population growth rate in per time units
N, N_0, N_t	Bacterial concentrations	Subscript 0 refers to initial concentrations, though in many cases this is implied so the subscript is not always present; subscript t refers to the concentration of unadsorbed bacteria following a previous time interval, t
N_F, N_T	Bacterial numbers	Subscript F refers to a "Final" number of unadsorbed bacteria; subscript T refers to "Total" and is used instead of N_0 to distinguish starting bacterial concentration (N_0) from starting bacterial numbers (N_T)
p	Probability	This is lower-case "p" without italicization
P, P_0, P_F, P_t	Phage titer	Subscripts are equivalent to those of N_0, N_F, N_t, with P in all cases referring to phage concentrations, i.e., phage titers
$P_{adsorbed}$	Prior titer of adsorbed virions	Number of previously free phages that have now adsorbed, divided by volume, as to be distinguished from P_0
P_K	Killing titer	Titer of phages required to reduce a bacterial population from a given starting number to a given ending number, assuming 100% adsorption
r	Poisson category	Here, e.g., 0 phages adsorbed, 1 phage adsorbed, etc., all per bacterium
$r!$	r factorial	For example, $3! = 1 \times 2 \times 3$; $2! = 1 \times 2$; $1! = 1$; $0! = 1$
t	Time	Generally, here, this is an interval over which adsorption occurs
$t_{0.1}, t_{0.01}$	Decimal reduction time(s)	Time it takes for 90% of unadsorbed bacteria to become adsorbed ($t_{0.1}$) or 99% ($t_{0.01}$)
$t_{0.5}$	Bacterial half-life	Time it takes for one-half of unadsorbed bacteria to become adsorbed
t_{MFT}	Mean free time	Average length of time it takes for a bacterium to become phage-adsorbed
V	Volume	Volume that targeted bacteria and targeting phages are suspended in during phage treatments
x	Fraction bacteria	As surviving following phage exposure ($-N_F/N_T$)

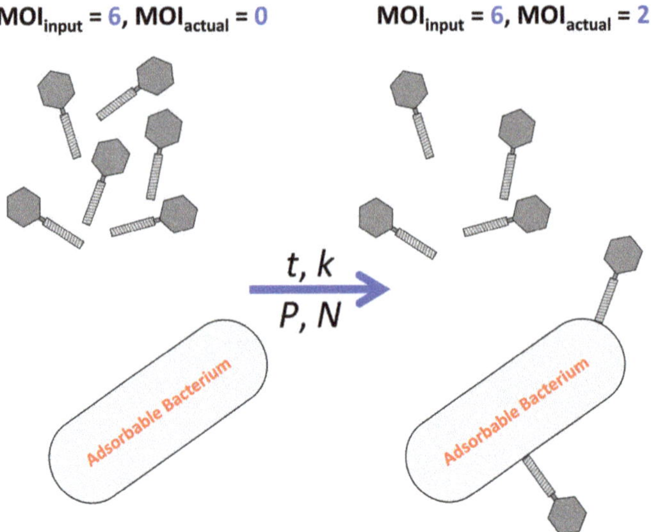

Figure 1. Comparing MOI$_{input}$ with MOI$_{actual}$. On both sides is the same MOI$_{input}$, whereas MOI$_{actual}$ to the right is equal to two vs. equal to zero on the left. Note that, generally, more than one adsorbable bacterium would be present and phages would adsorb over a Poisson distribution (Section 2.5), i.e., with the average number of virions adsorbed per bacterium equal to MOI$_{actual}$. In addition, keep in mind that the quantitative distinction between MOI$_{input}$ and MOI$_{actual}$ results from durations of adsorption periods (t) and the phage adsorption rate constant (k), the latter defined by a combination of the properties of the adsorbing phages, adsorbable bacteria, and adsorption environment. Phage (P) and bacterial (N) concentrations, however, also play important roles in determining MOI$_{actual}$, as considered below especially in Equation (5).

2.1.1. MOI$_{input}$ vs. MOI$_{actual}$

The two ways of defining MOI are MOI$_{input}$ vs. MOI$_{actual}$ [14,18]. The simplest as well as easiest to use—but the one that is also often misleading [12]—is MOI$_{input}$:

$$\text{MOI}_{input} = P/N, \tag{1}$$

where P is a starting phage titer and N is the initial concentration of targeted bacteria. This definition, in my opinion, is only useful to phage therapies to the extent that it can be contrasted with determinations as well as predictions of MOI$_{actual}$ (below). Phage therapy dosing based on MOI$_{input}$, in other words, at best should be viewed as "hopeful" since in many cases MOI$_{input}$ does not guarantee nor necessarily even approximate MOI$_{actual}$.

MOI$_{actual}$ instead is the more traditionally used meaning of MOI [19]. It is relevant to phage therapy first because it serves as the basis of Poisson distributions of adsorbed phages over susceptible bacteria and second because the extent of the impact of phages on bacteria also is Poissonal (Section 2.5).

Notwithstanding their distinctions, the definition of MOI$_{actual}$ is similar to that of MOI$_{input}$, though with a clear difference:

$$\text{MOI}_{actual} = P_{adsorbed}/N, \tag{2}$$

with $P_{adsorbed}$ not the initial phage titer but instead the concentration, such as per mL, of phages that have adsorbed bacteria, especially as seen after some interval of incubation of free phages with those phage-susceptible bacteria. That is, whereas MOI$_{input}$ is defined in terms of the total number of phages added to bacteria (again, such as per mL), MOI$_{actual}$ is based only on those virions that succeed in adsorbing and, importantly regarding phage therapy, generally only adsorbed phages have an impact on targeted bacteria.

2.1.2. Predicting $\text{MOI}_{\text{actual}}$

Though not as simple as for $\text{MOI}_{\text{input}}$, nevertheless $\text{MOI}_{\text{actual}}$ still can be fairly easy to determine in vitro as

$$\text{MOI}_{\text{actual}} = (P_0 - P_F)/N. \tag{3}$$

Here, P_0 is the starting concentration (titer) of free phages and P_F is the number free phages remaining unadsorbed following some interval of time (F standing for "Final"), assuming that all free phage losses are due to virion adsorption of targeted bacteria. Unfortunately, determining P_F can be impractical in vivo. Consequently, it can be helpful instead to be able to predict $\text{MOI}_{\text{actual}}$. In particular, it can be useful to possess some appreciation of the extent to which targeted bacteria may be impacted by treatment phages, with that impact, for a given phage type, generally being a function of $\text{MOI}_{\text{actual}}$ (e.g., Section 2.5).

An approximation of the suggested estimation [12,14] can be made based solely on initial phage titers (here shown as just P), the phage adsorption rate constant (k), and time (t):

$$\text{MOI}_{\text{actual}} = Pkt. \tag{4}$$

That approximation, however, is useful only at lower bacterial concentrations, e.g., such as below $10^7/\text{mL}$, and/or over shorter adsorption intervals, such as over a few minutes rather than over many tens of minutes. In contrast, at all bacterial concentrations or adsorption intervals, one can instead employ

$$\text{MOI}_{\text{actual}} = P\left(1 - e^{-Nkt}\right)/N. \tag{5}$$

Equation (5) differs from Equation (4) particularly in that it does not assume phage adsorption with replacement; that is, newly adsorbed phages are conceptually replaced with new free phages (Figure 2). Instead, in Equation (5) numbers of free phages are allowed to decline over time as those phages adsorb bacteria, i.e., as is expected in real systems. However, that consideration, as noted, may be qualitatively relevant only when bacterial concentrations are higher or adsorption intervals are longer.

Note in any case that e^{-Nkt} goes to zero as Nkt becomes larger, i.e., given higher concentrations of targeted bacteria, higher rates of phage adsorption to individual targeted bacteria, and/or longer incubation and thereby longer adsorption times. In that case, to the extent that e^{-Nkt} trends toward zero, then $\text{MOI}_{\text{actual}}$ will in fact come to approximate $\text{MOI}_{\text{input}}$.

2.1.3. Running the Calculator

The calculation that is presented in Equation (5) is solved via the online multiplicity of infection calculator found at moi.phage.org, there along with solutions to Equations (1) and (4) (with those latter equations solved by the calculator for the sake of comparison). Entering 1×10^7 phages/mL, 5×10^6 bacteria/mL, a 10 min adsorption period, and an adsorption rate constant [12,15] of 2.5×10^{-9} mL^{-1} min^{-1} [20] yields an $\text{MOI}_{\text{input}}$ (as equivalent to $\text{MOI}_{\text{addition}}$) of 2 but an $\text{MOI}_{\text{actual}}$ based on Equation (5) instead of 0.25, or 8-fold lower. Additionally, a total of only 1.1×10^6 phages of that original 1×10^7 will be expected to have adsorbed over that interval, while roughly 4×10^6 bacteria/mL will be expected to have remained unadsorbed out of that original 5×10^6, i.e., about 80% of bacteria targeted will not have been phage adsorbed in this example.

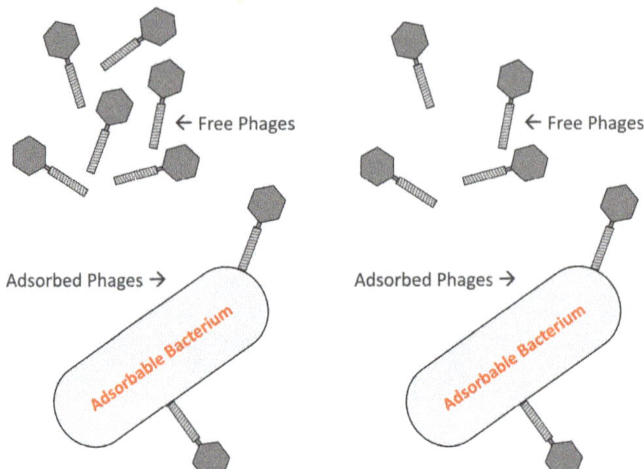

Figure 2. Adsorption with and without replacement of free phages. The mathematically simplified perspective is adsorption with replacement (**left**) since the result is a constant free phage concentration over time. Depending on circumstances, however, that assumption may or may not be realistic. It *may* be realistic, though, if free phage numbers are replaced as a consequence of in situ phage replication or if bacterial numbers are small, thereby resulting in few free phage losses due to adsorption. Alternatively, free phage adsorption without replacement (**right**) explicitly takes into account free phage losses that result from bacterial adsorptions, that is, with free phage concentrations thereby declining over time.

2.2. Bacterial Likelihood of Being Phage Adsorbed

Related to MOI_{actual}, and also solved using moi.phage.org, is simply the likelihood that a targeted bacterium will become phage adsorbed per unit of time, such as per min [15]. An appreciation of this likelihood can be helpful in gaining a better understanding of what may be accomplished upon achieving a given in situ phage titer during treatments. Here, I start with a model of phage adsorption over time and use this to derive the probability of adsorption to a single bacterium over a single unit of time.

2.2.1. Predicting Bacterial Adsorption Likelihood: $p(A_c)$

The number of adsorptions predicted to occur per unit time, particularly per unit of volume, such as per mL (A_t), is as follows:

$$A_t = NPkt. \tag{6}$$

If we are considering just a single bacterium, then the average number of adsorptions expected (A_c, with the "c" standing for "cell") can be found simply by setting N, the bacterial concentration, to 1 (again keeping in mind that this is all considered as occurring within 1 mL; see Appendix A of [15] for additional detail):

$$A_c = 1Pkt. \tag{7}$$

This is equivalent to our calculations of MOI_{actual} (Equation (4)). We can then approximate the probability of a single bacterium becoming adsorbed per mL and per min as

$$p(A_c) \approx 1Pk1, \tag{8}$$

and this is particularly so if the number of adsorptions expected per min, *Pk*, is somewhat less than one. If that is not the case, i.e., if the average number of adsorptions per min approaches or exceeds 1, then the probability that a bacterium will become adsorbed by a least one phage over the course of one min can instead be defined as

$$\mathrm{p}(A_c) = 1 - \mathrm{e}^{-P(1-\mathrm{e}^{-Nk})/N}, \tag{9}$$

that is, one minus e raised to the opposite of $\mathrm{MOI_{actual}}$ as calculated over one min (see Equation (5) for the latter). Note in Equations (8) and (9) that the lowercase "p" stands for "probability" vs. the uppercase, italicized "*P*", which stands for phage concentration, i.e., phage titer. In those equations, P is also implicitly equivalent to P_0 as to is N with N_0.

2.2.2. Running the Calculator

The online calculator can be found at adsorptions.phage-therapy.org. By way of example, if we again set k to 2.5×10^{-9} mL^{-1} min^{-1}, for 10^6 phages/mL (=P) the probability that a given bacterium ($N = 1$) will become phage adsorbed over one min, $\mathrm{p}(A_c)$, will be 0.0025. For $P = 10^7$ phages/mL, $\mathrm{p}(A_c)$ is instead raised to 0.025. At $P = 10^8$ phages/mL, the probability is instead 0.25. This is all assuming that phages are adsorbing with replacement, i.e., as specified by Equation (8). If we assume that phages are not adsorbing with replacement, then bacterial concentration (N) will come to matter somewhat more. Thus, with $P = 10^8$ and $N = 10^7$, the number of adsorptions per bacterium that are expected to occur over one min, which is the exponent in Equation (9), is 0.2469, while for $N = 10^8$ it is 0.2212, and for $N = 10^9$, it is 0.0918. These correspond to $\mathrm{p}(A_c)$ values of 0.2188, 0.1984, and 0.0877, respectively. The declines seen with greater bacterial numbers in turn are due to substantial losses of free phages to adsorption to the now substantial numbers of bacteria (the Nk term in Equation (9)) in combination with there simply being more bacteria for a given number of phages to adsorb (N as found in the exponent's denominator).

These latter calculations come to matter somewhat more if we assume both phage adsorption without replacement and longer adsorption intervals. Thus, for $P = 10^8, N = 10^8$, and $t = 60$ min, we have an expectation (Equation (5)) of a total (on average) of 1 phage adsorption per bacterium (i.e., in this case 1 = P/N vs. the 0.2212 indicated in the previous paragraph and $\mathrm{p}(A_c) = 0.3679$). With replacement of free phages following adsorption, however, the expectation (from Equations (4) or (7)) is instead an average of 15 phage adsorptions per bacterium over that same 60 min interval with $\mathrm{p}(A_c) = 0.0000$! Thus, unless phage concentrations can be sustained at high levels—e.g., by adding more phages, targeting smaller numbers of bacteria, or if phages are able sustain their numbers on their own such as due to in situ replication (Section 2.6)—then Equation (7)-type estimations can grossly overestimate expected per-bacterium levels of phage adsorption.

It is important in any case to recognize how dependent these outputs are on the magnitude of k [15]. If k is smaller, i.e., if we are working with a phage that has a lower potential to adsorb, then $\mathrm{p}(A_c)$ too will be smaller. Alternatively, with phages that adsorb faster, the resulting $\mathrm{p}(A_c)$ will be larger. These various ideas can be translated directly into what can be described as bacterial half-lives and related decimal reduction times (next section).

2.3. Bacterial Reduction Times

One measure of the susceptibility of a microorganism to an antimicrobial agent is what is described as decimal reduction time [14]. This is how long it takes for a given concentration of antimicrobial agent to reduce target numbers by 90% (here, abbreviated as $t_{0.1}$). Nearly equivalent mathematically, we can speak of half-lives, which is the time it takes to reduce a target bacterial population by 50% ($t_{0.5}$). Alternatively, we can consider reductions by 99% ($t_{0.01}$), and so on. In addition, and also similar mathematically, is mean free time (t_{MFT}), which for our purposes is the amount of time on average that it takes until a given bacterium becomes phage adsorbed. Overall, these constructs, as with likelihoods of bacteria being adsorbed by phages (above), can provide insight into the antibacterial utility of a given in situ phage titer.

Note that generally speaking, $t_{0.5} < t_{\mathrm{MFT}} < t_{0.1} < t_{0.01}$. This means that half of a bacterial population will become phage adsorbed faster than the average for single bacterium in a population to become phage adsorbed, and in turn it will take even longer for 90% of bacteria to become adsorbed, or indeed for 99% of bacteria to succumb to phage adsorption. In any case, for all of the presented equations in Sections 2.3.1 and 2.3.2, it is assumed that phages adsorb with replacement, with the without-replacement case addressed instead in Section 2.3.3.

2.3.1. Bacterial Half-Lives: $t_{0.5}$, and Also t_{MFT}

The bacterial mean free time in the presence of phages is simply the inverse of the likelihood of a bacterium being phage adsorbed per unit time, as seen with Equation (8), i.e.,

$$t_{\mathrm{MFT}} = 1/Pk. \tag{10}$$

The time it takes for one-half of a bacterial population to become phage adsorbed is slightly shorter, owing to the exponential decline associated with phage adsorption, i.e., where more adsorptions in absolute terms by a given population of free phages occur early during adsorption periods rather than later (assuming for that assertion that free phage adsorption is, again, without replacement) while later can be much later. Specifically, we multiply t_{MFT} by $-\ln(0.5)$ (the 0.5 for half-life), which is equivalent to $\ln(2)$. Thus,

$$t_{0.5} = \ln(2)/Pk = 0.69/Pk. \tag{11}$$

2.3.2. Decimal Reduction Times: $t_{0.1}$, plus $t_{0.01}$

As noted, decimal reduction times simply extend the bacterial reduction to 90% declines, up from the above 50%. The viable (unadsorbed) bacterial population has thus been reduced to 1/10th of its previous size. This can be calculated as

$$t_{0.1} = \ln(10)/Pk = 2.3/Pk. \tag{12}$$

The time it takes to reduce bacterial numbers 100-fold, i.e., to 0.01 of its original number, can be calculated instead as

$$t_{0.01} = \ln(100)/Pk = 4.6/Pk. \tag{13}$$

2.3.3. Phage Adsorption without Replacement

Considering phage adsorption without replacement complicates these formulae somewhat [14], with in the following x being equal to the resulting reduction, i.e., such as the above 0.5, 0.1, or 0.01 (and also adding explicitly the zero subscripts for consistency with the following section):

$$t_x = t(x) = -\ln\left(1 - \ln\left(\frac{1}{x}\right)\frac{N_0}{P_0}\right)/N_0 k. \tag{14}$$

Thus, for $x = 0.5$, then $1/x = 2$; for $x = 0.1$, then $1/x = 10$, etc. Note, though, that for this equation to be valid then sufficient numbers of phages must be initially present to achieve the indicated reduction, e.g., one must start with $P_0 > \ln(10)N_0$ to achieve decimal reduction or $P_0 > \ln(2)N_0$ phages to reduce unadsorbed bacterial numbers by half.

2.3.4. Running the Calculators

A dedicated, online bacterial half-life calculator can be found at b-half-life.phage.org. Starting with a phage concentration of $10^6/\mathrm{mL}$, and an adsorption rate constant as above, then t_{MFT} is calculated as 400 min vs. 277 min for $t_{0.5}$. Raise the phage titer to $10^7/\mathrm{mL}$ and these numbers are reduced to 40 and 28 min, respectively, or 4 and 2.8 min given

10^8 phages/mL (all holding phage titers constant over time). An equivalent calculator, but instead determining phage half-lives as a function of bacterial concentrations, can be found at p-half-life.phage.org. The latter can be used to gain an appreciation of how rapidly a given titer of supplied phages will be expected, as a function of bacterial concentrations (N), to become explicitly antibacterial as they adsorb, e.g., such as 50% of those phages adsorbing per min vs. instead 50% per hour. See also Bull and Regoes [21] for an extension of phage half-life calculations to also include phage losses for reasons other than adsorption to phage-infected bacteria.

A decimal reduction, etc., online calculator can be found at decimal.phage-therapy.org. This provides calculations not only for 10- and 100-fold declines in bacterial numbers but also makes this determination both with and without taking starting bacterial concentrations into account. That is, considering phage adsorption both without and with free phage replacement, respectively. The default settings are phage titers of 10^8/mL and bacterial concentrations of 10^6/mL. With no decline in phage numbers over time, output is $t_{0.1}$ = 9.2 min while $t_{0.01}$ = 18.4 min. At such a low bacterial concentration, the equivalent numbers, if assuming instead phage losses to adsorption, are only 9.3 min and 18.9 min, respectively, keeping in mind that total reductions in numbers of unadsorbed bacteria is ten times that for the latter ($t_{0.01}$) vs. the former ($t_{0.1}$). Raise bacterial concentrations to 10^7/mL and the equivalent numbers again assuming phage adsorption without replacement instead are 10.5 min and 24.7 min. Then, raise phage titers to 10^9 (while keeping N at 10^7/mL) and we find that $t_{0.1}$ = 0.9 min while $t_{0.01}$ = 1.8 or 1.9 min (these latter two values are without losses due to phage adsorption and with losses due to phage adsorption, respectively).

2.4. Inundative Phage Quantities

A slightly more sophisticated way of thinking about degrees of phage impact on bacteria is to consider not just durations of treatments in combination with how fast phages are adsorbing, but also how large a reduction in numbers of a bacterial population is desired [12]. This differs from the above bacterial reduction times (Section 2.3) because the sought end points are not fractional declines in bacterial numbers but, instead, are absolute declines. Thus, rather than, for example, a 99% reduction, a reduction to, e.g., 10^3 bacteria in total is sought. In terms of required starting phage titers, I have dubbed this an "inundative phage density" (IPD_{min}), with "density" and "titer" here being used synonymously. Alternatively, there is an "inundative phage number" (IPN_{min}), which is the starting absolute number of phages required, that is, rather than starting phage concentrations (the latter again equivalent to "titer" and "density"). As with the other calculations already considered, an implicit assumption is that all targeted bacteria are equally available to phages for adsorption.

In all of these cases, these are minimum values ("min") because it is assumed that bacterial losses are occurring as calculated whereas less-than-ideal phage adsorption and infection circumstances likely would result in a requirement for more phages than this "min", such as $IPN_{actual} > IPN_{min}$. Thus, a failure to successfully predict the extent of reductions in bacterial viability in the presence of predicted inundative quantities of phages can be used to indicate the presence of additional phenomena not considered by models. For example, less bacteria killing than expected can be due to not all targeted bacteria being equally available to phages, such as due to the presence of spatial or physiological refuges from phage attack [22]. Lower levels of killing than expected can also be a consequence of outright genetic bacterial resistance to phages and/or instead underestimations of phage adsorption rate constants. Alternatively, greater bacteria killing than expected can be due to the presence of additional antibacterial mechanisms and/or because new phages have been generated in situ (for the latter, see "Active treatment", below; Section 2.6). In any case, calculations of inundative phage quantities can provide an appreciation of what phage titers should be required to reduce phage-susceptible bacteria to a given total number of remaining bacteria, over a desired length time, particularly as based on the antibacterial action of dosed phages alone.

2.4.1. Inundative Phage Densities: $\text{IPD}_{\min}$

The minimum titer of phages required to reduce a volume of bacteria to a given amount over a specific span of time, or $\text{IPD}_{\min}$, can be calculated either assuming or not assuming that these titers remain constant over time (Figure 2). As with the approaches considered above, assuming a constant phage titer simplifies calculations but becomes less valid the higher bacterial concentrations or the longer the time frame over which adsorption is allowed to occur. In any case, both phage and bacterial replication are ignored for these $\text{IPD}_{\min}$, or $\text{IPN}_{\min}$, determinations.

The total starting number of bacteria is equal to the volume of the relevant environment (V) multiplied by the starting concentration of bacteria (N_0). The final number of bacteria is independent of volume. That is, often when reducing bacterial presence, you want to reduce the number of bacteria to a given lower amount (N_F) rather than to a given lower concentration. Thus, the fraction of bacteria that are expected to survive given the application of some inundative titer of phages will be N_F/VN_0 (total ending bacterial numbers divided by total starting numbers of bacteria) and this fraction, or at least its inverse, is used in the same manner as for, e.g., decimal reduction time calculations (Section 2.3.2). If phage titers can be held more or less constant over time, then the minimum titer of phages required to achieve that fraction of surviving bacteria can be descried as

$$\text{IPD}_{\min} = \frac{\ln(VN_0/N_F)}{kt}. \tag{15}$$

This is the natural log of the fold-decrease in bacterial concentrations, i.e., as equal to $1/x$ in Equation (14), divided by the product of the phage adsorption rate constant and time, with $\text{IPD}_{\min}$ representing some phage concentration, i.e., P. For a 10-fold decline in bacterial numbers—a decimal reduction and thus $x = 0.1$—this would be $P = \ln(10)/kt$. With rearranging and modifying the abbreviation for time, this is equivalent to the $t_{0.1} = \ln(10)/Pk$, as seen above in Equation (12). The quantity $\ln(10)$ in turn is equal to the $\text{MOI}_{\text{actual}}$ required to achieve this 10-fold reduction in concentrations of viable bacteria, i.e., 2.3.

Taking into account phage losses due to adsorption to bacteria has the effect of requiring higher starting phage titers, and this can be described instead as

$$\text{IPD}_{\min} = \frac{N_0 \cdot \ln(VN_0/N_F)}{\left(1 - e^{-N_0 kt}\right)}. \tag{16}$$

This is equivalent to [starting numbers of bacteria] × [$\text{MOI}_{\text{actual}}$ required to achieve the desired degree of reduction in bacterial numbers] divided by [fraction of added phages which succeed in adsorbing over time, t].

2.4.2. Inundative Phage Number: $\text{IPN}_{\min}$

An alternative perspective is just how many phages are needed to similarly reduce numbers of bacteria as seen for $\text{IPD}_{\min}$, but without prior knowledge of bacterial concentrations. This approach can be relevant if numbers of bacteria are known or at least can be estimated, but where treatment volumes are less easily determined. There are two ways of going about this. One is to assume that phage titers are known and remain more or less constant or, alternatively, that 100% adsorption of added free phages can be assumed. Missing is the case where phage numbers are instead declining to some intermediate extent, due to phage adsorptions of bacteria, as that extent cannot be calculated without knowledge of bacterial concentrations.

The first case looks simply like

$$\text{IPD}_{\min} = \frac{\ln(N_T/N_F)}{kt}, \tag{17}$$

where N_T is initial, unadsorbed bacterial numbers ("T" standing for "Total"), and this is rather than initial bacterial concentrations. As indicted though, this is again a calculation of IPD_{min}, rather than of a minimum inundative phage number (IPN_{min}), and this is because required phage titers rather than just phage numbers would be calculated; note also that the numerator again is equivalent to $\ln(1/x)$. If phage titers are not easily predicted, i.e., as due to phage application to volumes that are not well defined, we need to resort to assuming instead the noted 100% adsorption of added free phages:

$$IPN_{min} = \frac{N_T \cdot \ln(N_T/N_F)}{1}. \tag{18}$$

IPN_{min} is thus the total number of phages that need to be supplied, but again assuming 100% adsorption. Note that $\ln(N_T/N_F)$ is equal to that MOI_{actual} (Section 2.1) required to reduce bacterial numbers from N_T to N_F, which, in turn, is an N_T/N_F-fold reduction, and equivalently this is $(1/x)$-fold. For example, with a 10-fold reduction, $\ln(N_T/N_F) = 2.3 = MOI_{actual}$. IPN_{min}, as described by this equation, is therefore equal to that MOI_{actual} multiplied by the total number of bacteria targeted, i.e., by N_T.

If MOI_{actual} should fail to approximate MOI_{input}, and the degree of discrepancy is known, then one can modify the previous equation as

$$IPN_{min} = \frac{N_T \cdot \ln(N_T/N_F)}{MOI_{actual}/MOI_{input}} \tag{19}$$

We expect, in any case, for $MOI_{input}/MOI_{actual} \geq 1$ to hold under all circumstances, since it is impossible to adsorb more phages than there are phages (as above, assuming no in situ phage replication). Therefore, the less extensively that phage adsorption occurs, e.g., $MOI_{input} \gg MOI_{actual}$, even assuming ideal adsorption conditions, then the more phages that will be required to reduce bacterial numbers to an equivalent extent.

2.4.3. Running the Calculator

An online calculator is available for determining inundative phage quantities, as found at inundative.phage-therapy.org. If we start with 10^6 bacteria/mL ($=N_0$), and consider only 1 mL of volume (V), then reductions to 10^3 bacteria in total (N_F) over one hour (t) requires 4.5×10^7 phages/mL, assuming via Equation (15) that there are no phage losses ($=IPD_{min}$). This changes to 5.0×10^7 phages/mL given phage losses to adsorption, as per Equation (16) ($=IPD_{min}$). Alternatively, via Equation (18), a starting number of only 6.9×10^6 phages ($=IPN_{min}$) is required if 100% phage adsorption is assumed. (Note in the example that 10^6 is both the starting bacterial concentration and starting bacterial number since only 1 mL is being considered.)

Additional examples of IPD_{min} determinations are found in Table 3, all assuming a value for k of 2.5×10^{-9} mL^{-1} min^{-1}. Notice how nearly the same numbers of phages are required to reduce bacterial numbers to the same amount, e.g., 1 ($=10^0$), regardless of starting bacterial numbers. Thus, starting with 10^6 bacteria/mL in 100 mL requires 1.2×10^8 phages per mL (assuming no phage losses over time) but still half as many phages starting with only 10^2 bacteria/mL despite the 10,000-fold difference in numbers of starting bacteria. Thus, reducing bacterial populations to a substantial extent requires relatively high phage titers and this is so even if starting bacterial concentrations are relatively low. The explanation for why this is the case has to do with the statistics of Poisson distributions (next section).

Table 3. Calculating inundative phage quantities for one-hour treatments *.

$N_T \rightarrow$ $VN_T \rightarrow$ $N_F \downarrow$	10^{10} 10^{12}	10^9 10^{11}	10^8 10^{10}	10^7 10^9	10^6 10^8	10^5 10^7	10^4 10^6	10^3 10^5	10^2 10^4	
10^{-3}	2.3×10^8	2.1×10^8	2.0×10^8	1.8×10^8	1.7×10^8	1.5×10^8	1.4×10^8	1.2×10^8	1.1×10^8	Eq. (15)
10^{-3}	3.5×10^{11}	3.2×10^{10}	3.0×10^9	3.6×10^8	1.8×10^8	1.5×10^8	1.4×10^8	1.2×10^8	1.1×10^8	Eq. (16)
10^{-3}	3.0×10^{11}	2.8×10^{10}	2.5×10^9	2.3×10^8	2.1×10^7	1.8×10^6	1.6×10^5	1.4×10^4	1.2×10^3	Eq. (18)
10^{-2}	2.1×10^8	2.0×10^8	1.8×10^8	1.7×10^8	1.5×10^8	1.4×10^8	1.2×10^8	1.1×10^8	9.2×10^7	Eq. (15)
10^{-2}	3.2×10^{11}	3.0×10^{10}	2.8×10^9	3.3×10^8	1.7×10^8	1.4×10^8	1.2×10^8	1.1×10^8	9.2×10^7	Eq. (16)
10^{-2}	2.8×10^{11}	2.5×10^{10}	2.3×10^9	2.1×10^8	1.8×10^7	1.6×10^6	1.4×10^5	1.2×10^4	9.2×10^2	Eq. (18)
10^{-1}	2.0×10^8	1.8×10^8	1.7×10^8	1.5×10^8	1.4×10^8	1.2×10^8	1.1×10^8	9.2×10^7	7.7×10^7	Eq. (15)
10^{-1}	3.0×10^{11}	2.8×10^{10}	2.5×10^9	3.0×10^8	1.5×10^8	1.2×10^8	1.1×10^8	9.2×10^7	7.7×10^7	Eq. (16)
10^{-1}	2.5×10^{11}	2.3×10^{10}	2.1×10^9	1.8×10^8	1.6×10^7	1.4×10^6	1.2×10^5	9.2×10^3	6.9×10^2	Eq. (18)
10^0	1.8×10^8	1.7×10^8	1.5×10^8	1.4×10^8	1.2×10^8	1.1×10^8	9.2×10^7	7.7×10^7	6.1×10^7	Eq. (15)
10^0	2.8×10^{11}	2.5×10^{10}	2.3×10^9	2.7×10^8	1.3×10^8	1.1×10^8	9.2×10^7	7.7×10^7	6.1×10^7	Eq. (16)
10^0	2.3×10^{11}	2.1×10^{10}	1.8×10^9	1.6×10^8	1.4×10^7	1.2×10^6	9.2×10^4	6.9×10^3	4.6×10^2	Eq. (18)
10^1	1.7×10^8	1.5×10^8	1.4×10^8	1.2×10^8	1.1×10^8	9.2×10^7	7.7×10^7	6.1×10^7	4.6×10^7	Eq. (15)
10^1	2.5×10^{11}	2.3×10^{10}	2.1×10^9	2.4×10^8	1.2×10^8	9.3×10^7	7.7×10^7	6.1×10^7	4.6×10^7	Eq. (16)
10^1	2.1×10^{11}	1.8×10^{10}	1.6×10^9	1.4×10^8	1.2×10^7	9.2×10^5	6.9×10^4	4.6×10^3	2.3×10^2	Eq. (18)
10^2	1.5×10^8	1.4×10^8	1.2×10^8	1.1×10^8	9.2×10^7	7.7×10^7	6.1×10^7	4.6×10^7	3.1×10^7	Eq. (15)
10^2	2.3×10^{11}	2.1×10^{10}	1.8×10^9	2.1×10^8	9.9×10^7	7.7×10^7	6.1×10^7	4.6×10^7	3.1×10^7	Eq. (16)
10^2	1.8×10^{11}	1.6×10^{10}	1.4×10^9	1.2×10^8	9.2×10^6	6.9×10^5	4.6×10^4	2.3×10^3		Eq. (18)
10^3	1.4×10^8	1.2×10^8	1.1×10^8	9.2×10^7	7.7×10^7	6.1×10^7	4.6×10^7	3.1×10^7	1.5×10^7	Eq. (15)
10^3	2.1×10^{11}	1.8×10^{10}	1.6×10^9	1.8×10^8	8.3×10^7	6.2×10^7	4.6×10^7	3.1×10^7	1.5×10^7	Eq. (16)
10^3	1.6×10^{11}	1.4×10^{10}	1.2×10^9	9.2×10^7	6.9×10^6	4.6×10^5	2.3×10^4			Eq. (18)
10^4	1.2×10^8	1.1×10^8	9.2×10^7	7.7×10^7	6.1×10^7	4.6×10^7	3.1×10^7	1.5×10^7		Eq. (15)
10^4	1.8×10^{11}	1.6×10^{10}	1.4×10^9	1.5×10^8	6.6×10^7	4.6×10^7	3.1×10^7	1.5×10^7		Eq. (16)
10^4	1.4×10^{11}	1.2×10^{10}	9.2×10^8	6.9×10^7	4.6×10^6	2.3×10^5				Eq. (18)
10^5	1.1×10^8	9.2×10^7	7.7×10^7	6.1×10^7	4.6×10^7	3.1×10^7	1.5×10^7			Eq. (15)
10^5	1.6×10^{11}	1.4×10^{10}	1.2×10^9	1.2×10^8	5.0×10^7	3.1×10^7	1.5×10^7			Eq. (16)
10^5	1.2×10^{11}	9.2×10^9	6.9×10^8	4.6×10^7	2.3×10^6					Eq. (18)

* Arrows are used to indicate what values the upper-left abbreviations are describing. N_T refers to starting bacterial numbers within 1 mL, VN_T refers to starting bacterial numbers here within 100 mL, and N_F refers to ending bacterial numbers, with the value N_T in its two instances being used equivalent to N_0. Stacked quantities from top to bottom are IPD_{min} assuming constant phage titers over time (Equation (15)), IPD_{min} not assuming constant phage titers over time (Equation (16)), and IPN_{min} (Equation (18)). "Equation" in the last column has been abbreviated as "Eq."

2.5. Poisson Distributions

A Poisson distribution is similar to a normal distribution except that the x axis, defining the independent variable, r, consists solely of integers that cannot fall below 0. Thus, $r = 0$, 1, 2, 3, etc. Furthermore, what are varied on the y axis are the probabilities associated with each of those integers, i.e., $y = p(r)$. The magnitude of $p(r)$ is defined as follows

$$p(r) = \frac{n^r e^{-n}}{r!}. \tag{20}$$

For our purposes, r represents categories of phage adsorptions to bacteria, i.e., the unadsorbed fraction ($r = 0$), the fraction adsorbed by only a single phage ($r = 1$), the fraction adsorbed by two phages ($r = 2$), and so on. In contrast, the variable, n, is MOI_{actual} (Section 2.1). Thus, the fraction of bacteria expected within each of the r categories is defined for a given MOI_{actual} by a Poisson distribution [16].

2.5.1. Predicting Bacterial Survival

If we set r to zero, then Equation (20) is reduced to

$$p(0) = e^{-n}, \tag{21}$$

keeping in mind that $0! = 1$, as is also the case for any number raised to zero, i.e., n^0. Rearranging, then $n = \text{MOI}_{\text{actual}} = -\ln(p(0)) = -\ln(N_F/N_T) = -\ln(x)$, keeping in mind that $\ln(1/x) = -\ln(x)$. N_T is the starting number of unadsorbed bacteria, i.e., as found prior to phage addition, while N_F is the ending or "Final" number of unadsorbed bacteria. $\text{MOI}_{\text{actual}}$ is thus equal to the negative natural log of the fraction of bacteria remaining unadsorbed following some extent of phage exposure, or the positive natural log of the fold-decrease in bacterial numbers.

2.5.2. Killing Titers: P_K

Killing titer (P_K) calculations [12] take the above prediction of bacterial survival and literally rearrange it. This, in contrast to much of the above, is therefore a phage titer determination that is based on bacterial survival rather than a prediction of bacterial survival that is determined, at least in part, by knowledge of initial phage titers. As equivalently seen with Equation (18), $\text{MOI}_{\text{actual}}$ is multiplied by the initial bacterial concentration, but here with $\text{MOI}_{\text{actual}}$ calculated based on the fraction of bacteria that have survived, assuming that all added phages have adsorbed:

$$P_K = -\ln(p(0))N_0, \tag{22}$$

recalling that $\text{MOI}_{\text{actual}} = -\ln(p(0))$. Thus, if 10^8 bacteria per mL are reduced to 10^7, then the calculated killing titer is $-\ln(0.1) \times 10^8 = 2.3 \times 10^8$. This would be equal to P_0, i.e., the starting phage concentration, assuming that all free phages initially present adsorbed (and that no phage replication has occurred).

Note, though, that the requirement that all free phages must adsorb means that, for this calculation, $\text{MOI}_{\text{actual}}$ must equal $\text{MOI}_{\text{input}}$, that is, in order for P_K to be an actual phage titer determination. If insufficient time is allowed for adsorption, however, then $\text{MOI}_{\text{actual}}$ will be lower than $\text{MOI}_{\text{input}}$, resulting in the calculated P_K being less than P_0. Consequently, killing titer determinations will always underestimate starting phage titers unless complete phage adsorption is allowed to occur, keeping in mind though that often a small fraction of phages will fail to adsorb seemingly no matter what [23–26]. Of course, for killing titer calculations to hold true, then bacterial replication also must be insubstantial during phage application. Nevertheless, killing titers can provide at least an approximation of what phage titers would have been necessary to achieve the amount of bacteria killing observed, which can in turn be compared with what phage titers actually had been present at the start of phage treatments of a bacterial population.

2.5.3. Running the Calculators

A Poisson frequency calculator is presented at Poisson.phage.org, requiring a single input, that of $\text{MOI}_{\text{actual}}$. Note that this need not be an integer. For example, for $\text{MOI}_{\text{actual}} = 1.5$, the app indicates that the fraction of bacteria expected to not have been phage adsorbed is 0.22 (or 0.37 for $\text{MOI}_{\text{actual}} = 1$). Additionally, relevant for certain phage biology experiments is the fraction of bacteria which are singly vs. multiply adsorbed [12]. For $\text{MOI}_{\text{actual}} = 1.5$, these fractions are 0.33 and 0.44, respectively, such that, though with rounding error, $1 = 0.22 + 0.33 + 0.44$. Also calculated are the fraction of bacteria, of those that have been adsorbed at all, which have been singly vs. multiply adsorbed. For this same example ($\text{MOI}_{\text{actual}} = 1.5$), those fractions are 0.43 and 0.57, respectively, which also add up to 1. That is, 43% of bacteria that have been adsorbed in this example are predicted to have been singly adsorbed.

The killing titer calculator can be found at killingtiter.phage-therapy.org. Entered here are concentrations of still-viable bacteria as found both before and after phage adsorption, keeping in mind (again) that for this to be an actual titer determination, then phage adsorption must go effectively to completion, thus requiring sufficient time though also an absence of bacterial replication during that time. If for example you were to start with 10^7 bacteria/mL and end up with 10^4 bacteria/mL, then your calculated killing titers would be 6.9×10^7 phages/mL. Additionally, MOI_{actual} is calculated, which in this example, would be 6.9. The greatest utility of such killing titer determinations is for use toward establishing the titers of phages—or other agents such as phage tail-like bacteriocins—which, for whatever reason, are unable to form plaques on the bacterial strain being targeted, while still possessing single-hit kinetics of those bacteria [21]. Nevertheless, it is also useful to compare calculated killing titers (P_K) with actual titers (P_0) to assess treatments, with $P_K < P_0$ implying a less-than-ideal phage impact while $P_K > P_0$ would imply instead a greater-than-expected phage impact.

2.6. Active Treatments

All of the above-discussed approaches ignore both phage and bacterial population growth. In my opinion, ignoring bacterial population growth is reasonably justified, particularly (1) if one is treating bacterial populations which already are somewhat mature in terms of not displaying substantial additional bacterial population growth or (2) if phage impact is fast relative to rates of bacterial growth. The latter generally can be achieved by supplying phages in high concentrations, i.e., inundative densities acting over relatively short periods.

Phage in situ replication, on the other hand, is less easily ignored, except in a limited number of circumstances, e.g., such as application of overwhelming phage numbers, so-called passive treatments [27–30], or when phages are used which are unable to replicate [21,31]. Such phage replication, giving rise to in situ phage population growth during phage therapies, can result in what have been described as active treatments [27–30], with "Active" referring to a relevance of virion-productive phage infections of bacteria, again in situ, toward enhancing phage therapeutic efficacy. Importantly, however, we can also differentiate phage in situ population growth into that associated with high vs. low overall bacterial concentrations and also that phage population growth occurring in association with vs. without bacterial clumping or clustering, that is clumping or clustering of bacteria such as into biofilm microcolonies. Thus, for example:

1. Low bacterial concentrations *without* clumping and lower starting phage titers. In the case of low bacterial concentrations and no bacterial clumping, phage population growth likely is mostly irrelevant, since in situ phage replication will not be expected to have a substantial impact on more "global" phage titers. That is, bacteria are present in insufficient quantities to produce relatively large concentrations of new phages across environments. Still, these circumstances, given sufficient environmental mixing, are easily modelled mathematically.
2. Low bacterial concentrations *with* clumping and lower starting phage titers. With spatial structure in combination with bacteria being found in clonal clusters—but bacteria nonetheless overall found at low concentrations—phage in situ replication could in fact be relevant, though not globally, and the mathematics portraying such situations is not straightforward. I describe this latter scenario as a *locally* active treatment [32].

In other words, for the latter, once a bacterial microcolony has been infected by a single phage, it is not unlikely that other bacteria found in the same microcolony or cellular arrangement will be impacted by resulting locally produced phage progeny [33,34]. It is just that those newly generated phages, if amplified in number from only sparsely available bacterial microcolonies, may be unlikely to easily find other bacterial microcolonies to infect, due to those phages not achieving relatively high titers across treated environments.

Higher bacterial concentrations with bacterial clumping, such as existing as biofilms, again greatly complicate the necessary mathematics and are therefore not straightforward to model. Without clumping, though, we still may describe two basic scenarios when considering the treatment of higher bacterial concentrations, distinguishing passive from active treatments. These are:

3. Higher bacterial concentrations without clumping and *higher* starting phage titers. First is the noted passive treatment in which phage in situ replication is not required to achieve desired levels of bacterial eradication, e.g., as due to the employment of inundative phage concentrations (Section 2.4). This is because sufficient quantities of phages have been supplied via phage dosing alone.

4. Higher bacterial concentrations without clumping and *lower* starting phage titers. Second is what I have described as *globally* active treatment [32]. Here, the assumption is that phage virions are free to diffuse relatively rapidly about environments or otherwise be readily moved about, such as within blood. Therefore, phages produced in one location can give rise to sufficient increases in phage titers, i.e., to inundative densities (Section 2.4.1) throughout a phage-treated environment.

It is this globally active treatment that is the primary focus of this section and indeed, this is how active treatments typically are envisioned, at least from modeling perspectives [27–29].

As follows, I first take into account phage population growth (Section 2.6.1) and then bacterial population growth (Section 2.6.2), both in terms of such globally active treatment. In Section 2.6.3, the two approaches are brought together, with an online calculator for running the resulting model introduced. Section 2.6.4 then extends the idea of modeling phage active treatments but reviewing in vitro experimentation in particular, rather than in silico modeling.

2.6.1. Considering Phage Population Growth

Modeling of phage population growth as well as bacterial population growth has typically been performed within a context of chemostat-based phage–bacteria community dynamics, e.g., [35,36]. Here, for simplicity as well as because it likely is at least equivalently relevant to phage therapy, only batch-culture-type scenarios are considered, e.g., as modeled in Abedon et al. [37]. Batch- vs. chemostat-based modeling is equivalent, except that inflow of new media and outflow of culture media along with bacteria and phages is not considered with batch growth. Additionally, here no bacterial-concentration-associated constraints on phage or bacterial growth rates are considered [36,38].

Though models of phage population growth are often presented based on calculus (as reviewed in Stopar and Abedon [13]), in reality their numerical solutions will typically employ discrete iterations, e.g., advancing simulations in one-minute intervals. Therefore, the relevant equations are presented here explicitly as these iterated equations. Thus:

$$P_{t+1} = P_t + BkP_{t-L}N_{t-L}e^{-LI_N} - kP_tN_t - I_PP_t. \tag{23}$$

This can be expressed in words as follows: The phage concentration found one interval later (P_{t+1}) is equal to the just-previous phage concentration (P_t) plus new phages generated upon phage-induced bacterial lysis (B meaning burst size) of those bacteria infected one latent period (L) earlier ($BkP_{t-L}N_{t-L}$). Subtracted from this are those phages lost to adsorption (kP_tN_t) along with any free phages lost for any additional reasons (I_PP_t). In addition is the construct, e^{-LI_N}, which has the effect of removing phage-infected bacteria that have been lost to non-phage-related decay over the course of one latent period. I_N is defined as the rate of loss of bacterial cells for non-phage-related reasons, as is also employed in the following section.

2.6.2. Considering Bacterial Population Growth

Bacterial growth is introduced as an additional iterated equation, one which feeds into the equation modeling phage population growth (Equation (23)) and vice versa. The growth itself is modeled using what is known as the Malthusian parameter (μ), which basically is the rate of exponential growth of the population as occurs over a single interval, such as one minute. Thus,

$$N_{t+1} = N_t + \mu N_t - kP_t N_t - I_N N_t \tag{24}$$

This—similar to the phage equation, Equation (23)—is the concentration of unadsorbed bacteria one interval later (N_{t+1}) as equal to the just-previous unadsorbed bacterial concentration (N_t) but also with new bacteria being added due to bacterial binary fission (μN_t). Bacteria are lost to phage adsorption ($kP_t N_t$) as well as to phage-unrelated forms of inactivation ($I_N N_t$). As with modeling phage population dynamics, inactivation can be ignored (that is, by setting the parameter, I_N, to zero). Alternatively, by setting I_N and I_P to the same value, then a chemostat-like system can be modeled, with both parameters thereby describing outflow. Inflow in any case can be ignored because, as noted, nutrient concentrations are not being considered.

2.6.3. Running the Calculator

Running a simulation based on the above two equations involves simply employing appropriate parameter values along with starting conditions and then stepping through both equations one interval at a time. There is increasing imprecision the longer the incrementation interval, but I have found that resulting error is minor given use of one min intervals. I have had a tendency to employ a spreadsheet, such as Microsoft Excel®, to run these sorts of simulations [13], which involves stepping through the equations vertically in columns, with each row corresponding to one interval. Alternatively, an online calculator for modeling globally active treatment can be found at active.phage-therapy.org, though there this is described simply as "Active treatment".

The default, though otherwise fully adjustable parameters entered in the online calculator are latent period (15 min), burst size (100), initial phage titer (10^7/mL), a phage inactivate rate (0.00001 as a per min fractional loss), an initial bacterial concentration (10^3/mL), the Malthusian parameter (0.013), a phage-independent rate of bacterial loss (also set to 0.00001 and which, as also for phages, is set there deliberately small by default), and a simulation duration (60 min). Running the calculator using those inputs yields an only minor, $\log_{10}$ 0.003 increase in phage titers, owing to the very small starting bacterial concentration. This is roughly a 1% increase in phage numbers. In contrast, bacterial concentrations over this span are reduced by $\log_{10}$ 0.317, which corresponds to a 52% reduction. Change the starting bacterial concentration to 10^6/mL and over that hour, phage concentrations increase by 1300% (1.146 $\log_{10}$, which is from 10^7 to ~1.4 × 10^8 phages/mL) while bacterial concentrations decline by 3.834 logs (down to 1.5 × 10^2/mL, or nearly a 100% decline). Thus, in this latter case, there are sufficient bacteria present to support substantial phage population growth, and this in turn results in more substantial declines in numbers of bacteria, i.e., considerably effective active treatment is occurring. The caveats, however, are that it is difficult to determine in situ phage latent periods or burst sizes as well as bacterial rates of replication, and indeed, determining in situ phage adsorption rate constants as well. Furthermore, it is difficult to assume that in situ environments are homogeneous, or necessarily well mixed, both as required implicitly by the simulation. Still, this calculator allows one to easily play a number "what if?" scenarios regarding starting phage and bacterial concentrations.

2.6.4. Additional Approaches to Predicting In Situ Efficacy, from In Vitro Characteristics

A major issue with such mathematical modeling of active treatments, toward prediction of in situ phage behavior as presented above, is that it is labor intensive to obtain the needed parameter values, even in vitro, especially those of phage latent periods, phage burst sizes, phage adsorption rate constants, and bacterial growth rates. Consequently, efforts have been made to model active treatments experimentally, also in vitro, again toward predicting phage abilities during actual treatments, but without going through the struggle of obtaining those individual measures. These efforts, such as found in [39–41], particularly involve determinations of phage impacts on in vitro optical densities (i.e., turbidity) of broth bacterial cultures over time, as summarized in terms of areas under the curve (AUCs); see also [42,43] for review of this general broth-culture approach. Smaller AUCs (less overall bacterial culture density over time) are indicative of higher phage antibacterial virulence—due to sooner, faster, or more complete phage-induced lysis of bacterial populations—and larger AUCs indicate lower phage antibacterial virulence.

An issue with these optical density-based approaches is, unfortunately, the occurrence of lysis inhibition [44], which has the effect of retaining and even boosting the turbidity of bacterial cultures despite the productive lytic phage infection of most or all phage-sensitive bacteria present. Importantly, though, only a subset of phage types display this phenotype. Nevertheless, see panel B in Figure 1 of [39], where the phage T4 studied there is historically well known for displaying the lysis inhibition phenotype [45–47], and see also, e.g., [48–50] for examples of lysis inhibition as displayed by other phages. This issue of lysis inhibition was a constraint also on our own work, studying phage broth performance using an optical density approach, which limited what phages we were able to analyze [51].

Optical-density-based in vitro modeling approaches, to a degree, build on earlier work where in vitro determined phage population growth rates, presumably a key measure of active treatment effectiveness, were found to correlate with in vivo phage therapy efficacy [52,53]. Those efforts involve measuring increases in phage prevalence over time, an approach that should in fact mostly not be impacted by lysis inhibition, and this is instead of the above-noted optical-density measures of decreases in bacterial prevalence. To a degree, though, contrast those correlations between in vitro phage growth rates and in vivo efficacy with the in vivo observations of Bull et al. [54], which clearly indicate that in vivo phage population growth is not necessarily always a robust predictor of phage antibacterial effectiveness. Indeed, phage population growth rates presumably are far more relevant toward active treatment efficacy than toward instead passive treatment efficacy, since the latter, by definition, does not require phage population growth (i.e., as discussed at the start of Section 2.6). Nonetheless, determining rates of phage population growth is simpler to accomplish than determining separately phage latent periods, burst sizes, and adsorption rates. (For protocols determining the latter, see [55] along with adsorption rate-determination citations found in [15].) Bacterial turbidity measurements are, in turn, somewhat less labor intensive to obtain than phage population growth rates, especially given the use of kinetic microplate readers vs. plaque-based measures of changes in phage titers over time.

A limitation for all of these approaches, including the mathematical modeling emphasized here, is that resulting predictions of subsequent phage performance during therapies will only be as useful as experimental in vitro environments are representative of subsequent in situ conditions [56–58]. Nonetheless, the goal with all of these methods is to gain a more robust appreciation of what at least might be achievable by a given phage during an actual treatment rather than choosing phages for phage therapy based solely on more simplistic measures of host range such as just spotting or just plaquing abilities [43,59,60], or instead relying upon genome sequence-based methods for phage host-range determination, the latter as currently are under development [61,62].

In any case, when models of phage treatments—whether in vitro, in silico, or indeed in vivo—indicate a lower likelihood of phage therapy success, then that should bode less well, ultimately, for treatment effectiveness than if these models instead suggest a higher potential for attainment of antibacterial efficacy.

3. Discussion

The strengths of the various approaches provided here stem not just from their simplicity but also from their bases in mechanistic modeling. Specifically, a phage life cycle consists of virion infection and release which is then followed by virion movement and then adsorption. All of these processes are well studied, including at the whole-organism levels that are considered here. That is, phage populations adsorb following well-studied models of exponential decline in free phage numbers, models which were first published on in the early 1930s [63,64]. The duration of phage infections and resulting burst sizes were first studied quantitatively later in that same decade, now over 80 years ago [65], and the use of Poisson distributions to describe phage adsorptions has been around for almost as long [16]. Models of the type, as utilized here under the heading of "Active treatments" (Section 2.6), can be traced at least to Campbell in 1961 [38] and have been exploited extensively by Levin and colleagues [66–69], the latter starting in 1977 [35] (see also, e.g., [21,27,28,56,70–83]). An important goal of those studies is the use of models and parameter values that allow some predictive power despite the complexity of the phage–bacteria community dynamics that these studies have sought to emulate.

I have been especially involved in analysis of the earliest work of Bohannon and Lenski [36]. They used *Escherichia coli* B and bacteriophage T4 in a minimal-medium chemostat experiment, which they followed for 200 h prior to the takeover of their cultures by phage-resistant bacteria. Notably, their model provided predictions that were more qualitative than quantitative, that is, consistent with trends but less consistent with actual phage titers and bacterial concentrations, even prior to phage-resistant bacteria coming to dominate populations. Particularly since the phage they employed (T4) displays the complicating phenotype of lysis inhibition [44] (see also Section 2.6.4), I sought to improve the quantitative predictive power of their model [84]. The result of a number of modifications (Appendix A) was a substantial increase in predictive power through the first 100 h of their actual chemostat. (See also Abedon et al. [37], their Figure 2, for explicit validation of the ability of models such as those presented here to predict phage population growth rates again in vitro. See also Figure 1 of Weld et al. [71].)

These comparisons between experiments and models, in combination with the long history of study of these sorts of mechanistic phage–bacteria community dynamics modeling, or simply of phage population dynamics, suggests that though the approaches provided in this article may not be 100% predictive, they are likely as close in their predictive power as the precision that phage therapy experiments themselves will tend to be monitored. This, though, comes with the caveat that modeling outputs are only as good as modeling inputs, meaning that knowledge of actual phage adsorption rates as well as latent periods and burst sizes for active treatments can be relevant to predicting phage therapy outcomes. Alternatively, failures of models to accurately predict outcomes can be suggestive of a less-than-ideal appreciation of the magnitudes of those phage growth parameters in situ.

4. Conclusions

Though I suggest that modeling can have a place in gaining a better understanding of the pharmacology of phage therapy, more sophisticated phage therapy models [77,79,82,83] may be less accessible to the typical phage therapy practitioner or less useful in terms of application to novel circumstances. Alternatively, and at the other extreme, dismissing mathematical descriptions of phage treatments altogether seems as though it can, if my reading of the phage therapy literature is any indication, result in reduced understanding of phage treatments and their outcomes than should otherwise be possible. Explicitly, I typically employ simple mathematical constructs to better understand the underlying pharmacology, particularly pharmacodynamics, of published phage-bacterial interactions, e.g., [85]. It is my feeling that such applications might be as useful *prior to* the publication of phage therapy studies as they can be to me when analyzing studies *following* their publication, hence the emphasis of this review.

A different consideration is the utility of these various mathematical approaches to clinical phage therapy. My suspicion, in fact, is that in explicit terms, this math may be less useful than can be the case for preclinical studies, if only because there is less opportunity to make the detailed measurements that many of these models require, e.g., such as of bacterial concentrations, phage titers in association with targeted bacteria, phage adsorption rate constants, phage burst sizes, etc. These are all as found in situ while treating infections caused by what are typically somewhat uncharacterized bacterial strains and, in many cases, also in combination with antibiotics [41,57,86–89], which can have antagonistic impacts on phage infection abilities [41,51,85,90]. In particular for the latter, note that of 18 clinical phage therapy studies that I was able to obtain—published in 2023 or, at the time of writing, which are published but still online ahead of print—at least 16 indicate treatments using phages in combination with antibiotics [57,91–108]. See also [109], where 79 of the 114 clinical phage treatments reported "were administered in combination with standard-of-care antibiotics".

Notwithstanding the greater modeling imprecision which inevitably results when transitioning from pre-clinical studies to real-world phage therapy implementation, it is unlikely to be productive for clinical phage therapists to be unaware of the various presented models and especially their outputs. Thus, my suggestion, effectively for all phage therapists—whether or not they choose to explicitly apply these methods to specific phage therapies—is nonetheless to "play" with these models. That is, to run the described online calculators (Table 1) using different input values, e.g., by varying in situ phage titers or targeted bacterial concentrations, and to do so simply to gain an appreciation for how changing treatment approaches or conditions might impact treatment effectiveness. The goal should be to gain greater understanding especially of what phage doses can be more or less likely to result in sought phage treatment efficacies.

Funding: Funding was provided by Public Health Service grants R21AI156304 (Stephen T. Abedon, PI) and R01AI169865 (Daniel Wozniak, PI; Stephen T. Abedon, co-investigator).

Institutional Review Board Statement: Not applicable.

Informed Consent Statement: Not applicable.

Data Availability Statement: Not applicable.

Acknowledgments: Thank you to Aleksandra Petrovic Fabijan for her invitation to me to speak at the 2022 ESCMID Education Course, which served as the inspiration for writing this piece.

Conflicts of Interest: S.T.A. has consulted for and served on advisory boards for companies with phage therapy interests, has held equity stakes in a number of these companies, and maintains the websites phage.org and phage-therapy.org. No additional competing financial interests exist. The text presented represents the perspectives of the author alone, and no outside help was received in its writing.

Appendix A. Improving the Realism of Phage–Bacteria Chemostat Modeling

In this appendix, I note the key modifications that I made [84] to the Bohannon and Lenski [36] model toward greater predictive power, both for the sake of improved future modeling realism and to indicate where these modifications could be substantive. A first key modification was to assume that the Bohannon and Lenski chemostat was initiated with stationary phase rather than using log phase bacteria, thus requiring bacteria to go through an initial lag phase prior to their start of exponential growth. This is unlikely to impact the approaches presented here (Section 2.6), unless other chemostats that have been initiated with stationary phase bacteria are being modeled.

The second key modification was to assume that the robustness of both phage and bacterial population growth, when occurring, was less than had been determined initially, outside of chemostats. This is highly relevant to the approaches presented here to the extent that laboratory-determined phage growth parameters—adsorption rates, latent period, and burst size—might be optimistic relative to real-world circumstances. In other words, if you are predicting, using the various calculations presented here, that your treatments may just barely work in terms of sufficiently removing targeted bacteria, then in actuality, those treatments simply may not work. It is always important with phages, however, to recognize that their ability to replicate in situ can cover up what would otherwise be dosing insufficiencies, though this is without guarantee.

A last point has to do instead with the robustness of phage–virion survival. Based on the analysis of a number of published phage-containing chemostats beyond just that of Bohannon and Lenski [36] I found in many cases that rates of free phage decay could be greater than rates of chemostat outflow would suggest. In other words, something otherwise unaccounted for appeared to be either inactivating or removing phage virions in or from these various chemostats. I thus included a bacterium-independent free-phage decay parameter not only in the chemostat modeling found in Abedon [84] but also in the batch-culture phage–bacteria population dynamics model described in Equation (23). Similarly, and quite relevant to phage therapy, in many cases phage titers may decline in situ faster than simply free phage adsorption to bacteria would suggest. This likely would be beyond just considerations of the familiar (to phage therapists) expected declines in phage numbers over time during circulation in blood [110,111].

References

1. Górski, A.; Międzybrodzki, R.; Borysowski, J. *Phage Therapy: A Practical Approach*; Springer Nature Switzerland AG: Cham, Switzerland, 2019.
2. Pirnay, J.-P.; Ferry, T.; Resch, G. Recent progress toward the implementation of phage therapy in Western medicine. *FEMS Microbiol. Rev.* **2022**, *46*, fuab040. [CrossRef] [PubMed]
3. Baral, B. Phages against killer superbugs: An enticing strategy against antibiotics-resistant pathogens. *Front. Pharmacol.* **2023**, *14*, 1036051. [CrossRef]
4. Cahan, E. As superbugs flourish, bacteriophage therapy recaptures researchers' interest. *J. Am. Med. Assoc.* **2023**, *329*, 781–784. [CrossRef] [PubMed]
5. Diallo, K.; Dublanchet, A. A century of clinical use of phages: A literature review. *Antibiotics* **2023**, *12*, 751. [CrossRef] [PubMed]
6. Hitchcock, N.M.; Devequi Gomes, N.D.; Shiach, J.; Valeria Saraiva, H.K.; Dantas Viana, B.J.; Alencar Pereira, R.L.; Coler, B.S.; Botelho Pereira, S.M.; Badaro, R. Current clinical landscape and global potential of bacteriophage therapy. *Viruses* **2023**, *15*, 1020. [CrossRef]
7. Jones, J.D.; Trippett, C.; Suleman, M.; Clokie, M.R.J.; Clark, J.R. The future of clinical phage therapy in the United Kingdom. *Viruses* **2023**, *15*, 721. [CrossRef] [PubMed]
8. Petrovic Fabijan, A.; Iredell, J.; Danis-Wlodarczyk, K.; Kebriaei, R.; Abedon, S.T. Translating phage therapy into the clinic: Recent accomplishments but continuing challenges. *PLoS Biol.* **2023**, *21*, e3002119. [CrossRef] [PubMed]
9. Strathdee, S.A.; Hatfull, G.F.; Mutalik, V.K.; Schooley, R.T. Phage therapy: From biological mechanisms to future directions. *Cell* **2023**, *186*, 17–31. [CrossRef] [PubMed]
10. García-Cruz, J.C.; Huelgas-Méndez, D.; Jiménez-Zúñiga, J.S.; Rebollar-Juárez, X.; Hernández-Garnica, M.; Fernández-Presas, A.M.; Husain, F.M.; Alenazy, R.; Alqasmi, M.; Albalawi, T.; et al. Myriad applications of bacteriophages beyond phage therapy. *PeerJ* **2023**, *11*, e15272. [CrossRef] [PubMed]
11. Harper, D.R. Biological control by microorganisms. In *eLS*; John Wiley & Sons, Ltd: Chichester, UK, 2013. [CrossRef]
12. Abedon, S.T. Further considerations on how to improve phage therapy experimentation, practice, and reporting: Pharmacodynamics perspectives. *Phage* **2022**, *3*, 98–111. [CrossRef]
13. Stopar, D.; Abedon, S.T. Modeling bacteriophage population growth. In *Bacteriophage Ecology*; Abedon, S.T., Ed.; Cambridge University Press: Cambridge, UK, 2008; pp. 389–414.
14. Abedon, S. Phage therapy pharmacology: Calculating phage dosing. *Adv. Appl. Microbiol.* **2011**, *77*, 1–40. [PubMed]
15. Abedon, S.T. Bacteriophage adsorption: Likelihood of virion encounter with bacteria and other factors affecting rates. *Antibiotics* **2023**, *12*, 723. [CrossRef]
16. Dulbecco, R. Appendix: On the reliability of the Poisson distribution as a distribution of the number of phage particles infecting individual bacteria in a population. *Genetics* **1949**, *34*, 122–125.
17. Abedon, S.T. Phage therapy dosing: The problem(s) with multiplicity of infection (MOI). *Bacteriophage* **2016**, *6*, e1220348. [CrossRef] [PubMed]
18. Kasman, L.M.; Kasman, A.; Westwater, C.; Dolan, J.; Schmidt, M.G.; Norris, J.S. Overcoming the phage replication threshold: A mathematical model with implications for phage therapy. *J. Virol.* **2002**, *76*, 5557–5564. [CrossRef]
19. Benzer, S.; Hudson, W.; Weidel, W.; Delbrück, M.; Stent, G.S.; Weigle, J.J.; Dulbecco, R.; Watson, J.D.; Wollman, E.L. A syllabus on procedures, facts, and interpretations in phage. In *Viruses 1950*; Delbrück, M., Ed.; California Institute of Technology: Pasadena, CA, USA, 1950; pp. 100–147.
20. Stent, G.S. *Molecular Biology of Bacterial Viruses*; WH Freeman and Co.: San Francisco, CA, USA, 1963.
21. Bull, J.J.; Regoes, R.R. Pharmacodynamics of non-replicating viruses, bacteriocins and lysins. *Proc. R. Soc. Lond. B Biol. Sci.* **2006**, *273*, 2703–2712. [CrossRef]
22. Abedon, S.T. Ecology and evolutionary biology of hindering phage therapy: The phage tolerance vs. phage resistance of bacterial biofilms. *Antibiotics* **2023**, *12*, 245. [CrossRef]

23. Gallet, R.; Lenormand, T.; Wang, I.N. Phenotypic stochasticity protects lytic bacteriophage populations from extinction during the bacterial stationary phase. *Evolution* **2012**, *66*, 3485–3494. [CrossRef]

24. Storms, Z.J.; Sauvageau, D. Evidence that the heterogeneity of a T4 population is the result of heritable traits. *PLoS ONE* **2014**, *9*, e116235. [CrossRef] [PubMed]

25. Storms, Z.J.; Smith, L.; Sauvageau, D.; Cooper, D.G. Modeling bacteriophage attachment using adsorption efficiency. *Biochem. Eng. J.* **2012**, *64*, 22–29. [CrossRef]

26. Storms, Z.J.; Arsenault, E.; Sauvageau, D.; Cooper, D.G. Bacteriophage adsorption efficiency and its effect on amplification. *Bioprocess. Biosyst. Eng* **2010**, *33*, 823–831. [CrossRef]

27. Payne, R.J.H.; Jansen, V.A.A. Phage therapy: The peculiar kinetics of self-replicating pharmaceuticals. *Clin. Pharmacol. Ther.* **2000**, *68*, 225–230. [CrossRef] [PubMed]

28. Payne, R.J.H.; Jansen, V.A.A. Understanding bacteriophage therapy as a density-dependent kinetic process. *J. Theor. Biol.* **2001**, *208*, 37–48. [CrossRef] [PubMed]

29. Payne, R.J.H.; Jansen, V.A.A. Pharmacokinetic principles of bacteriophage therapy. *Clin. Pharmacokinet.* **2003**, *42*, 315–325. [CrossRef]

30. Nang, S.C.; Lin, Y.W.; Petrovic Fabijan, A.; Chang, R.Y.K.; Rao, G.G.; Iredell, J.; Chan, H.K.; Li, J. Pharmacokinetics/pharmacodynamics of phage therapy: A major hurdle to clinical translation. *Clin. Microbiol. Infect.* **2023**, *29*, 702–709. [CrossRef] [PubMed]

31. Hagens, S.; Habel, A.; von Ahsen, U.; von Gabain, A.; Bläsi, U. Therapy of experimental *Pseudomonas* infections with a nonreplicating genetically modified phage. *Antimicrob. Agents Chemother.* **2004**, *48*, 3817–3822. [CrossRef] [PubMed]

32. Abedon, S.T. Active bacteriophage biocontrol and therapy on sub-millimeter scales towards removal of unwanted bacteria from foods and microbiomes. *AIMS Microbiol.* **2017**, *3*, 649–688. [CrossRef] [PubMed]

33. Abedon, S.T. Spatial vulnerability: Bacterial arrangements, microcolonies, and biofilms as responses to low rather than high phage densities. *Viruses* **2012**, *4*, 663–687. [CrossRef]

34. Eriksen, R.S.; Mitarai, N.; Sneppen, K. On phage adsorption to bacterial chains. *Biophys. J.* **2020**, *119*, 1896–1904. [CrossRef]

35. Levin, B.R.; Stewart, F.M.; Chao, L. Resource limited growth, competition, and predation: A model and experimental studies with bacteria and bacteriophage. *Am. Nat.* **1977**, *111*, 3–24. [CrossRef]

36. Bohannan, B.J.M.; Lenski, R.E. Effect of resource enrichment on a chemostat community of bacteria and bacteriophage. *Ecology* **1997**, *78*, 2303–2315. [CrossRef]

37. Abedon, S.T.; Herschler, T.D.; Stopar, D. Bacteriophage latent-period evolution as a response to resource availability. *Appl. Environ. Microbiol.* **2001**, *67*, 4233–4241. [CrossRef] [PubMed]

38. Campbell, A. Conditions for the existence of bacteriophages. *Evolution* **1961**, *15*, 153–165. [CrossRef]

39. Storms, Z.J.; Teel, M.R.; Mercurio, K.; Sauvageau, D. The virulence index: A metric for quantitative analysis of phage virulence. *Phage* **2020**, *1*, 27–36. [CrossRef]

40. Konopacki, M.; Grygorcewicz, B.; Dołęgowska, B.; Kordas, M.; Rakoczy, R. PhageScore: A simple method for comparative evaluation of bacteriophages lytic activity. *Biochem. Eng. J.* **2020**, *161*, 107652. [CrossRef]

41. Grygorcewicz, B.; Roszak, M.; Rakoczy, R.; Augustyniak, A.; Konopacki, M.; Jabłońska, J.; Serwin, N.; Cecerska-Heryć, E.; Kordas, M.; Galant, K.; et al. PhageScore-based analysis of Acinetobacter baumannii infecting phages antibiotic interaction in liquid medium. *Arch. Microbiol.* **2022**, *204*, 421. [CrossRef] [PubMed]

42. Gelman, D.; Yerushalmy, O.; Alkalay-Oren, S.; Rakov, C.; Ben-Porat, S.; Khalifa, L.; Adler, K.; Abdalrhman, M.; Coppenhagen-Glazer, S.; Aslam, S.; et al. Clinical phage microbiology: A suggested framework and recommendations for the in-vitro matching steps of phage therapy. *Lancet Microbe* **2021**, *2*, e555–e563. [CrossRef]

43. Glonti, T.; Pirnay, J.-P. In vitro techniques and measurements of phage characteristics that are important for phage therapy success. *Viruses* **2022**, *14*, 1490. [CrossRef] [PubMed]

44. Abedon, S.T. Look who's talking: T-even phage lysis inhibition, the granddaddy of virus-virus intercellular communication research. *Viruses* **2019**, *11*, 951. [CrossRef]

45. Hershey, A.D. Mutation of bacteriophage with respect to type of plaque. *Genetics* **1946**, *31*, 620–640. [CrossRef]

46. Doermann, A.H. Lysis and lysis inhibition with Escherichia coli bacteriophage. *J. Bacteriol.* **1948**, *55*, 257–275. [CrossRef] [PubMed]

47. Moussa, S.H.; Kuznetsov, V.; Tran, T.A.; Sacchettini, J.C.; Young, R. Protein determinants of phage T4 lysis inhibition. *Protein Sci.* **2012**, *21*, 571–582. [CrossRef]

48. Gromkoua, R.H. T-related bacteriophage isolated from *Shigella sonnei*. *J. Virol.* **1968**, *2*, 692–694. [CrossRef] [PubMed]

49. Schito, G.C. Dvelopment of coliphage N4: Ultrastructural studies. *J. Virol.* **1974**, *13*, 186–196. [CrossRef] [PubMed]

50. Hays, S.G.; Seed, K.D. Dominant *Vibrio cholerae* phage exhibits lysis inhibition sensitive to disruption by a defensive phage satellite. *eLife* **2020**, *9*, e53200. [CrossRef]

51. Danis-Wlodarczyk, K.M.; Cai, A.; Chen, A.; Gittrich, M.R.; Sullivan, M.B.; Wozniak, D.J.; Abedon, S.T. Friends or foes? Rapid determination of dissimilar colistin and ciprofloxacin antagonism of *Pseudomonas aeruginosa* phages. *Pharmaceuticals* **2021**, *14*, 1162. [CrossRef]

52. Henry, M.; Lavigne, R.; Debarbieux, L. Predicting *in vivo* efficacy of therapeutic bacteriophages used to treat pulmonary infections. *Antimicrob. Agents Chemother.* **2013**, *57*, 5961–5968. [CrossRef]

53. Lindberg, H.M.; McKean, K.A.; Wang, I.-N. Phage fitness may help predict phage therapy efficacy. *Bacteriophage* **2014**, *4*, e964081. [CrossRef]
54. Bull, J.J.; Otto, G.; Molineux, I.J. *In vivo* growth rates are poorly correlated with phage therapy success in a mouse infection model. *Antimicrob. Agents Chemother.* **2012**, *56*, 949–954. [CrossRef]
55. Kropinski, A.M. Practical advice on the one-step growth curve. *Methods Mol. Biol.* **2018**, *1681*, 41–47.
56. Bull, J.J.; Gill, J.J. The habits of highly effective phages: Population dynamics as a framework for identifying therapeutic phages. *Front. Microbiol.* **2014**, *5*, 618. [CrossRef] [PubMed]
57. Fungo, G.B.N.; Uy, J.C.W.; Porciuncula, K.L.J.; Candelario, C.M.A.; Chua, D.P.S.; Gutierrez, T.A.D.; Clokie, M.R.J.; Papa, D.M.D. "Two is better than one": The multifactorial nature of phage-antibiotic combinatorial treatments against ESKAPE-induced infections. *Phage* **2023**, *4*, 55–67. [CrossRef]
58. Marchi, J.; Zborowsky, S.; Debarbieux, L.; Weitz, J.S. The dynamic interplay of bacteriophage, bacteria and the mammalian host during phage therapy. *iScience* **2023**, *26*, 106004. [CrossRef] [PubMed]
59. Kutter, E. Phage host range and efficiency of plating. *Methods Mol. Biol.* **2009**, *501*, 141–149. [PubMed]
60. Khan Mirzaei, M.; Nilsson, A.S. Isolation of phages for phage therapy: A comparison of spot tests and efficiency of plating analyses for determination of host range and efficacy. *PLoS ONE* **2015**, *10*, e0118557. [CrossRef] [PubMed]
61. Aggarwal, S.; Dhall, A.; Patiyal, S.; Choudhury, S.; Arora, A.; Raghava, G.P.S. An ensemble method for prediction of phage-based therapy against bacterial infections. *Front. Microbiol.* **2023**, *14*, 1148579. [CrossRef]
62. Bajiya, N.; Dhall, A.; Aggarwal, S.; Raghava, G.P.S. Advances in the field of phage-based therapy with special emphasis on computational resources. *Brief. Bioinform.* **2023**, *24*, bbac574. [CrossRef]
63. Schlesinger, M. Ueber die Bindung des Bakteriophagen an homologe Bakterien. II. Quantitative, Untersuchungen über die Bindungsgeschwindigkeit und die Sättigung. Berechnung der Teilchengröße des Bakteriophagen aus deren Ergebnissen. [Adsorption of bacteriophages to homologous bacteria. II Quantitative investigations of adsorption velocity and saturation. Estimation of particle size of the bacteriophage]. *Z. Fur Hygenie Immunitiitsforschung* **1932**, *114*, 149–160.
64. Schlesinger, M. Adsorption of bacteriophages to homologous bacteria. II Quantitative investigations of adsorption velocity and saturation. Estimation of particle size of the bacteriophage [translation]. In *Bacterial Viruses*; Stent, G.S., Ed.; Little, Brown and Co.: Boston, MA, USA, 1960; pp. 26–36.
65. Ellis, E.L.; Delbrück, M. The growth of bacteriophage. *J. Gen. Physiol.* **1939**, *22*, 365–384. [CrossRef]
66. Levin, B.R.; Lenski, R.E. Coevolution in bacteria and their viruses and plasmids. In *Coevolution*; Futuyma, D.J., Slatkin, M., Eds.; Sinauer Associates, Inc.: Sunderland, MA, USA, 1983; pp. 99–127.
67. Levin, B.R.; Lenski, R.E. Bacteria and phage: A model system for the study of the ecology and co-evolution of hosts and parasites. In *Ecology and Genetics of Host-Parasite Interactions*; Rollinson, D., Anderson, R.M., Eds.; Academic Press: London, UK, 1985; pp. 227–242.
68. Levin, B.R.; Bull, J.J. Phage therapy revisited: The population biology of a bacterial infection and its treatment with bacteriophage and antibiotics. *Am. Nat.* **1996**, *147*, 881–898. [CrossRef]
69. Levin, B.R.; Bull, J.J. Population and evolutionary dynamics of phage therapy. *Nat. Rev. Microbiol.* **2004**, *2*, 166–173. [CrossRef] [PubMed]
70. Lenski, R.E. Dynamics of Interactions between Bacteria and Virulent Bacteriophage. In *Advances in Microbial Ecology*; Springer: Boston, MA, USA, 1988; Volume 10, pp. 1–44.
71. Weld, R.J.; Butts, C.; Heinemann, J.A. Models of phage growth and their applicability to phage therapy. *J. Theor. Biol.* **2004**, *227*, 1–11. [CrossRef] [PubMed]
72. Kysela, D.T.; Turner, P.E. Optimal bacteriophage mutation rates for phage therapy. *J. Theor. Biol.* **2007**, *249*, 411–421. [CrossRef] [PubMed]
73. Cairns, B.J.; Payne, R.J.H. Bacteriophage therapy and the mutant selection window. *Antimicrob. Agents Chemother.* **2008**, *52*, 4344–4350. [CrossRef] [PubMed]
74. Gill, J.J. Modeling of bacteriophage therapy. In *Bacteriophage Ecology*; Abedon, S.T., Ed.; Cambridge University Press: Cambridge, UK, 2008; pp. 439–464.
75. Cairns, B.J.; Timms, A.R.; Jansen, V.A.; Connerton, I.F.; Payne, R.J. Quantitative models of *in vitro* bacteriophage-host dynamics and their application to phage therapy. *PLoS Pathog.* **2009**, *5*, e1000253. [CrossRef]
76. Schmerer, M.; Molineux, I.J.; Bull, J.J. Synergy as a rationale for phage therapy using phage cocktails. *PeerJ* **2014**, *2*, e590. [CrossRef]
77. Leung, C.Y.J.; Weitz, J.S. Modeling the synergistic elimination of bacteria by phage and the innate immune system. *J. Theor. Biol.* **2017**, *429*, 241–252. [CrossRef]
78. Roach, D.R.; Leung, C.Y.; Henry, M.; Morello, E.; Singh, D.; Di Santo, J.P.; Weitz, J.S.; Debarbieux, L. Synergy between the host immune system and bacteriophage is essential for successful phage therapy against an acute respiratory pathogen. *Cell Host Microbe* **2017**, *22*, 38–47. [CrossRef]
79. Sinha, S.S.; Grewal, R.K.; Roy, S. Modeling bacteria–phage interactions and its implications for phage therapy. *Adv. Appl. Microbiol.* **2018**, *103*, 103–141.
80. Li, G.; Leung, C.Y.; Wardi, Y.; Debarbieux, L.; Weitz, J.S. Optimizing the timing and composition of therapeutic phage cocktails: A control-theoretic approach. *Bull. Math Biol.* **2020**, *82*, 75. [CrossRef]

81. Rodriguez-Gonzalez, R.A.; Leung, C.Y.; Chan, B.K.; Turner, P.E.; Weitz, J.S. Quantitative models of phage-antibiotic combination therapy. *mSystems* **2020**, *5*, e00756-19. [CrossRef] [PubMed]

82. Styles, K.M.; Brown, A.T.; Sagona, A.P. A review of using mathematical modeling to improve our understanding of bacteriophage, bacteria, and eukaryotic interactions. *Front. Microbiol.* **2021**, *12*, 724767. [CrossRef] [PubMed]

83. Kyaw, E.E.; Zheng, H.; Wang, J.; Hlaing, H.K. Stability analysis and persistence of a phage therapy model. *Math. Biosci. Eng.* **2021**, *18*, 5552–5572. [CrossRef]

84. Abedon, S.T. Deconstructing chemostats towards greater phage-modeling precision. In *Contemporary Trends in Bacteriophage Research*; Adams, H.T., Ed.; Nova Science Publishers: Hauppauge, NY, USA, 2009; pp. 249–283.

85. Abedon, S.T. Phage-antibiotic combination treatments: Antagonistic impacts of antibiotics on the pharmacodynamics of phage therapy? *Antibiotics* **2019**, *8*, 182. [CrossRef] [PubMed]

86. Abedon, S.T.; Danis-Wlodarczyk, K.; Alves, D.R. Phage therapy in the 21st Century: Is there modern, clinical evidence of phage-mediated clinical efficacy? *Pharmaceuticals* **2021**, *14*, 1157. [CrossRef]

87. Kebriaei, R.; Lehman, S.M.; Shah, R.M.; Stamper, K.C.; Kunz Coyne, A.J.; Holger, D.; El Ghali, A.; Rybak, M.J. Optimization of phage-antibiotic combinations against *Staphylococcus aureus* biofilms. *Microbiol. Spectr.* **2023**, *11*, e0491822. [CrossRef]

88. Osman, A.H.; Kotey, F.C.N.; Odoom, A.; Darkwah, S.; Yeboah, R.K.; Dayie, N.T.K.D.; Donkor, E.S. The potential of bacteriophage-antibiotic combination therapy in treating infections with multidrug-resistant bacteria. *Antibiotics* **2023**, *12*, 1329. [CrossRef]

89. Santamaría-Corral, G.; Senhaji-Kacha, A.; Broncano-Lavado, A.; Esteban, J.; García-Quintanilla, M. Bacteriophage-antibiotic combination therapy against *Pseudomonas aeruginosa*. *Antibiotics* **2023**, *12*, 1089. [CrossRef]

90. Pons, B.J.; Dimitriu, T.; Westra, E.R.; van Houte, S. Antibiotics that affect translation can antagonize phage infectivity by interfering with the deployment of counter-defenses. *Proc. Natl. Acad. Sci. USA* **2023**, *120*, e2216084120. [CrossRef]

91. Bhartiya, S.K.; Prasad, R.; Sharma, S.; Shukla, V.; Nath, G.; Kumar, R. Biological therapy on infected traumatic wounds: A case-control study. *Int. J. Low. Extrem. Wounds* **2022**. [CrossRef]

92. Blasco, L.; López-Hernández, I.; Rodríguez-Fernández, M.; Pérez-Florido, J.; Casimiro-Soriguer, C.S.; Djebara, S.; Merabishvili, M.; Pirnay, J.-P.; Rodríguez-Baño, J.; Tomás, M.; et al. Case report: Analysis of phage therapy failure in a patient with a *Pseudomonas aeruginosa* prosthetic vascular graft infection. *Front Med.* **2023**, *10*, 1199657. [CrossRef] [PubMed]

93. Bradley, J.S.; Hajama, H.; Akong, K.; Jordan, M.; Stout, D.; Rowe, R.S.; Conrad, D.J.; Hingtgen, S.; Segall, A.M. Bacteriophage therapy of multidrug-resistant *Achromobacter* in an 11-year-old boy with cystic fibrosis assessed by metagenome analysis. *Pediatr. Infect. Dis. J.* **2023**, *42*, 754–759. [CrossRef] [PubMed]

94. Cesta, N.; Pini, M.; Mulas, T.; Materazzi, A.; Ippolito, E.; Wagemans, J.; Kutateladze, M.; Fontana, C.; Sarmati, L.; Tavanti, A.; et al. Application of phage therapy in a case of a chronic hip-prosthetic joint infection due to *Pseudomonas aeruginosa*: An Italian real-life experience and *in vitro* analysis. *Open Forum Infect. Dis.* **2023**, *10*, ofad051. [CrossRef] [PubMed]

95. Dedrick, R.M.; Smith, B.E., Cristinziano, M.; Freeman, K.G.; Jacobs-Sera, D.; Belessis, Y.; Whitney, B.A.; Cohen, K.A.; Davidson, R.M.; Van, D.D.; et al. Phage therapy of *Mycobacterium Infections*: Compassionate-use of phages in twenty patients with drug-resistant mycobacterial disease. *Clin. Infect. Dis.* **2023**, *76*, 103–112. [CrossRef]

96. Doub, J.B.; Chan, B.; Johnson, A.J. Salphage: Salvage bacteriophage therapy for a chronic *Enterococcus faecalis* prosthetic joint infection. *IDCases* **2023**, *33*, e01854. [CrossRef]

97. Fedorov, E.; Samokhin, A.; Kozlova, Y.; Kretien, S.; Sheraliev, T.; Morozova, V.; Tikunova, N.; Kiselev, A.; Pavlov, V. Short-term outcomes of phage-antibiotic combination treatment in adult patients with periprosthetic hip joint infection. *Viruses* **2023**, *15*, 499. [CrossRef]

98. Gainey, A.B.; Daniels, R.; Burch, A.K.; Hawn, J.; Fackler, J.; Biswas, B.; Brownstein, M.J. Recurrent ESBL *Escherichia coli* urosepsis in a pediatric renal transplant patient treated with antibiotics and bacteriophage therapy. *Pediatr. Infect. Dis. J.* **2023**, *42*, 43–46. [CrossRef]

99. Green, S.I.; Clark, J.R.; Santos, H.H.; Weesner, K.E.; Salazar, K.C.; Aslam, S.; Campbell, W.J.; Doernberg, S.B.; Blodget, E.; Morris, M.I.; et al. A retrospective, observational study of 12 cases of expanded access customized phage therapy: Production, characteristics, and clinical outcomes. *Clin. Infect. Dis.* **2023**, ciad335. [CrossRef]

100. Hahn, A.; Sami, I.; Chaney, H.; Koumbourlis, A.C.; Del Valle, M.C.; Cochrane, C.; Chan, B.K.; Koff, J.L. Bacteriophage therapy for pan-drug-resistant *Pseudomonas aeruginosa* in two persons with cystic fibrosis. *J. Investig. Med. High Impact. Case. Rep.* **2023**, *11*, 23247096231188243. [CrossRef]

101. Haidar, G.; Chan, B.K.; Cho, S.T.; Hughes, K.K.; Nordstrom, H.R.; Wallace, N.R.; Stellfox, M.E.; Holland, M.; Kline, E.G.; Kozar, J.M.; et al. Phage therapy in a lung transplant recipient with cystic fibrosis infected with multidrug-resistant *Burkholderia multivorans*. *Transpl. Infect. Dis.* **2023**, *25*, e14041. [CrossRef]

102. Köhler, T.; Luscher, A.; Falconnet, L.; Resch, G.; McBride, R.; Mai, Q.A.; Simonin, J.L.; Chanson, M.; Maco, B.; Galiotto, R.; et al. Personalized aerosolised bacteriophage treatment of a chronic lung infection due to multidrug-resistant *Pseudomonas aeruginosa*. *Nat. Commun.* **2023**, *14*, 3629. [CrossRef] [PubMed]

103. Le, T.; Nang, S.C.; Zhao, J.; Yu, H.H.; Li, J.; Gill, J.J.; Liu, M.; Aslam, S. Therapeutic potential of intravenous phage as standalone therapy for recurrent drug resistant urinary tract infections. *Antimicrob. Agents Chemother.* **2023**, *67*, e0003723. [CrossRef] [PubMed]

104. Li, L.; Zhong, Q.; Zhao, Y.; Bao, J.; Liu, B.; Zhong, Z.; Wang, J.; Yang, L.; Zhang, T.; Cheng, M.; et al. First-in-human application of double-stranded RNA bacteriophage in the treatment of pulmonary *Pseudomonas aeruginosa* infection. *Microb. Biotechnol.* **2023**, *16*, 862–867. [CrossRef] [PubMed]

105. Onallah, H.; Hazan, R.; Nir-Paz, R.; Yerushalmy, O.; Rimon, A.; Braunstein, R.; Gelman, D.; Alkalay, S.; Abdalrhman, M.; Stuczynski, D.; et al. Compassionate use of bacteriophages for failed persistent infections during the first 5 years of the Israeli phage therapy center. *Open Forum Infect. Dis.* **2023**, *10*, ofad221. [CrossRef] [PubMed]

106. Racenis, K.; Lacis, J.; Rezevska, D.; Mukane, L.; Vilde, A.; Putnins, I.; Djebara, S.; Merabishvili, M.; Pirnay, J.-P.; Kalnina, M.; et al. Successful bacteriophage-antibiotic combination therapy against multidrug-resistant *Pseudomonas aeruginosa* left ventricular assist device driveline infection. *Viruses* **2023**, *15*, 1210. [CrossRef]

107. Samaee, H.R.; Eslami, G.; Rahimzadeh, G.; Saeedi, M.; Davoudi, B.A.; Asare-Addo, K.; Nokhodchi, A.; Roozbeh, F.; Moosazadeh, M.; Ghasemian, R.; et al. Inhalation phage therapy as a new approach to preventing secondary bacterial pneumonia in patients with moderate to severe COVID-19: A double-blind clinical trial study. *J. Drug Deliv. Sci. Technol.* **2023**, *84*, 104486. [CrossRef]

108. Young, M.J.; Hall, L.M.L.; Merabishvilli, M.; Pirnay, J.-P.; Clark, J.R.; Jones, J.D. Phage therapy for diabetic foot infection: A case series. *Clin. Ther.* **2023**, *45*, 797–801. [CrossRef]

109. Jean-Paul, P.; Sarah, D.; Griet, S.; Johann, G.; Christel, C.; Steven, D.S.; Tea, G.; An, S.; Emily, V.B.; Sabrina, G.; et al. Retrospective, observational analysis of the first one hundred consecutive cases of personalized bacteriophage therapy of difficult-to-treat infections facilitated by a Belgian consortium. *medRxiv* **2023**. [CrossRef]

110. Merril, C.R.; Biswas, B.; Carlton, R.; Jensen, N.C.; Creed, G.J.; Zullo, S.; Adhya, S. Long-circulating bacteriophage as antibacterial agents. *Proc. Natl. Acad. Sci. USA* **1996**, *93*, 3188–3192. [CrossRef]

111. Vitiello, C.L.; Merril, C.R.; Adhya, S. An amino acid substitution in a capsid protein enhances phage survival in mouse circulatory system more than a 1000-fold. *Virus Res.* **2005**, *114*, 101–103. [CrossRef]

Case Report

Local Treatment of Driveline Infection with Bacteriophages

Anja Püschel [1,*], Romy Skusa [1,2], Antonia Bollensdorf [1] and Justus Gross [1]

1 Klinik für Allgemein-, Viszeral-, Gefäß- und Transplantationschirurgie, Universitätsmedizin Rostock, 18057 Rostock, Germany
2 Institut für Medizinische Mikrobiologie, Virologie und Hygiene, Universitätsmedizin Rostock, 18057 Rostock, Germany
* Correspondence: anja.pueschel@med.uni-rostock.de

Abstract: Drive line infections (DLI) are common infectious complications after left ventricular assist devices (LVAD) implantation. In case of severe or persistent infections, when conservative management fails, the exchange of the total LVAD may become necessary. We present a case of successful treatment of DL infection with a combination of antibiotics, debridement and local bacteriophage treatment.

Keywords: bacteriophages; phage therapy; application of phages; antibiotic resistance; antibacterial therapy

Citation: Püschel, A.; Skusa, R.; Bollensdorf, A.; Gross, J. Local Treatment of Driveline Infection with Bacteriophages. *Antibiotics* **2022**, 11, 1310. https://doi.org/10.3390/antibiotics11101310

Academic Editors: Aneta Skaradzińska and Anna Nowaczek

Received: 31 August 2022
Accepted: 20 September 2022
Published: 27 September 2022

Publisher's Note: MDPI stays neutral with regard to jurisdictional claims in published maps and institutional affiliations.

1. Introduction

Left ventricular assist devices (LVAD) are either used as a bridging until heart transplantation or as a final therapy (destination therapy) for advanced heart failure.

With an increase in the number of cases and a longer duration of LVAD support, a new spectrum of long-term complications has opened up. In addition to coagulation disorders, infections are an increasing problem.

Infections occur in 18 to 59% of the cases and may be local, with pocket infection and drive line infection (DLI), or they may be systemic with involvement of the bloodstream and endocarditis. Among them, DLI is the most frequent, with a prevalence of 14–28% [1,2]; it is defined as an infection affecting the soft tissues around the driveline outlet, accompanied by redness, warmness and purulent discharge. Despite it being a confined infection, it has the potential to become systemic with serious consequences.

Medical therapy with broad-spectrum antibiotics is the cornerstone of every treatment. In case of severe infection or ineffectiveness of the primary antibiotic therapy, driveline unroofing and debridement may become necessary.

Reinfections occur frequently and do not seem to be prevented by the use of long-term antibiotics as bacteria embedded in the tissue around the driveline form a surface adherent biofilm leading to a 1000-fold greater tolerance to antibiotics [3–5]. Moreover, an increasing occurrence of antibiotic resistance, in particular multi-resistance of Gram-negative germs, poses a challenge for the treatment of implant-associated infections.

The last resource, if severe or persistent infections are present, is the exchange of the total LVAD—an intervention that is associated with high morbidity and mortality [6]. Therefore, less invasive approaches should conservative treatments fail are urgently needed to reduce long-term morbidity and mortality. In this context, bacteriophages and their bacteriolytic activities represent promising therapeutic options.

Bacteriophages are viruses that infect bacteria. They are ubiquitous and highly specific to their bacterial host but are 10 times more numerous, making them the most abundant life forms on Earth, with an estimated 10^{31} bacteriophages on the planet [7]. Already, a decade before the discovery of penicillin, bacteriophages were successfully used to treat bacterial infections. However, at least in the Western world, the initial success was short-lived. The newly emerging antibiotics led to phage treatment being mostly ignored with the exception of some countries of the former Soviet Union and Eastern Europe [8].

Despite longstanding experience in those countries, evidence for the therapeutic application of bacteriophages is scarce and limited to case studies, few preclinical studies and animal models.

2. Case Report

In August 2021, a 57-year-old male patient was admitted to the hospital with a local infection on the drive line insertion site. He had no fever, a leukocyte count of $7.65 \times 10/1$ and a serum C reactive protein of 13 mg/L. Blood cultures were negative.

The indication for LVAD implantation in 2018 was a dilative cardiomyopathy and end-stage heart failure. The patient had a history of prior DLI that was treated with antibiotics and surgical debridement. Risk factors for infection were obesity (BMI 36.3 kg/m^2) and diabetes mellitus. After admission, empiric antibiotic therapy with piperacillin/tazobactam (3×4.5 g/day) was initiated. At the site of infection, local wound swabs were taken. The results revealed a mixed driveline infection with *Proteus mirabilis* and *Staphylococcus aureus* that were both sensitive to antibiotic therapy (Table 1).

Table 1. Susceptibility profile of *Proteus mirabilis* and *Staphylococcus aureus* (S: sensitive, I: intermediate resistant, R: resistant).

Antibiotics	*Proteus mirabilis*	*Staphylococcus aureus*
Oxacillin	R	S
Ampicillin	S	S
Ampicillin/Sulbactam	S	S
Piperacillin/Tazobactam	S	S
Cefuroxime	I	S
Cefotaxime	S	S
Ceftazidime	S	S
Imipenem	I	S
Meropenem	S	S
Gentamicin	S	S
Tetracycline	R	S
Cotimoxazole	S	S
Erythromycin	R	S
Clindamycin	R	S
Vancomycin	R	S
Fosfomycin	S	S
Fusidic acid	R	S
Rifampin	R	S
Linezolid	R	S
Daptomycin	R	S
Tigecycline	R	S

Additionally, the patient underwent a (F18) Fluordeoxyglucose-PET-CT scan 90 min after injection of approximately 220 MBq F-18-FDG on a Sensation 16 Biograph PET/CT scanner (Siemens-Healthineers, Erlangen, Germany). A standardized F18-FDG PET/CT protocol was used including 6 h of fasting, blood glucose levels less than 150 mg/dL, diluted oral contrast (Telebrix, 300 mg; Guerbet, Sulzbach, Germany) and low-dose CT (26 mAs, 120 kV, 0.5 s per rotation, 5 mm slice thickness) from base of the skull to mid-thigh for attenuation correction. Semiquantitative analysis was performed using a circular region of interest (ROI) (diameter 1.5 cm) with TrueD software (Siemens Medical Solutions, Siemens, Germany) and was normalized for injected dose and patient's body weight. The scan revealed pathologically increased local metabolic activity of the driveline from the exit point to the entire surrounding subcutaneous adipose tissue up to the abdominal wall muscles (Figure 1). The infection was strictly isolated to the DL exit site without expansion to the pump.

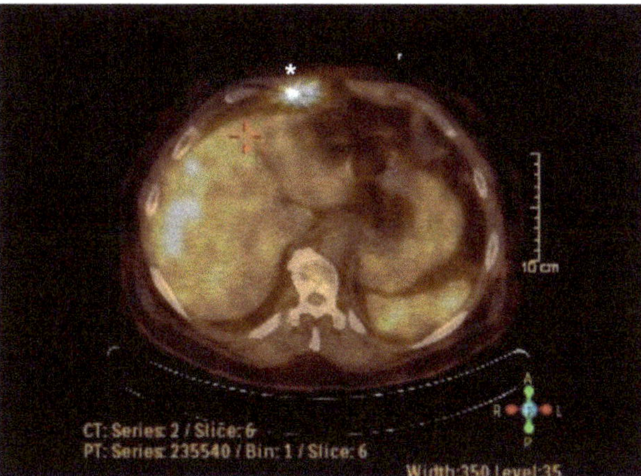

Figure 1. PET-CT scan of the upper abdomen before complex wound treatment. * PET positive driveline of the LVAD-System.

Since the patient had already been unsuccessfully treated surgically for a driveline infection, further local and systemic expansion was to be prevented as a chronic infection is a contraindication to a possible heart transplantation. Therefore, we decided to use an experimental approach with local bacteriophage therapy in addition to renewed surgical therapy.

Local bacteriophage application was planned according to Article 37 of the Declaration of Helsinki (to treat an individual patient for which there are no proven interventions or other known interventions are ineffective, the physician may use an unproven intervention with the patient's informed consent) in accordance with our ethics committee (A 2021-0132).

After obtaining further local swab specimens for microbiological analysis, local debridement of infected tissue, jet lavage with antiseptic (Lavanox, Serag-Wiesner, Naila, Germany) and driveline coating with Gore® Synecore were performed. For this purpose, the Synecore was cut into shape and put around the driveline. Then 20 mL SniPha 360 (1×10^7 CFU) (SniPha 360, Sanubiom GmbH, Fritzens, Austria, Phage 24.com) was diluted in saline and polysaccharide (Starsil, Hemostat Manufacturing GmbH, Velen Germany). The resulting viscous phage-containing fluid was applied between the driveline and the Synecore coating that was further secured by sutures (Figure 2). Additionally, a part of the bacteriophage galenic was applied to the subcutaneous tissue surrounding the driveline before the wound was closed.

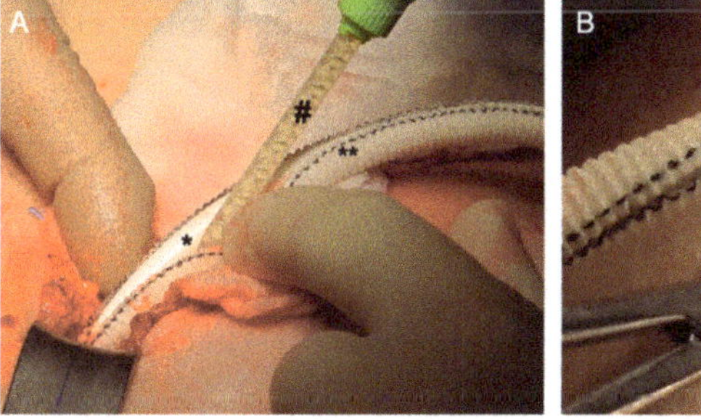
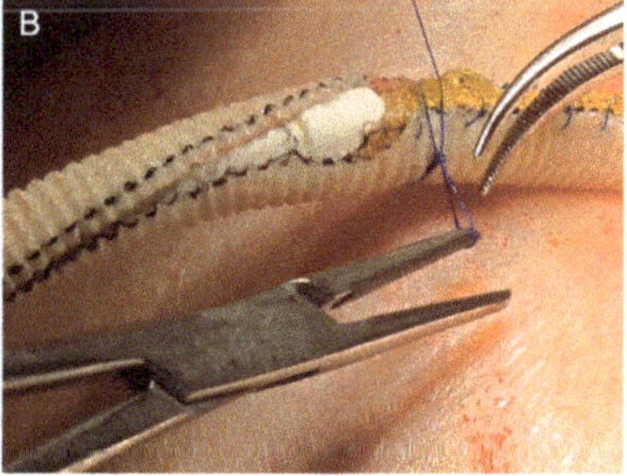

Figure 2. Intraoperative setup: (**A**) the viscous bacteriophage-rich galenic (#) was applied between the driveline (*) and the Gore® Synecore (**). (**B**) The Gore® Synecore was secured with sutures around the driveline before closing the wound.

SniPha 360 is a commercially available cocktail of lytic bacteriophages against *Escherichia coli*, *Staphylococcus aureus*, *Pseudomonas aeruginosa*, *Streptococcus pyogenes*, *Proteus vulgaris* and *Proteus mirabilis*.

Microbiological analysis of intraoperative samples confirmed infection with *Staphylococcus aureus* and *Proteus mirabilis*.

Susceptibility testing of the patient's bacterial isolates was performed by spot test. Following an overnight incubation in LB Medium at 37 °C, 200 µL of the bacteria strain suspension was added to the soft agar, mixed and immediately poured on the bottom agar plates. After the bacteria containing top agar solidified, 50 µL of phage suspension was randomly spotted onto the surface of the plates and allowed to dry. The inoculated plates were incubated overnight (18 h) at 37 °C under ambient atmosphere, followed by inspection for lysis zones. A spot test of SniPha 360 on the patient's strains showed no lysis zone on *S. aureus* and substantial turbidity throughout the cleared lysis zone on *P. mirabilis*.

Because of an uncomplicated postoperative wound healing, we decided against surgical revision and renewal of local phage therapy and continued the conservative treatment. Antibiotic therapy was switched to oral application of cotrimoxazole and the patient was discharged after 20 days with primary wound healing.

Two months later, he was readmitted with a mild local infection at the drive line exit. The microbiological analysis detected only *Staphylococcus aureus*, but not *Proteus mirabilis*. That suggests that the bacteriophages contributed to the treatment success and it is a good example for the correlation of in vitro testing and in vivo results. Long time calculated antibiotic therapy with flucloxacillin was initiated and the patient was discharged after 14 days.

In a follow-up examination 8 months later, no sign of a local or systemic infection was found.

3. Discussion

The present case demonstrates the multimodal treatment of an LVAD driveline infection with a combination of antibiotics, surgical debridement and local bacteriophage treatment.

DLI are common infectious complications after LVAD implantation. Besides host risk factors like immunosuppression, diabetes mellitus and obesity, other factors for recurrent or chronic infections are the duration of implants, poor tissue penetration by antibiotics, poor vascular supply, multidrug-resistant microbes and biofilm formation.

Biofilms associated with implanted medical devices represent a significant clinical problem, and often, the total removal of these implants is the only therapeutic option. In the case of LVAD infections, however, this is associated with high morbidity and mortality [6].

As an alternative, less invasive approach, we used a local bacteriophage application.

Bacteriophages, as a non-antibiotic technique for treating bacterial infections, have recently gained popularity. They have been used successfully in humans and other animal species [9,10]. Compared to antibiotics, bacteriophages have a completely different antibacterial mechanism. Phages cannot infect mammalian cells, have no cytotoxic effects on vascular cells and are highly specific to their respective bacterial hosts, thereby protecting the physiological host flora and reducing the risk of secondary infections [11,12].

Furthermore, bacteriophages have been shown to penetrate poorly vascularized tissues and to cross the blood–brain barrier [13]. Many Gram-negative and Gram-positive bacterial infections have been effectively treated in this manner either by local or systemic application [14,15]. Aslam et al. described systemic phage therapy for the first time as an adjuvant to antibiotics to treat left ventricular assist device infection. Since then, only a few case reports or small case studies have been published on the subject of DLI [9,16]. Additionally, the use of specialized and individualized phage mixtures has shown to be an alternative in the fight against multi-drug resistant bacteria as well as in persistent transplant or implant-related infections [11,16,17]. Moreover, in vitro studies have demonstrated that bacteriophages are able to disrupt certain biofilm matrices by exopolysaccharide degradation, bacterial cell infection and subsequent cell lysis [18]. Due to the increasing emergence of biofilm-associated infections such as DLI, there is a need for a therapeutic alternative to antibiotics that could be satisfied by phage therapy [19].

Despite these promising results, bacteriophage treatment is still not common and not officially recommended in the western hemisphere. There is a lack of real clinical trials–the known treatments rely mostly on case reports or small case studies (summarized in Plumet et al. and Aslam et al. [17,20]).

In the present case report, a recurrent local infection with biofilm-forming bacteria was treated with local bacteriophage application after unsuccessful surgical and antibiotic therapy.

Our results are somewhat mixed. Despite the broad host range of the bacteriophage cocktail we used, further bacteriophage in vitro testing revealed no lytic activity on the patient's *Staphylococcus aureus* strain. This correlated with our clinical observations of reinfection.

As bacteria have co-evolved over the last 3–4 billion years with phages, they developed a variety of mechanisms for preventing viral infections [21]. But with a wide phenotypic variability among phages, even closely related ones, Phage resistant bacteria remain susceptible to other phages of a similar target range [12,22]. Mixing several bacteriophages also helps to minimize the likelihood of bacteria acquiring resistance as well as synergistic antibiotic treatment or higher initial doses [12].

In the present case, however, using a commercially available phage cocktail, the exact composition of the phages and their concentration remain unknown to us. Since primary uncomplicated wound healing occurred, we decided against surgical revision and renewal of local phage therapy and continued the conservative treatment with empiric broad-spectrum antibiotics.

Antibiotics and bacteriophages can have additive or synergistic effects, as demonstrated in vitro and in vivo [22,23].

4. Conclusions

In summary, this case report describes the multimodal treatment of a driveline infection. The microbiological data suggest that bacteriophages contributed considerably to the treatment success, and it is also a good example of the correlation of in vitro bacteriophage testing and in vivo results. This case further demonstrates that "one fits all" bacteriophage cocktails, although readily available for immediate clinical use, do not render preceding in vitro testing indispensable.

Many questions, however, remain about the potential of bacteriophage therapy. Further studies are needed to prove and optimize safety, to focus on pharmacokinetics and pharmacodynamics and modes of application as well as on bacteriophage resistance and immune response.

Author Contributions: A.P. draft manuscript preparation; R.S., A.B. and J.G., review and editing. All authors have read and agreed to the published version of the manuscript.

Funding: This research received no external funding.

Institutional Review Board Statement: Local bacteriophage application was planned according to Article 37 of the Declaration of Helsinki in accordance with our ethics committee (A 2021-0132).

Informed Consent Statement: Informed consent was obtained from all subjects involved in the study.

Conflicts of Interest: The authors declare no conflict of interest.

References

1. Gordon, R.J.; Quagliarello, B.; Lowy, F.D. Ventricular assist device-related infections. *Lancet Infect. Dis.* **2006**, *6*, 426–437. [CrossRef]
2. Pereda, D.; Conte, J.V. Left Ventricular Assist Device Driveline Infections. *Cardiol. Clin.* **2011**, *29*, 515–527. [CrossRef] [PubMed]
3. Leuck, A.-M. Left ventricular assist device driveline infections: Recent advances and future goals. *J. Thorac. Dis.* **2015**, *7*, 2151–2157. [CrossRef]

4. Nienaber, J.J.C.; Kusne, S.; Riaz, T.; Walker, R.C.; Baddour, L.M.; Wright, A.J.; Park, S.J.; Vikram, H.R.; Keating, M.R.; Arabia, F.A.; et al. Clinical Manifestations and Management of Left Ventricular Assist Device-Associated Infections. *Clin. Infect. Dis.* **2013**, *57*, 1438–1448. [CrossRef]

5. Macia, M.; Rojo-Molinero, E.; Oliver, A. Antimicrobial susceptibility testing in biofilm-growing bacteria. *Clin. Microbiol. Infect.* **2014**, *20*, 981–990. [CrossRef]

6. Levy, D.; Guo, Y.; Simkins, J.; Puius, Y.; Muggia, V.; Goldstein, D.; D'Alessandro, D.; Minamoto, G. Left ventricular assist device exchange for persistent infection: A case series and review of the literature. *Transpl. Infect. Dis.* **2014**, *16*, 453–460. [CrossRef]

7. Comeau, A.M.; Hatfull, G.F.; Krisch, H.M.; Lindell, D.; Mann, N.H.; Prangishvili, D. Exploring the prokaryotic virosphere. *Res. Microbiol.* **2008**, *159*, 306–313. [CrossRef]

8. Hesse, S.; Adhya, S. Phage Therapy in the Twenty-First Century: Facing the Decline of the Antibiotic Era; Is It Finally Time for the Age of the Phage? *Annu. Rev. Microbiol.* **2019**, *73*, 155–174. [CrossRef] [PubMed]

9. Rubalskii, E.; Ruemke, S.; Salmoukas, C.; Boyle, E.C.; Warnecke, G.; Tudorache, I.; Shrestha, M.; Schmitto, J.D.; Martens, A.; Rojas, S.V.; et al. Bacteriophage Therapy for Critical Infections Related to Cardiothoracic Surgery. *Antibiotics* **2020**, *9*, 232. [CrossRef]

10. Greene, W.; Chan, B.; Bromage, E.; Grose, J.H.; Walsh, C.; Kortright, K.; Forrest, S.; Perry, G.; Byrd, L.; Stamper, M.A. The Use of Bacteriophages and Immunological Monitoring for the Treatment of a Case of Chronic Septicemic Cutaneous Ulcerative Disease in a Loggerhead Sea Turtle *Caretta caretta. J. Aquat. Anim. Health* **2021**, *33*, 139–154. [CrossRef]

11. Herten, M.; Idelevich, E.A.; Sielker, S.; Becker, K.; Scherzinger, A.S.; Osada, N.; Torsello, G.B.; Bisdas, T. Vascular Graft Impregnation with Antibiotics: The Influence of High Concentrations of Rifampin, Vancomycin, Daptomycin, and Bacteriophage Endolysin HY-133 on Viability of Vascular Cells. *Med Sci. Monit. Basic Res.* **2017**, *23*, 250–257. [CrossRef] [PubMed]

12. Sulakvelidze, A.; Alavidze, Z.; Morris, J.G., Jr. Bacteriophage Therapy. *Antimicrob. Agents Chemother.* **2001**, *45*, 649–659. [CrossRef] [PubMed]

13. Alisky, J.; Iczkowski, K.; Rapoport, A.; Troitsky, N. Bacteriophages show promise as antimicrobial agents. *J. Infect.* **1998**, *36*, 5–15. [CrossRef]

14. Kortright, K.E.; Chan, B.K.; Koff, J.L.; Turner, P.E. Phage Therapy: A Renewed Approach to Combat Antibiotic-Resistant Bacteria. *Cell Host Microbe* **2019**, *25*, 219–232. [CrossRef]

15. Luong, T.; Salabarria, A.-C.; Roach, D.R. Phage Therapy in the Resistance Era: Where Do We Stand and Where Are We Going? *Clin. Ther.* **2020**, *42*, 1659–1680. [CrossRef]

16. Aslam, S.; Pretorius, V.; Lehman, S.M.; Morales, S.; Schooley, R.T. Novel bacteriophage therapy for treatment of left ventricular assist device infection. *J. Hear. Lung Transplant.* **2019**, *38*, 475–476. [CrossRef]

17. Plumet, L.; Ahmad-Mansour, N.; Dunyach-Remy, C.; Kissa, K.; Sotto, A.; Lavigne, J.-P.; Costechareyre, D.; Molle, V. Bacteriophage Therapy for Staphylococcus Aureus Infections: A Review of Animal Models, Treatments, and Clinical Trials. *Front. Cell. Infect. Microbiol.* **2022**, *12*, 907314. [CrossRef]

18. Hughes, K.A.; Sutherland, I.W.; Jones, M.V. Biofilm susceptibility to bacteriophage attack: The role of phage-borne polysaccharide depolymerase. *Microbiology* **1998**, *144*, 3039–3047. [CrossRef]

19. Spellberg, B.; Gilbert, D.N. The Future of Antibiotics and Resistance: A Tribute to a Career of Leadership by John Bartlett. *Clin. Infect. Dis.* **2014**, *59* (Suppl. 2), S71–S75. [CrossRef]

20. Aslam, S.; Lampley, E.; Wooten, D.; Karris, M.; Benson, C.; Strathdee, S.; Schooley, R.T. Lessons Learned From the First 10 Consecutive Cases of Intravenous Bacteriophage Therapy to Treat Multidrug-Resistant Bacterial Infections at a Single Center in the United States. *Open Forum Infect. Dis.* **2020**, *7*, ofaa389. [CrossRef]

21. Seed, K.D. Battling Phages: How Bacteria Defend against Viral Attack. *PLOS Pathog.* **2015**, *11*, e1004847. [CrossRef]

22. Yilmaz, C.; Colak, M.; Yilmaz, B.C.; Ersoz, G.; Kutateladze, M.; Gozlugol, M. Bacteriophage Therapy in Implant-Related Infections. *J. Bone Jt. Surg.* **2013**, *95*, 117–125. [CrossRef] [PubMed]

23. Chaudhry, W.N.; Concepción-Acevedo, J.; Park, T.; Andleeb, S.; Bull, J.J.; Levin, B.R. Synergy and Order Effects of Antibiotics and Phages in Killing Pseudomonas aeruginosa Biofilms. *PLoS ONE* **2017**, *12*, e0168615. [CrossRef] [PubMed]

Review

Bacteriophage–Antibiotic Combination Therapy against *Pseudomonas aeruginosa*

Guillermo Santamaría-Corral [1], Abrar Senhaji-Kacha [1,2], Antonio Broncano-Lavado [1], Jaime Esteban [1,2,*] and Meritxell García-Quintanilla [1,2]

[1] Department of Clinical Microbiology, IIS-Fundación Jiménez Díaz, Av. Reyes Católicos 2, 28040 Madrid, Spain; guillermo.shcb@gmail.com (G.S.-C.); abrarsen7@gmail.com (A.S.-K.); antonio.broncano@gmail.com (A.B.-L.)
[2] CIBERINFEC—Infectious Diseases CIBER, 28029 Madrid, Spain
* Correspondence: jesteban@fjd.es

Abstract: Phage therapy is an alternative therapy that is being used as the last resource against infections caused by multidrug-resistant bacteria after the failure of standard treatments. *Pseudomonas aeruginosa* can cause pneumonia, septicemia, urinary tract, and surgery site infections mainly in immunocompromised people, although it can cause infections in many different patient profiles. Cystic fibrosis patients are particularly vulnerable. In vitro and in vivo studies of phage therapy against *P. aeruginosa* include both bacteriophages alone and combined with antibiotics. However, the former is the most promising strategy utilized in clinical infections. This review summarizes the recent studies of phage-antibiotic combinations, highlighting the synergistic effects of in vitro and in vivo experiments and successful treatments in patients.

Keywords: *Pseudomonas aeruginosa*; treatment; bacteriophage; combination

check for **updates**

Citation: Santamaría-Corral, G.; Senhaji-Kacha, A.; Broncano-Lavado, A.; Esteban, J.; García-Quintanilla, M. Bacteriophage–Antibiotic Combination Therapy against *Pseudomonas aeruginosa*. *Antibiotics* **2023**, *12*, 1089. https://doi.org/10.3390/antibiotics12071089

Academic Editors: Aneta Skaradzińska and Anna Nowaczek

Received: 4 April 2023
Revised: 19 June 2023
Accepted: 20 June 2023
Published: 22 June 2023

1. Introduction

Pseudomonas aeruginosa is a Gram-negative bacillus, widely distributed in the environment, that causes important infections, generally as an opportunistic pathogen [1–3]. This bacterium is associated with high morbidity and mortality in patients with underlying pathology, such as cystic fibrosis (CF), chronic obstructive pulmonary disease (COPD), and bronchiectasis, among other pulmonary diseases [2,3]. This opportunistic pathogen is one of the causal agents of ventilator-associated pneumonia (VAP), since it can colonize hospital equipment, saline solution, and soap. The treatment of this pathogen is very diverse depending on the pathology in which it is involved. Among the drugs that are most used against this bacterium are ceftazidime, amikacin, colistin, and meropenem. Monotherapy and dual therapy regimens are established based on the antibiotic susceptibility that is present and the type of pathology. The resistance mechanisms presented by this bacterium make it challenging to establishing an effective treatment for *P. aeruginosa* [4]. In recent decades there has been a notable increase in infections by this bacterium, especially by multidrug-resistant (MDR) and extensively drug-resistant (XDR) clones [5,6]. The continued use of antibiotics, the increase in the prevalence of chronic diseases (especially respiratory), and the use of immunosuppressants have resulted in an increase in infections by this pathogen and adverse consequences in terms of the morbidity and mortality of these patients [7].

MDR and XDR *P. aeruginosa* bacteria exhibit varied resistance mechanisms acquired by chromosomal mutations or by horizontal transmission of genetic material. Among these resistance mechanisms, we can find the production of β-lactamases of the AmpC type, in addition to the fact that the outer membrane of this bacterium has extremely low permeability; this confers innate resistance to many antimicrobials and, in turn, the membrane itself is capable of expressing porins and active expulsion systems such as MexAB-OprM, MexCD-OprJ, MexEF-OprN and MexXY-OprM [8,9]. These systems can affect antimicrobials such as beta-lactams, fluoroquinolones, and aminoglycosides. Regarding the mechanisms of horizontal transmission, this bacterium can encode different modifying enzymes that can act against many antimicrobials; among them, we can mention the production of β-lactamases of extended-spectrum (BLEE), carbapenemases, and aminoglycoside-modifying enzymes [10]. Other species of the genus *Pseudomonas* and also of the genus *Acinetobacter* act as environmental reservoirs of antimicrobial resistance systems of *P. aeruginosa* that may be encoded in integron cassettes and be transmitted to *P. aeruginosa* [3]. Regarding its virulence factors, *P. aeruginosa* presents a wide variety, which contribute to its pathogenicity. In its external membrane, there is the lipopolysaccharide LPS common with other Gram-negative bacilli with endotoxic activity [11] and several porins, such as the OprF, OprH, and OprD Superfamilies [12]. On the other hand, it is important to highlight their high capacity to form biofilms, which increase the resistance not only to antimicrobial but also to inhospitable environmental factors [13]. Among the mucoid substances they produce in their biofilms, the most studied is alginate [14]. Finally, other factors such as type IV pili, flagellum, and numerous secretion systems as well as secondary metabolites such as pyocyanin can be highlighted, the latter being responsible for aggravating pulmonary diseases due to its pro-inflammatory activity and oxidative damage mediated by the formation of oxygen free radicals [15]. All these virulence factors allow *P. aeruginosa* to persist in vulnerable patients with various chronic pathologies, worsening their prognosis and causing persistent infections or recurrences whose treatment is a clinical challenge [7].

All these factors contribute to the need to search for new therapeutic alternatives, among which phage therapy can be highlighted. Bacteriophages capable of infecting *P. aeruginosa* have been isolated since the mid-20th century, and this therapeutic tool, combined with antimicrobials, may be extremely useful in the treatment of unresolved *P. aeruginosa* infections.

Several studies have demonstrated the efficacy of bacteriophages against *P. aeruginosa*, especially in combination with antibiotics. In vitro studies show that the use of temperate phages (such as HK97) in combination with suboptimal concentrations of ciprofloxacin can drastically reduce the population of *P. aeruginosa*; these studies have been carried out using assays in agar plates as well as in microtiter plates, in which the bacteria were exposed against serial concentrations of antibiotics as well as different dilutions of phage [16]. In turn, lytic bacteriophages can serve as adjuvants in conjunction with antibiotics by reducing the MICs of antibiotics in such a way as to increase susceptibility to antibiotics that might previously have been unsuitable, although this phenomenon is highly dependent on the mechanisms of action of the antibiotics administered in conjunction with the phages. These assays are carried out in microtiter plates in which inocula with a determined concentration of bacteria were added and exposed to both the presence of bacteriophages and antibiotics, after which successive absorbance measurements were taken at 600 nm every 15 min for 24 h at 37 °C under shaking conditions [17].

The use of phages combined with antibiotics is the most realistic way of applying this therapy to patients to avoid the appearance of resistance and achieve greater therapeutic success. Phages have recently regained interest in the fight against antibiotic multi-resistance. They are safe, although their effectiveness is highly dependent on the strain against which they are applied [18]. There remains a lack of knowledge regarding the use of bacteriophages, including information on their influence on the immune response of patients as well as their production and processing for administration in; thus, their use is currently limited to clinical trials and compassionate use situations [18]. In order to be useful for clinical practice, in this paper we review the most relevant work of the last five years based on the combination of bacteriophages and antimicrobials against *P. aeruginosa* infections. We describe in vitro and in vivo studies as well as case reports.

2. In Vitro Models

Bacteriophage–antibiotic combination in vitro therapy against multidrug-resistant (MDR) *P. aeruginosa* has been demonstrated with almost all antibiotics commercially available: aminoglycosides (amikacin, gentamycin, streptomycin, and tobramycin), β-lactams (ceftazidime, ceftriaxone, meropenem, and piperacillin), fluoroquinolones (ciprofloxacin), fosfomycin, macrolides (erythromycin), polymyxins (colistin and polymyxin B) and tetracyclines. Furthermore, several researchers have tried to optimize a checkerboard method to determine bacteriophage–antibiotic interactions and to determine whether synergy can be obtained with both simultaneous and successive application of these antibacterial agents [19].

In vitro treatment with bacteriophages and antibiotics has been able to significantly increase susceptibility and re-sensitization of MDR *P. aeruginosa* strains to antibiotics [20–31]. All these studies were performed by conducting assays to determine the minimum inhibitory concentration (MIC) for antibiotics and bacteriophages according to the Clinical and Laboratory Standards Institute (CLSI) broth microdilution protocol and the fractional inhibitory concentrations (FIC) of antibiotics in the presence of phages using the checkerboard method [32]. However, this bacteriophage–antibiotic synergy can be inhibited by the competition between temperate and chronic viruses, as Landa et al. demonstrated [33]. Chronic viruses can trigger the production and release of new viruses by the host cell without killing it but also have a latent cycle in which its genetic material is embedded into the bacterium's genome [34,35]. Meanwhile, temperate viruses have a lytic cycle, in which virus production by bacteria bursts out of the host cell, and a latent cycle, where the virus remains inactive in the bacteria until induced to replicate. The authors modeled the synergy between antibiotics and these two viral types in controlling bacterial infections. While combinations of antibiotic and temperate viruses exhibited synergy, the combination of temperate and chronic viruses inhibited antibiotic control of bacteria. Antibiotics had the highest effect on the bacterial population infected with temperate viruses, since latent bacteria were induced by antibiotics into the lytic cycle. When chronic and temperate viruses were present in the absence of antibiotics, the temperate viruses could still lyse the bacteria. Otherwise, when the concentration of antibiotics was low, the presence of both viruses had a larger negative impact on the bacterial population than when only chronic viruses were present. Nevertheless, at higher concentrations of antibiotics the bacterial populations become equivalent to the effect when only chronic phages are present. The two populations converge because chronic viruses out-compete temperate viruses due to the stress-induced chronic virus production rate increase.

Ciprofloxacin has been the most utilized antibiotic combined with bacteriophages [36–39]. The impact of this antibiotic (as colistin) on the bactericidal, bacteriolytic, and new virion production of *P. aeruginosa* bacteriophages was assessed in a recent study by optical density-based "lysis profile" assays in the presence and absence of antibiotics [40]. Lysis profiles require the addition of bacteriophages at high bacterial densities, as the impact on the bacteria population is observed as a reduction of the turbidity of the bacterial culture. Colistin was shown to substantially interfere with bacteriolytic and virion-production activities; the bacteriophage utilizing LPS as its surface receptor could be a contributor to the observed antagonism of this bacteriophage infection activity by colistin, as LPS is directly disrupted by that antibiotic. In adsorption experiments, phage virion-attachment antagonism was observed in the presence of colistin (MIC). In contrast to the colistin results, negative impacts on lysis-profile kinetics are minimal with ciprofloxacin. Ciprofloxacin had no negative impact on phage adsorption rates even at high concentrations. These results suggest that ciprofloxacin could be useful as a concurrent phage therapy co-treatment, especially when phage replication is required for treatment success.

Bacteriophage PEV20 synergistic effects with ciprofloxacin has been proven in several studies [41–43] to enhance eradication of *P. aeruginosa* biofilm associated with cystic fibrosis and wound patients [43]. In addition, reducing the antibiotic concentration required to fight against these bacterial infections is associated with biofilms in these patients. These results were assessed by quantification of biofilm biomass, viability, and determination of minimum biofilm inhibitory concentration (MBIC). Furthermore, the antimicrobial effect of nebulized PEV20 with ciprofloxacin was determined against *P. aeruginosa* strains isolated from sputum CF patients by assessing bacterial killing and performing time-kill studies [42].

Conversely, a recent study demonstrated that combined bacteriophage and antibiotic pretreatment with ciprofloxacin and two phages (vB_PaeP_4024 and vB_PaeS_4069) prevents *P. aeruginosa* infection of wild type and CFTR epithelial cells and the emergence of bacteriophage-resistant mutants without inducing an inflammatory response, while administration of single bacteriophages, phage cocktails, or ciprofloxacin led to development of bacterial regrowth due to phage-resistant mutants [44]. In an innovative study, Ferran et al. simulated oral treatment with ciprofloxacin and phage-inhaled administration in *P. aeruginosa* respiratory infections [45]. Antibiotic in vitro treatment reproduced a maximum concentration of 1.5 µg/mL and a half-life of 4 h. Ciprofloxacin and bacteriophage single treatment generated resistant bacteria in less than 30 h. However, the combination of bacteriophages with ciprofloxacin was able to prevent the growth of resistant bacteria as simultaneous and delayed treatment. To assess the robustness of the combined treatment, the Hollow Fiber Infection Model (HFIM) was inoculated with a 1000-fold higher bacterial inoculum, while the regimen of either ciprofloxacin and phages at a Multiplicity of Infection (MOI) of 0.1 was the same. Simultaneous administration of combined treatment quickly decreased bacterial density below the limit of detection (LOD) but increased again after reducing susceptibility to ciprofloxacin (16- to 32-fold higher MIC) and bacteriophages compared to the naïve population. In contrast, in the delayed treatment, the initial reduction of bacteria was slower, with bacterial density falling below the LOD at 1 h for one replicate and 6 h for the other. However, after this decline, no increase in the bacterial density was observed, and gain no colony could be recovered on samples taken during the next 72 h. The authors concluded that when phages reduce the size of the bacterial population, the remaining population is not sufficient to include less-susceptible mutants to ciprofloxacin.

The synergistic action of bacteriophages and antibiotics has also been studied against dual-species biofilm, such as *P. aeruginosa–S. aureus* biofilm. Akturk et al. described the synergistic action of phages and antibiotics (ciprofloxacin, gentamicin, and meropenem) on 48 h *P. aeruginosa–S. aureus* biofilm when treated in simultaneous or sequential combination [46]. Phage or antibiotic single treatment developed a moderate effect on biofilm; however, when applied simultaneously, the effect was extensively improved. In addition, when gentamicin and ciprofloxacin were administered sequentially 6 h after phage treatment, a remarkable biofilm diminution was noticed, exhibiting even eradication of the biofilm. Furthermore, it was determined that the antibiofilm effect depends only on antibiotic concentration, not on its type: almost complete biofilm eradication was observed only when antibiotic concentration was higher or equal to MIC. Otherwise, achieving a similar gentamicin antibacterial effect on *P. aeruginosa–S. aureus* biofilm required increase of the antibiotic concentration: bacteriophage–gentamicin 8xMIC sequential administration nearly eradicated the *P. aeruginosa* population and was the most effective treatment on the *S. aureus* population.

Tkhilaishvili et al. demonstrated the potential use of combined bacteriophages Sb-1 and PYO with antibiotics for killing dual-species biofilm formed by *P. aeruginosa* and methicillin-resistant *S. aureus* (MRSA) [47]. They also investigated the effect of either simultaneous or staggered application of commercially available bacteriophages (Pyophage and Staphylococcal bacteriophage) and ciprofloxacin against dual-species biofilm in vitro. In this experiment, biofilms were formed in porous glass beads, and different techniques (microcalorimetry, sonication, and electron microscopy) were applied for assessing the anti-biofilm properties of treatments. Antibiotics tested alone against biofilms required high concentrations ranging from 256 to 512 µg/mL to show an inhibitory effect, whereas bacteriophage alone showed good and moderate activity against MRSA biofilms and dual-species biofilms, respectively, but low activity against *P. aeruginosa* biofilms. The combination of antibiotics and bacteriophages showed a remarkable improvement in the anti-biofilm activity of both antimicrobials with complete eradication of dual-species after staggered exposure to Pyophage or Pyophage+Staphylococcal phage for 12 h followed by 1 µg/mL of ciprofloxacin, a dose achievable by intravenous or oral antibiotic administration.

Lastly, Monahar et al. established the first approach to study the potential therapeutic approach of using bacteriophage–antibiotic combinations for treating infections caused by *P. aeruginosa* and *Candida albicans* [39]. Bacteriophage–fluconazole treatment was effective against 6-h-old dual-species biofilm, but not against 24-h-old biofilms. Likewise, the combination of antibiotics with the bacteriophage showed no synergistic effect on dual-biofilm.

3. In Vivo Models

The in vivo models described in the scientific literature testing the effect of combinations of bacteriophages and antibiotics are scarce (Table 1).

Table 1. Summary of in vivo studies using phage–antibiotic combination against MDR *P. aeruginosa*.

Infection Model	Bacteria	Phage Therapy	Antibiotic Combination	Outcome	Reference
Lung infection, mouse	*P. aeruginosa* MDR	PEV20 (10^6 PFU/mg)	Ciprofloxacin (0.33 mg)	Reduced bacterial load by 5.9 log	[45]
Acute immunocompromised, mouse	*P. aeruginosa* MDR	Three-phage cocktail (10^9 PFU/mL)	Alone or with Meropenem	Enhanced therapeutic protection against pulmonary infection	[43]
Cystic fibrosis zebrafish	*P. aeruginosa* (PA01)	Four-phage cocktail (300–500 PFU/embryo)	Ciprofloxacin (100 µL)	Reduced embryos lethality	[44]
Dorsal wound, mouse	*P. aeruginosa* (PA01)	PAM2H cocktail (10^8 PFU/mL)	Ceftazidime	Synergistic reduction in bacterial burden	[18]

Regarding the lung infection in vivo models, Lin et al. demonstrated the in vivo effect of an inhalable powder of co-spray drying *Pseudomonas aeruginosa* phage PEV20 with ciprofloxacin using a neutropenic model of acute lung infection [48]. Firstly, the clinical *P. aeruginosa* (resistant to ciprofloxacin, aztreonam, and amikacin) was sprayed directly into the trachea using a micro sprayer. The powders (1 mg) of single ciprofloxacin (0.33 mg), single PEV20 (10^6 PFU/mg), and the combination were aerosolized into the trachea of anesthetized mice using a dry powder insufflator. Intratracheally treatment with PEV20–ciprofloxacin combination powder significantly reduced the bacterial load in mice lungs by 5.9 $\log_{10}$, whereas single treatments with phage and antibiotics failed to reduce the burden. The efficacy was synergistic, as the observed killing effect for the combination powder was statistically higher than the additive effect of single treatments, with both showing nil effect at 24 h. Assessment of immunological responses in the lungs showed reduced inflammation associated with the bactericidal effect of PEV20–ciprofloxacin powder. This study represents the first proof-of-concept study demonstrating the synergistic efficacy of combined phage–antibiotic powder treatment in a mouse lung infection model.

In addition, Duplessis et al. described the IP administration of a three-bacteriophage cocktail with/without meropenem in an acute immunocompromised mouse model of MDR *P. aeruginosa* pulmonary infection [49]. Firstly, they assessed the potential therapeutic IP administration of the bacteriophage cocktail (10^9 PFU/mL) alone for 120 h, delayed relative to bacterial inoculation by 3 h. IP administration of phage cocktails did not protect mice from death. Lastly, they assessed if subcutaneous administration of meropenem at subinhibitory concentrations could enhance bacteriophage efficacy, delayed by 3 h relative to bacterial inoculation. The combined treatment of meropenem and phage significantly enhanced therapeutic protection against pulmonary infection and significantly reduced bacterial burden in the lungs and spleen. These data support that phage-administered IP can penetrate the pulmonary tissues and, in combination with a sub-efficacious dose of antibiotic, can slow bacterial proliferation but not protect against a lethal outcome.

Cafora et al. tested the effects of combining bacteriophage therapy (four-phage cocktail) and antibiotic treatment (ciprofloxacin) against *P. aeruginosa* infections in an innovative cystic fibrosis zebrafish model [50]. Zebrafish CFTR channels present a similar structure to human CFTR. Additionally, zebrafish CFTR knockdown presents susceptibility to *P. aeruginosa* infections. As bacteriophage therapy, the authors injected 300–500 PFU/embryo of phage cocktail into the yolk sac of CF+PAO1-infected embryos. In the case of antibiotic treatment, it was done by incubation of CF+PAO1 embryos in fish water containing 100 μL of ciprofloxacin. Antibiotic treatment reduced lethality in comparison to CF+PAO1 embryos. Interestingly, combined treatment with phages and ciprofloxacin enhanced the reduction of lethality compared to every single treatment.

Finally, Engeman et al. described the synergistic killing of *P. aeruginosa* by phage–antibiotic combination treatment in a mouse dorsal wound model [20]. Mice were wounded dorsally, infected with PA01::lux, and treated with a PAM2H bacteriophage cocktail (10^8 PFU/mL topically on the wound once a day and PBS intraperitoneally twice a day), ceftazidime (CAZ) (PBS topically on the wound once a day and CAZ intraperitoneally twice a day), or PAM2H and CAZ in combination (10^8 PFU/mL topically on the wound once a day and CAZ intraperitoneally twice a day). Treatment with PAM2H in combination with CAZ resulted in a synergistic reduction in bacterial burden in vivo. Reduced virulence was noticed in the bacteria recovered from post-treated mice wounds in a larvae model.

4. Case Reports

In the majority of clinical cases of *P. aeruginosa* infections, bacteriophages have not been administered as a single treatment they have been applied concomitantly with antibiotics as an adjuvant treatment. Hereafter, we outline a series of clinical cases in which compassionate use with bacteriophage or a cocktail of phages were administered (intravenously, locally, or nebulized) concomitantly with antibiotics as an adjuvant treatment against *P. aeruginosa* with clinical resolution of different infections, mainly chronic (Table 2).

Ferry et al. described several cases in which adjuvant bacteriophage therapy was necessary to treat *P. aeruginosa* infections. One of them was an 88-year-old male patient with prosthetic joint infection (PJI) of the knee caused by ceftazidime and ciprofloxacin susceptible to *P. aeruginosa* [51]. As conventional treatment with antibiotics (IV ceftazidime and oral ciprofloxacin) was not effective and prosthesis explantation or exchange was not suitable, phage therapy was established as an adjuvant treatment to try to control the infection. As bacteriophage therapy was utilized, three phages in a cocktail (10^9 PFU/mL) were administered through the arthroscope after conventional arthroscopy. After receiving bacteriophages and antibiotics, the patient rapidly showed signs of improvement.

They also described the case of a 74-year-old man with melanoma who experienced catheter-related bacteremia due to multidrug-resistant *P. aeruginosa* in December 2017, treated successfully with colistin and meropenem [52]. He was diagnosed with a spinal abscess in December 2018, and the aspiration revealed a pandrug-resistant *P. aeruginosa*, resistant to all antibiotics. Antibiotic treatment (colistin and rifampicin) rapidly ceased as a consequence of nephrotoxicity and ineffectiveness. The medical team proposed phage–antibiotic treatment combined with a surgical staged strategy. The first stage consisted of a spinal surgical procedure with local administration of a three-phage cocktail (10^6 PFU/mL) and IV cefiderocol for 6 weeks. Two weeks after the end of the first stage and 2 weeks after the withdrawal of cefiderocol, the second stage was performed, with local administration of a phage cocktail before inserting the intersomatic cages at L2–L3 and L3–L4 levels. Cefiderocol was started again intravenously, pending the culture results. However, *P. aeruginosa* still grew in cultures from the bone biopsy, with a small colony variant phenotype susceptible to bacteriophage cocktail and cefiderocol. Although the strain had become resistant to this antibiotic, colistin was added intravenously to potentially synergize with cefiderocol. As the cultures still revealed the persistence of *P. aeruginosa*, a phage cocktail was also added intravenously over 3-hour infusions every day for 21 days. Antibiotics were stopped at 3 months. The outcome of the patient was favorable during the follow-up of 21 months, without implant loosening nor clinical signs of infection, and the patient was walking without pain.

Table 2. Summary of clinical case reports using phage–antibiotic combination against MDR *P. aeruginosa*.

Disease	Bacteria	Phage Therapy	Antibiotic Combination	Outcome	Reference
Prosthetic joint infection (PJI)	*P. aeruginosa*	Three-phage cocktail (10^9 PFU/mL)	Ciprofloxacin Ceftazidime	Rapid improvement of patient's health	[46]
Catheter-related bacteremia	Pandrug-resistant *P. aeruginosa*	Personalized three-phage cocktail (10^6 PFU/mL) IV 3 h for 21 days	IV Cefiderocol 2 weeks later IV Colistin	Favorable to patient after 21 months follow-up	[47]
Catheter-related bacteremia	*P. aeruginosa* XDR	Phage cocktail (10^8 PFU/mL) by direct contact with the infected bone for 4 h	Colistin (local) IV Ceftolozane/Tazobactam	Favorable, with no bacterial growth and rapid healing of bone	[48]
Liver infection	*P. aeruginosa* XDR	IV BFC1 cocktail (10^7 PFU/mL)	IV Gentamycin, Colistin and Aztreonam	Controlled the bloodstream infection, and retransplantation was possible after 72 days	[22]
Cystic fibrosis	*P. aeruginosa* MDR	IV AB-PA01 (10^9 PFU/mL) every 6 h for 8 weeks	Ciprofloxacin and Piperaciclin-tazobactam for 3 weeks; added Doripenem	No *P. aeruginosa* recurrence or CF exacerbation	[49]
Pneumonia	*P. aeruginosa* MDR	1) Nebulized AB-PA01 (10^9 PFU/mL) for 2 weeks 2) AB-PA01-m1 and Navy-1 phage cocktail (10^9 PFU/mL)	Piperaciclin-Tazobactam and Colistin	No active *P. aeruginosa* pneumonia after 3 months	[50]
Recurrent infections post-transplant	*P. aeruginosa* MDR	IV AB-PA01 for 4 weeks (10^6 PFU/mL)	Inhaled Colistin Piperaciclin-Tazobactam from day 60 to 90	No additional *P. aeruginosa* was cultured	[50]
Pneumonia	Carbapenem-resistant *P. aeruginosa*	Personalized two-phage cocktail preparations (10^8 PFU/mL). Nebulized administration and intrapleural for 24 days	IV Amikacin, Azhitromycin, Imipenem, and Ceftazidime-Avibactam	Clearance of the pathogen and clinical improvement	[51]
Graft infection, bacteremia	*P. aeruginosa*	OMK01 (10^7 PFU/mL)	Ceftazidime	General clinical improvement	[52]
Wound infection	*P. aeruginosa*	PA5 and PA10 (10^{10} PFU/mL)	IV Ceftazidime-Avibactam and Colistin	The wound completely healed, with no *P. aeruginosa* detection	[53]
Relapsing bacteremia	*P. aeruginosa* MDR	Local application of BFC 1.10 (10^7 PFU/mL) cocktail	IV Ceftazidime-Avibactam	Bacterial eradication	[23]
Bacteremia	*P. aeruginosa* MDR	Local application (10^8 PFU/mL) during surgery every 8 h for 5 days	IV Colistin, Meropenem, and Ceftazidime	No *P. aeruginosa* detection	[54]

Another clinical case consisted of a male patient in his 60s with disseminated non-small cell lung cancer who underwent an external beam radiotherapy followed by cementoplasty performed for bone metastases located on the spine and the right sacroiliac joint [53]. Two months after surgery, a fistula occurred, with clinical evidence of infection of the cement located in the right sacroiliac joint. Surgery was required to remove the cement and to debride and abscess. The patient developed catheter-related bacteremia due to ceftazidime-resistant *P. aeruginosa*, and he received IV imipenem/cilastatin. Despite antibiotic treatment, the patient still had a fever with purulent local secretion, and a CT scan revealed persistent osteomyelitis caused by XDR *P. aeruginosa* only susceptible to polymyxins and ceftolozane/tazobactam. As an alternative treatment, during the surgical procedure, debridement of the necrotic bone was performed, and a bacteriophage cocktail (10^8 PFU/mL) was brought into contact with the bone in the cavity. The patient remained in ventral decubitus for 4 h to ensure that the phages remained in contact with the infected bone. As the patient had mild kidney injury, it was decided to use local administration of colistin. In addition, ceftolozane/tazobactam was given intravenously. At the time of surgical reconstruction, the macroscopic aspect was extremely favorable. After reconstruction, no bacteria grew in the culture and the healing was rapid.

Another scenario in which bacteriophage therapy has been widely used is in *P. aeruginosa* infections pre- and post-transplant. Nieuwenhuyse et al. described the case of a male toddler suffering from atresia with liver transplantation, with the nosocomial acquisition of extensively drug-resistant (XDR) *Pseudomonas aeruginosa* susceptible to colistin and intermediately susceptible to aztreonam [22]. The child presented multiple hepatic abscesses and severe septicemia. Despite intravenous (IV) antibiotic therapy, blood and abscess samples continued to grow XDR *P. aeruginosa*. Due to antibiotic therapy failure and the child's critical situation, the decision was to initiate adjuvant phage therapy with bacteriophage cocktail BFC1. A phage cocktail (10^7 PFU/mL) was administered intravenously combined with antibiotics (gentamycin, colistin, and high doses of aztreonam). Phage therapy combined with antibiotics controlled bloodstream infection and led to liver retransplantation after 72 days of combined treatment. More than two years after the second liver transplantation, total clearance of *P. aeruginosa* colonization was observed.

Law et al. described the case of a 26-year-old female with cystic fibrosis on the lung transplant waitlist with a pulmonary exacerbation leading to acute-on-chronic respiratory failure complicated by a pneumothorax [54]. She was colonized by two MDR *P. aeruginosa* strains: one non-mucoid susceptible to colistin and the other one mucoid susceptible to meropenem and piperacillin-tazobactam. She was treated with antibiotics for 4 weeks: the first two weeks with piperacillin-tazobactam, colistin, and azithromycin, and for the last two weeks, piperacillin-tazobactam was replaced by a carbapenem. At the end of the 4 weeks, the patient was transitioned to inhaled colistin. One week after discontinuation of IV antibiotics, the patient worsened and she was restarted on IV antibiotics (vancomycin, colistin, and meropenem, which were switched to piperacillin-tazobactam due to susceptibility profiles). Despite antibiotic treatment, the following week she experienced progressive respiratory and renal failure, attributed to colistin. At this time, they obtained approval for starting adjuvant phage therapy with AB-PA01, a cocktail of four bacteriophages. AB-PA01 was administered every 6 h (10^9 PFU/mL) intravenously for 8 weeks. The patient received concomitant ciprofloxacin and piperacillin-tazobactam for 3 weeks. Finally, ciprofloxacin was discontinued, and doripenem was added based on updated susceptibility profiles. After the end of bacteriophage therapy, she did not have a recurrence of *P. aeruginosa* pneumonia and CF exacerbation. She underwent successful bilateral lung transplantation 9 months later.

Aslam et al. described the cases of two lung transplant recipients that received bacteriophage therapy for complicated MDR *P. aeruginosa* infections [55]. The first one was a 67-year-old man who underwent a bilateral transplant for hypersensitivity to pneumonitis, complicated by multiple episodes of *P. aeruginosa* pneumonia. He developed chronic lung allograft dysfunction and progressive kidney failure. The patient suffered two distinct episodes of MDR *P. aeruginosa* pneumonia that were treated with bacteriophage therapy along with concomitant antibiotics. For the first episode, he received a 2-week course of IV and nebulized AB-PA01 (10^9 PFU/mL) as an adjunct to systemic antibiotics (piperacillin-tazobactam and colistin). After two weeks of treatment, he had significantly decreased inflammation and minimal respiratory secretions. Nebulized phage therapy was extended by an additional week without systemic antibiotics in an attempt to repopulate the airways with normal respiratory flora. By day 21 of treatment, bronchoalveolar lavage (BAL) cultures did not include *Pseudomonas* bacterial species, suggesting the reestablishment of respiratory flora. The patient completed inhaled phage therapy by day 29. However, on day 46, the patient suffered another episode characterized by clinical decompensation with respiratory failure and septic shock. In his respiratory cultures grew mucoid MDR *P. aeruginosa*; systemic antibiotics (piperacillin-tazobactam, tobramycin, and inhaled colistin) were restarted, and phage therapy was used again. In this case, bacteriophage therapy consisted of distinct courses of AB-PA01-m1 (prefixed cocktail of phages plus one new specific bacteriophage, 10^9 PFU/mL) and Navy phage cocktail 1 (personalized phage cocktail, 10^9 PFU/mL) with clinical resolution of pneumonia. After finishing treatment, the patient received suppressive bacteriophage therapy with Navy phage cocktail 1 and Navy phage cocktail 2 (10^9 PFU/mL) from day 93 to day 150. During this period and the following 3 months, there was no active *P. aeruginosa* pneumonia. In another case, a 57-year-old woman with non-CF bronchiectasis colonized by MDR *P. aeruginosa* was only susceptible to colistin; she experienced significant bilateral airway ischemic injury, and developed recurrent MDR *P. aeruginosa* infections post-transplant. She also developed *Mycobacterium abscessus* infection, initially treated with imipenem, tigecycline, and inhaled colistin. As a result of nephrotoxic antibiotic treatment, she had progressive renal failure. Due to the inability to clear *P. aeruginosa* from respiratory cultures and concern that the infection was preventing airway healing, bacteriophage therapy was initiated. The patient was treated with a 4-week IV AB-PA01 and continued only with inhaled colistin concomitantly. The patient clinically responded to treatment, and no additional *P. aeruginosa* was cultured since the start of phage therapy until 60 days after completion. The isolate grown at day 60 and subsequent strains showed improved antibiotic susceptibility. Additional infections were successfully treated with piperacillin-tazobactam, and by day 90 she was discharged from the hospital.

Chen et al. reported the case of a 68-year-old man who suffered broncho-pleural fistula (BPF)-associated empyema and pneumonia caused by carbapenem-resistant *P. aeruginosa* [56]. The patient's lung had been destroyed after tuberculosis and repeated hemoptysis for 2 years. A personalized lytic pathogen-specific two bacteriophage preparation was administered nebulized and injected intrapleurally to the patient continuously for 24 days in combination with conventional antibiotics IV (amikacin, azithromycin, imipenem, and ceftazidime-avibactam, among others). The combined treatment was well tolerated, resulting in clearance of the pathogen and improvement of clinical outcome.

Phage therapy has also been applied in the treatment of infections related to cardiothoracic surgery. A 76-year-old male patient with relapsing *P. aeruginosa* mediastinal and aortic graft infection was treated with moderately effective and indefinite IV ceftazidime [57]. The patient was an ideal candidate for bacteriophage therapy, so a procedure was proposed that comprised local administration of phage OMKO1 (10^7 PFU/mL) and ceftazidime solution into the mediastinal fistula. The day after the procedure, the patient showed signs of improvement and was discharged from IV ceftazidime; the patient returned home shortly thereafter. Rubalskii et al. also reported critical infections related to cardiothoracic surgery in which bacteriophage therapy was necessary [58], such as a 13-year-old male patient with *P. aeruginosa*-infected thoracotomy wound after lung transplantation, not eradicated after conventional treatment. The patient received local administration of PA5 and PA10 (10^{10} PFU/mL) bacteriophages concomitantly with IV colistin and ceftazidime-avibactam. After bacteriophage–antibiotic treatment, the cardiothoracic wound fully healed, and *P. aeruginosa* was not detected again.

Finally, for bone-related infections, administration of antibiotics and phages concomitantly has been applied. Racenis et al. depicted the case of a 21-year-old patient with persistent MDR *P. aeruginosa* femur osteomyelitis, regardless of extensive antibiotic treatment and surgical procedures [23]. The combination of IV ceftazidime-avibactam and local administration of a phage cocktail (10^7 PFU/mL) allowed for bacterial eradication and avoided leg amputation.

Tkhilaishvili et al. reported the case of an 80-year-old woman with metabolic syndrome (diabetes mellitus type 2, obesity, and hypertension), chronic kidney failure, diagnosis of relapsing right knee PJI, and chronic osteomyelitis of the femur after a gunshot injury [59]. One year earlier, the knee prosthesis was explanted, successfully treated, and reimplanted due to positive cultures of *Klebsiella pneumoniae* and *Providencia stuartii*. Three months after reimplantation, two morphologically distinct MDR *P. aeruginosa* isolates grew from the aspirated synovial fluid (one only susceptible to colistin and the other susceptible to ceftazidime and colistin). The knee prosthesis was explanted, and during surgery, an antibiotic-loaded cement spacer (containing 1 g gentamycin and 1 g clindamycin per 40 g poly(methyl methacrylate)) and four drainage tubes were placed. Adjunctive local bacteriophage therapy was applied during surgery (10^8 PFU/mL), followed by administration every 8 h through the drain tubes as a delivery system for 5 days. Moreover, after surgery, intravenous treatment with colistin, meropenem, and ceftazidime was started. The drainage fluid was collected for culture before bacteriophage instillation on days three, four, and five of phage treatment, and no *P. aeruginosa* was isolated.

Sinner et al. recently reported the case of a 25-year-old male with exposed calvarium in the left parietal–temporal region, due to accidental electrocution burn wounds, complicated by the development of skull osteomyelitis caused by *P. aeruginosa* [60]. After the failure of traditional (debridement and antibiotic) treatment, Whole Genome Sequencing (WGS) revealed increased MICs of all available β-lactams (except cefiderocol), likely due of the presence of blaGES-1, a β-lactamase gene, in combination with MDR efflux pumps MexD and MexX, in all six of the patient's isolates. After debridement of the infected scalp and bone, the patient was transitioned to cefiderocol but continued having relapses. Therefore, the patient received IV bacteriophage Pa14NPøPASA16 (1.7×10^{11} PFU) as adjuvant treatment for 6 weeks. The patient showed local wound improvement, with no further relapsing episodes and no abnormal laboratory values or findings on clinical exam suggesting toxicity. More than 12 months after ending antimicrobial treatment, the patient remained infection free.

5. Concluding Remarks

The worldwide spread of antibiotic resistance and the multiple failed antibiotic therapies against infectious diseases have made clear the urgent need to use an alternative or adjuvant to antibiotics. Phage therapy permits a specific union between the phage and the desired pathogen, becoming one of the most promising alternatives against infectious diseases produced by multi-drug resistant bacteria. The specificity of phages and the appearance of resistance against phages makes the use of cocktails more desirable in therapy, as shown in the in vivo and case-report studies. Despite the antibiotic–phage combination used in the mentioned case reports, not only against *P. aeruginosa* but also against most pathogens, there is a lack of in vivo studies with antibiotic–phage combinations. Interestingly, the scarce number of in vivo studies show a reduction of bacterial growth or eradication of the bacteria during and after phage therapy. The best antibiotic pairing should be chosen in consideration of the patient's sensitivity and the clinical presentation. Although nebulized phage administration is showing successful and promising results. Though the majority of clinical cases applied an intravenous treatment, this does not mean that this is best method of administration. To answer this question, a clinical trial should be performed to measure phage concentration and antiphage antibodies over time using both nebulized and intravenous routes.

The efficiency of phage therapy is still intrinsically related to the specific case of the patient. As reviewed here, the single use of antibiotic therapy did not eradicate the infection; however, the combination between antibiotics and bacteriophage cocktails did show promising results, with total eradication of the infection and no further relapses for the patient in some cases. Moreover, in all cases, the administration of phages combined with antibiotics achieved an improvement in the clinical case or a decrease of the bacterial load.

In the case reports reviewed here, there was no toxicity associated with phage administration, and no abnormal laboratory results were obtained nor significant clinical findings in the patient post-treatment that would suggest toxicity derived from the phage therapy.

The combination of phages with antibiotics could be a realistic way to eradicate infections caused by MDR/XDR *P. aeruginosa* strains using a personalized therapy, although more in vivo studies are needed to analyze the limitations.

Author Contributions: Conceptualization, M.G.-Q. and J.E.; writing—original draft preparation, A.B.-L., A.S.-K., G.S.-C. and J.E.; writing—review & editing M.G.-Q. and J.E.; funding acquisition, M.G.-Q. and J.E. All authors have read and agreed to the published version of the manuscript.

Funding: M.G.-Q. is supported by the Subprograma Miguel Servet from the Ministerio de Ciencia e Innovación of Spain (CP19/00104), Instituto de Investigación Carlos III (Plan Estatal de I+D+i 2017–2020), and is also co-funded by the European Social Fund, "El Fondo Social Europeoinvierteentufuturo".

Conflicts of Interest: The authors declare no conflict of interest.

References

1. Riquelme, S.A.; Liimatta, K.; Wong Fok Lung, T.; Fields, B.; Ahn, D.; Chen, D.; Lozano, C.; Sáenz, Y.; Uhlemann, A.C.; Kahl, B.C.; et al. *Pseudomonas aeruginosa* Utilizes Host-Derived Itaconate to Redirect Its Metabolism to Promote Biofilm Formation. *Cell Metab.* **2020**, *31*, 1091–1106.e6. [CrossRef] [PubMed]
2. Fernández-Barat, L.; Ferrer, M.; De Rosa, F.; Gabarrús, A.; Esperatti, M.; Terraneo, S.; Rinaudo, M.; Li Bassi, G.; Torres, A. Intensive care unit-acquired pneumonia due to *Pseudomonas aeruginosa* with and without multidrug resistance. *J. Infect.* **2017**, *74*, 142–152. [CrossRef]
3. Jurado-Martín, I.; Sainz-Mejías, M.; McClean, S. *Pseudomonas aeruginosa*: An Audacious Pathogen with an Adaptable Arsenal of Virulence Factors. *Int. J. Mol. Sci.* **2021**, *22*, 3128. [CrossRef] [PubMed]
4. Reynolds, D.; Kollef, M. The Epidemiology and Pathogenesis and Treatment of *Pseudomonas aeruginosa* Infections: An Update. *Drugs* **2021**, *81*, 2117–2131. [CrossRef] [PubMed]
5. Ho, J.; Tambyah, P.A.; Paterson, D.L. Multiresistant Gram-negative infections: A global perspective. *Curr. Opin. Infect. Dis.* **2010**, *23*, 546–553. [CrossRef]
6. Coates, A.R.; Halls, G.; Hu, Y. Novel classes of antibiotics or more of the same? *Br. J. Pharmacol.* **2011**, *163*, 184–194. [CrossRef]
7. Horcajada, J.P.; Montero, M.; Oliver, A.; Sorlí, L.; Luque, S.; Gómez-Zorrilla, S.; Benito, N.; Grau, S. Epidemiology and treatment of multidrug-resistant and extensively drug-resistant *Pseudomonas aeruginosa* infections. *Clin. Microbiol. Rev.* **2019**, *32*, e00031-19. [CrossRef]
8. Vila, J.; Marco, F. Interpretive reading of the non-fermenting gram-negative bacilli antibiogram. *Enferm. Infecc. Microbiol. Clin.* **2010**, *28*, 726–736. [CrossRef]
9. Lister, P.D.; Wolter, D.J.; Hanson, N.D. Antibacterial-resistant *Pseudomonas aeruginosa*: Clinical impact and complex regulation of chromosomally encoded resistance mechanisms. *Clin. Microbiol. Rev.* **2009**, *22*, 582–610. [CrossRef]
10. Miriagou, V.; Cornaglia, G.; Edelstein, M.; Galani, I.; Giske, C.G.; Gniadkowski, M.; Malamou-Lada, E.; Martinez-Martinez, L.; Navarro, F.; Nordmann, P.; et al. Acquired carbapenemases in Gram-negative bacterial pathogens: Detection and surveillance issues. *Clin. Microbiol. Infect.* **2010**, *16*, 112–122. [CrossRef]
11. King, J.D.; Kocíncová, D.; Westman, E.L.; Lam, J.S. Review: Lipopolysaccharide biosynthesis in *Pseudomonas aeruginosa*. *Innate Immun.* **2009**, *15*, 261–312. [CrossRef] [PubMed]
12. Chevalier, S.; Bouffartigues, E.; Bodilis, J.; Maillot, O.; Lesouhaitier, O.; Feuilloley, M.G.J.; Orange, N.; Dufour, A.; Cornelis, P. Structure, function and regulation of *Pseudomonas aeruginosa* porins. *FEMS Microbiol. Rev.* **2017**, *41*, 698–722. [CrossRef] [PubMed]
13. Lee, K.; Yoon, S.S. *Pseudomonas aeruginosa* Biofilm, a Programmed Bacterial Life for Fitness. *J. Microbiol. Biotechnol.* **2017**, *27*, 1053–1064. [CrossRef] [PubMed]
14. Cross, A.R.; Raghuram, V.; Wang, Z.; Dey, D.; Goldberg, J.B. Overproduction of the AlgT Sigma Factor Is Lethal to Mucoid *Pseudomonas aeruginosa*. *J. Bacteriol.* **2020**, *202*, e00445-20. [CrossRef]
15. Hall, S.; McDermott, C.; Anoopkumar-Dukie, S.; McFarland, A.J.; Forbes, A.; Perkins, A.V.; Davey, A.K.; Chess-Williams, R.; Kiefel, M.J.; Arora, D.; et al. Cellular Effects of Pyocyanin, a Secreted Virulence Factor of *Pseudomonas aeruginosa*. *Toxins* **2016**, *8*, 236. [CrossRef]

16. Al-Anany, A.M.; Fatima, R.; Hynes, A.P. Temperate phage-antibiotic synergy eradicates bacteria through depletion of lysogens. *Cell Rep.* **2021**, *35*, 109172. [CrossRef]
17. Liu, C.G.; Green, S.I.; Min, L.; Clark, J.R.; Salazar, K.C.; Terwilliger, A.L.; Kaplan, H.B.; Trautner, B.W.; Ramig, R.F.; Maresso, A.W. Phage-antibiotic synergy is driven by a unique combination of antibacterial mechanism of action and stoichiometry. *MBio* **2020**, *11*, e01462-20. [CrossRef]
18. Hatfull, G.F.; Dedrick, R.M.; Schooley, R.T. Phage Therapy for Antibiotic-Resistant Bacterial Infections. *Annu. Rev. Med.* **2022**, *73*, 197–211. [CrossRef]
19. Nikolic, I.; Vukovic, D.; Gavric, D.; Cvetanovic, J.; Aleksic Sabo, V.; Gostimirovic, S.; Narancic, J.; Knezevic, P. An Optimized Checkerboard Method for Phage-Antibiotic Synergy Detection. *Viruses* **2022**, *14*, 1542. [CrossRef]
20. Engeman, E.; Freyberger, H.R.; Corey, B.W.; Ward, A.M.; He, Y.; Nikolich, M.P.; Filippov, A.A.; Tyner, S.D.; Jacobs, A.C. Synergistic Killing and Re-Sensitization of *Pseudomonas aeruginosa* to Antibiotics by Phage-Antibiotic Combination Treatment. *Pharmaceuticals* **2021**, *14*, 184. [CrossRef]
21. Xuan, G.; Lin, H.; Kong, J.; Wang, J. Phage Resistance Evolution Induces the Sensitivity of Specific Antibiotics in *Pseudomonas aeruginosa* PAO1. *Microbiol. Spectr.* **2022**, *10*, e01356-22. [CrossRef] [PubMed]
22. Van Nieuwenhuyse, B.; Van der Linden, D.; Chatzis, O.; Lood, C.; Wagemans, J.; Lavigne, R.; Schroven, K.; Paeshuyse, J.; de Magnée, C.; Sokal, E.; et al. Bacteriophage-antibiotic combination therapy against extensively drug-resistant *Pseudomonas aeruginosa* infection to allow liver transplantation in a toddler. *Nat. Commun.* **2022**, *13*, 5725. [CrossRef] [PubMed]
23. Racenis, K.; Rezevska, D.; Madelane, M.; Lavrinovics, E.; Djebara, S.; Petersons, A.; Kroica, J. Use of Phage Cocktail BFC 1.10 in Combination With Ceftazidime-Avibactam in the Treatment of Multidrug-Resistant *Pseudomonas aeruginosa* Femur Osteomyelitis— A Case Report. *Front. Med.* **2022**, *9*, 851310. [CrossRef] [PubMed]
24. Damir, G.; Knezevic, P. Filamentous Pseudomonas Phage Pf4 in the Context of Therapy-Inducibility, Infectivity, Lysogenic Conversion, and Potential Application. *Viruses* **2022**, *14*, 1261. [CrossRef]
25. Holger, D.J.; Lev, K.L.; Kebriaei, R.; Morrisette, T.; Shah, R.; Alexander, J.; Lehman, S.M.; Rybak, M.J. Bacteriophage-antibiotic combination therapy for multidrug-resistant *Pseudomonas aeruginosa*: In vitro synergy testing. *J. Appl. Microbiol.* **2022**, *133*, 1636–1649. [CrossRef]
26. Wannasrichan, W.; Htoo, H.H.; Suwansaeng, R.; Pogliano, J.; Nonejuie, P.; Chaikeeratisak, V. Phage-resistant *Pseudomonas aeruginosa* against a novel lytic phage JJ01 exhibits hypersensitivity to colistin and reduces biofilm production. *Front. Microbiol.* **2022**, *13*, 1004733. [CrossRef]
27. Valappil, S.K.; Shetty, P.; Deim, Z.; Terhes, G.; Urbán, E.; Váczi, S.; Patai, R.; Polgár, T.; Pertics, Z.B.; Schneider, G.; et al. Survival Comes at a Cost: A Coevolution of Phage and Its Host Leads to Phage Resistance and Antibiotic Sensitivity of *Pseudomonas aeruginosa* Multidrug Resistant Strains. *Front. Microbiol.* **2021**, *12*, 783722. [CrossRef]
28. Fiscarelli, E.V.; Rossitto, M.; Rosati, P.; Essa, N.; Crocetta, V.; Giulio, A.D.; Lupetti, V.; Bonaventura, G.D.; Pompilio, A. In Vitro Newly Isolated Environmental Phage Activity against Biofilms Preformed by *Pseudomonas aeruginosa* from Patients with Cystic Fibrosis. *Microorganisms* **2021**, *9*, 478. [CrossRef]
29. Aghaee, B.L.; Mirzaei, M.K.; Alikhani, M.Y.; Mojtahedi, A.; Maurice, C.F. Improving the Inhibitory Effect of Phages against *Pseudomonas aeruginosa* Isolated from a Burn Patient Using a Combination of Phages and Antibiotics. *Viruses* **2021**, *13*, 334. [CrossRef]
30. Gurney, J.; Pradier, L.; Griffin, J.S.; Gougat-Barbera, C.; Chan, B.K.; Turner, P.E.; Kaltz, O.; Hochberg, M.E. Phage steering of antibiotic-resistance evolution in the bacterial pathogen, *Pseudomonas aeruginosa*. *Evol. Med. Public Health* **2020**, *2020*, 148–157. [CrossRef]
31. Uchiyama, J.; Shigehisa, R.; Nasukawa, T.; Mizukami, K.; Takemura-Uchiyama, I.; Ujihara, T.; Murakami, H.; Imanishi, I.; Nishifuji, K.; Sakaguchi, M.; et al. Piperacillin and ceftazidime produce the strongest synergistic phage–antibiotic effect in *Pseudomonas aeruginosa*. *Arch. Virol.* **2018**, *163*, 1941–1948. [CrossRef]
32. CLSI. *Methods dilution Antimicrobial Susceptibility Tests Bact That Grow Aerobically*, 11th ed.; CLSI standard M07; Clinical and Laboratory Standards Institute: Wayne, PA, USA, 2018; p. 91.
33. Landa, K.J.; Mossman, L.M.; Whitaker, R.J.; Rapti, Z.; Clifton, S.M. Phage-antibiotic synergy inhibited by temperate and chronic virus competition. *Bull. Math. Biol.* **2022**, *84*, 54. [CrossRef] [PubMed]
34. Lwoff, A. Lysogeny. *Bacteriol. Rev.* **1953**, *17*, 269–337. [CrossRef] [PubMed]
35. Weinbauer, M.G. Ecology of prokaryotic viruses. *FEMS Microbiol. Rev.* **2004**, *28*, 127–181. [CrossRef] [PubMed]
36. Rodriguez-Gonzalez, R.A.; Leung, Y.; Chan, B.K.; Turner, P.E.; Weitz, J.S. Quantitative Models of Phage-Antibiotic Combination Therapy. *mSystems* **2020**, *5*, e00756-19. [CrossRef] [PubMed]
37. Issa, R.; Chanishvili, N.; Caplin, J.; Kakabadze, E.; Bakuradze, N.; Makalatia, K.; Cooper, I. Antibiofilm potential of purified environmental bacteriophage preparations against early stage *Pseudomonas aeruginosa* biofilms. *J. Appl. Microbiol.* **2019**, *126*, 1657–1667. [CrossRef]
38. Henriksen, K.; Rørbo, N.; Rybtke, M.L.; Martinet, M.G.; Tolker-Nielsen, T.; Høiby, N.; Middelboe, M.; Ciofu, O.P. *P. aeruginosa* flow-cell biofilms are enhanced by repeated phage treatments but can be eradicated by phage-ciprofloxacin combination-monitoring the phage-*P. aeruginosa* biofilms interactions. *Pathog. Dis.* **2019**, *77*, ftz011. [CrossRef]
39. Manohar, P.; Loh, B.; Nachimuthu, R.; Leptihn, S.; Nachimuthu Assistant Professor, R. Phage-antibiotic combinations to control *Pseudomonas aeruginosa*-Candida two-species biofilms. *bioRxiv* **2022**, bioRxiv:2022.08.18.504394. [CrossRef]

40. Danis-Wlodarczyk, K.M.; Cai, A.; Chen, A.; Gittrich, M.R.; Sullivan, M.B.; Wozniak, D.J.; Abedon, S.T. Friends or Foes? Rapid Determination of Dissimilar Colistin and Ciprofloxacin Antagonism of *Pseudomonas aeruginosa* Phages. *Pharmaceuticals* **2021**, *14*, 1162. [CrossRef]

41. Lin, Y.; Chang, R.Y.K.; Britton, W.J.; Morales, S.; Kutter, E.; Li, J.; Chan, H.-K. Inhalable combination powder formulations of phage and ciprofloxacin for *P. aeruginosa* respiratory infections. *Eur. J. Pharm. Biopharm.* **2019**, *142*, 543–552. [CrossRef]

42. Lin, Y.; Yoon Kyung Chang, R.; Britton, W.J.; Morales, S.; Kutter, E.; Chan, H.-K. Synergy of nebulized phage PEV20 and ciprofloxacin combination against *Pseudomonas aeruginosa*. *Int. J. Pharm.* **2018**, *15*, 158–165. [CrossRef] [PubMed]

43. Yoon Kyung Chang, R.; Das, T.; Manos, J.; Kutter, E.; Morales, S.; Chan, H.-K. Bacteriophage PEV20 and ciprofloxacin combination treatment enhances removal of *P. aeruginosa* biofilm isolated from cystic fibrosis and wound patients. HHS Public Access. *AAPS J.* **2020**, *21*, 49. [CrossRef] [PubMed]

44. Luscher, A.; Simonin, J.; Falconnet, L.; Valot, B.; Hocquet, D.; Chanson, M.; Resch, G.; Köhler, T.; Delden, C. van Combined Bacteriophage and Antibiotic Treatment Prevents *Pseudomonas aeruginosa* Infection of Wild Type and *cftr*- Epithelial Cells. *Front. Microbiol.* **2020**, *11*, 1947. [CrossRef]

45. Ferran, A.A.; Lacroix, M.Z.; Gourbeyre, O.; Huesca, A.; Gaborieau, B.; Debarbieux, L.; Bousquet-Mélou, A. The Selection of Antibiotic- and Bacteriophage-Resistant *Pseudomonas aeruginosa* Is Prevented by Their Combination. *Microbiol. Spectr.* **2022**, *10*, e0287422. [CrossRef]

46. Akturk, E.; Oliveira, H.; Santos, S.B.; Costa, S.; Kuyumcu, S.; Melo, D.R.; Azeredo, J. Synergistic Action of Phage and Antibiotics: Parameters to Enhance the Killing E ffi cacy Against Mono and Dual-Species Biofilms. *Antibiotics* **2019**, *8*, 103. [CrossRef] [PubMed]

47. Tkhilaishvili, T.; Wang, L.; Perka, C.; Trampuz, A.; Moreno, M.G. Using Bacteriophages as a Trojan Horse to the Killing of Dual-Species Biofilm Formed by *Pseudomonas aeruginosa* and Methicillin Resistant *Staphylococcus aureus*. *Front. Microbiol.* **2020**, *11*, 695. [CrossRef]

48. Lin, Y.; Quan, D.; Chang, R.Y.K.; Chow, M.Y.T.; Wang, Y.; Li, M.; Morales, S.; Britton, W.J.; Kutter, E.; Li, J.; et al. Synergistic activity of phage PEV20-ciprofloxacin combination powder formulation—A proof-of-principle study in a *P. aeruginosa* lung infection model. *Eur. J. Pharm. Biopharm.* **2021**, *158*, 166–171. [CrossRef]

49. Duplessis, C.; Warawa, J.M.; Lawrenz, M.B.; Henry, M.; Biswas, B. Successful intratracheal treatment of phage and antibiotic combination therapy of a multi-drug resistant *Pseudomonas aeruginosa* murine model. *Antibiotics* **2021**, *10*, 946. [CrossRef]

50. Cafora, M.; Deflorian, G.; Forti, F.; Ferrari, L.; Binelli, G.; Briani, F.; Ghisotti, D.; Pistocchi, A. Phage therapy against *Pseudomonas aeruginosa* infections in a cystic fibrosis zebrafish model. *Sci. Rep.* **2019**, *9*, 1527. [CrossRef]

51. Ferry, T.; Kolenda, C.; Batailler, C.; Gaillard, R.; Gustave, C.A.; Lustig, S.; Fevre, C.; Petitjean, C.; Leboucher, G.; Laurent, F. Case Report: Arthroscopic "Debridement Antibiotics and Implant Retention" With Local Injection of Personalized Phage Therapy to Salvage a Relapsing *Pseudomonas aeruginosa* Prosthetic Knee Infection. *Front. Med.* **2021**, *8*, 569159. [CrossRef]

52. Ferry, T.; Kolenda, C.; Laurent, F.; Leboucher, G.; Merabichvilli, M.; Djebara, S.; Gustave, C.; Perpoint, T.; Barrey, C.; Pirnay, J.; et al. Personalized bacteriophage therapy to treat pandrug-resistant spinal *Pseudomonas aeruginosa* infection. *Nat. Commun.* **2022**, *13*, 4239. [CrossRef] [PubMed]

53. Ferry, T.; Boucher, F.; Fevre, C.; Perpoint, T.; Chateau, J.; Petitjean, C.; Josse, J.; Chidiac, C.; L'hostis, G.; Leboucher, G.; et al. Innovations for the treatment of a complex bone and joint infection due to XDR *Pseudomonas aeruginosa* including local application of a selected cocktail of bacteriophages. *J. Antimicrob. Chemother.* **2018**, *73*, 2901–2903. [CrossRef] [PubMed]

54. Law, N.; Logan, C.; Yung, G.; Furr, C.L.L.; Lehman, S.M.; Morales, S.; Rosas, F.; Gaidamaka, A.; Bilinsky, I.; Grint, P.; et al. Successful adjunctive use of bacteriophage therapy for treatment of multidrug-resistant *Pseudomonas aeruginosa* infection in a cystic fibrosis patient. *Infection* **2019**, *47*, 665–668. [CrossRef] [PubMed]

55. Aslam, S.; Courtwright, A.M.; Koval, C.; Lehman, S.M.; Morales, S.; Furr, C.L.L.; Rosas, F.; Brownstein, M.J.; Fackler, J.R.; Sisson, B.M.; et al. Early clinical experience of bacteriophage therapy in 3 lung transplant recipients. *Am. J. Transplant.* **2019**, *19*, 2631–2639. [CrossRef]

56. Chen, P.; Liu, Z.; Tan, X.; Wang, H.; Liang, Y.; Kong, Y.; Sun, W.; Sun, L.; Ma, Y.; Lu, H. Bacteriophage therapy for empyema caused by carbapenem-resistant *Pseudomonas aeruginosa*. *Biosci. Trends* **2022**, *16*, 158–162. [CrossRef]

57. Chan, B.K.; Turner, P.E.; Kim, S.; Mojibian, H.R.; Elefteriades, J.A.; Narayan, D. Phage treatment of an aortic graft infected with *Pseudomonas aeruginosa*. *Evol. Med. Public Health* **2018**, *2018*, 60–66. [CrossRef]

58. Rubalskii, E.; Ruemke, S.; Salmoukas, C.; Boyle, E.C.; Warnecke, G.; Tudorache, I.; Shrestha, M.; Schmitto, J.D.; Martens, A.; Rojas, S.V.; et al. Bacteriophage Therapy for Critical Infections Related to Cardiothoracic Surgery. *Antibiotics* **2020**, *9*, 232. [CrossRef]

59. Tkhilaishvili, T.; Winkler, T.; Müller, M.; Perka, C.; Trampuz, A. Bacteriophages as Adjuvant to Antibiotics for the Treatment of Periprosthetic Joint Infection Caused by Multidrug-Resistant *Pseudomonas aeruginosa*. *Antimicrob. Agents Chemother.* **2019**, *64*, e00924-19. [CrossRef]

60. Simner, P.J.; Cherian, J.; Suh, G.A.; Bergman, Y.; Beisken, S.; Fackler, J.; Lee, M.; Hopkins, R.J.; Tamma, P.D. Combination of phage therapy and cefiderocol to successfully treat *Pseudomonas aeruginosa* cranial osteomyelitis. *JAC Antimicrob. Resist.* **2023**, *4*, dlac046. [CrossRef]

 *antibiotics*

Article

Vibrio Phage VMJ710 Can Prevent and Treat Disease Caused by Pathogenic MDR *V. cholerae* O1 in an Infant Mouse Model

Naveen Chaudhary [1], Balvinder Mohan [1], Harpreet Kaur [1], Vinay Modgil [1], Vishal Kant [1], Alka Bhatia [2] and Neelam Taneja [1,*]

[1] Department of Medical Microbiology, Postgraduate Institute of Medical Education and Research, Chandigarh 160012, India
[2] Department of Experimental Medicine and Biotechnology, Postgraduate Institute of Medical Education and Research, Chandigarh 160012, India
* Correspondence: drneelampgi@yahoo.com

Abstract: Cholera, a disease of antiquity, is still festering in developing countries that lack safe drinking water and sewage disposal. *Vibrio cholerae*, the causative agent of cholera, has developed multi-drug resistance to many antimicrobial agents. In aquatic habitats, phages are known to influence the occurrence and dispersion of pathogenic *V. cholerae*. We isolated *Vibrio* phage VMJ710 from a community sewage water sample of Manimajra, Chandigarh, in 2015 during an outbreak of cholera. It lysed 46% of multidrug-resistant *V. cholerae* O1 strains. It had significantly reduced the bacterial density within the first 4–6 h of treatment at the three multiplicity of infection (MOI 0.01, 0.1, and 1.0) values used. No bacterial resistance was observed against phage VMJ710 for 20 h in the time–kill assay. It is nearest to an ICP1 phage, i.e., *Vibrio* phage ICP1_2012 (MH310936.1), belonging to the class *Caudoviricetes*. ICP1 phages have been the dominant bacteriophages found in cholera patients' stools since 2001. Comparative genome analysis of phage VMJ710 and related phages indicated a high level of genetic conservation. The phage was stable over a wide range of temperatures and pH, which will be an advantage for applications in different environmental settings. The phage VMJ710 showed a reduction in biofilm mass growth, bacterial dispersal, and a clear disruption of bacterial biofilm structure. We further tested the phage VMJ710 for its potential therapeutic and prophylactic properties using infant BALB/c mice. Bacterial counts were reduced significantly when phages were administered before and after the challenge of orogastric inoculation with *V. cholerae* serotype O1. A comprehensive whole genome study revealed no indication of lysogenic genes, genes associated with possible virulence factors, or antibiotic resistance. Based on all these properties, phage VMJ710 can be a suitable candidate for oral phage administration and could be a viable method of combatting cholera infection caused by MDR *V. cholerae* pathogenic strains.

Keywords: phage; genome; cholera; antibiotic-resistance; mice

check for **updates**

Citation: Chaudhary, N.; Mohan, B.; Kaur, H.; Modgil, V.; Kant, V.; Bhatia, A.; Taneja, N. *Vibrio* Phage VMJ710 Can Prevent and Treat Disease Caused by Pathogenic MDR *V. cholerae* O1 in an Infant Mouse Model. *Antibiotics* **2023**, *12*, 1046. https://doi.org/10.3390/antibiotics12061046

Academic Editors: Aneta Skaradzińska and Anna Nowaczek

Received: 28 May 2023
Revised: 3 June 2023
Accepted: 7 June 2023
Published: 14 June 2023

1. Introduction

Cholera causes around 1.4–4.3 million cases and over 21,000–143,000 deaths each year [1]. The disease is endemic and causes outbreaks in several parts of Southeast Asia and Africa. Seven cholera pandemics have been reported till now. Two serogroups, O1 and O139, of *V. cholerae* are mainly responsible for cholera. The serogroup O1 is further divided into two biotypes, El Tor and classical, each of which has Ogawa and Inaba serotypes. The main pathogenesis and virulence of *V. cholerae* is due to the production of cholera toxin (CT) encoded by a bacteriophage harbored by the pathogen [2]. The cornerstone of cholera treatment is fluid and electrolyte replacement therapy. Antibiotics are not compulsory for a successful treatment, but are used as an adjunct therapy. Antibiotics decrease the duration of disease, reduce the volume of diarrhea, and the duration of shedding of the infective organism in stools. Currently, doxycycline, ciprofloxacin, and azithromycin are effectively used for the treatment of cholera [1,2]. However, due to indiscriminate use, multi-drug resistant (MDR) *V. cholerae* have emerged, and the strains are showing resistance not only to first-line agents such as ampicillin, cotrimoxazole, nalidixic acid, and tetracycline, but also to fluoroquinolones such as ciprofloxacin, third-generation cephalosporins, and azithromycin [3]. Resistance has also emerged against ceftriaxone, and NDM-1 (New Delhi metallo-beta lactamase) gene-encoding carbapenem resistance has been reported in India [4,5]. Mass antibiotic prophylaxis is not recommended by the World Health Organization (WHO) as it carries the risk of the development and spread of antimicrobial resistance (AMR). WHO emphasizes that antibiotic treatment should be used only in severely dehydrated patients in conjunction with rehydration therapy and susceptible household contact cases of cholera patients [4]. Cholera is endemic in most parts of India, according to the study conducted by Ali et al. In India, 150 out of 641 districts have reported cholera, and some of those districts, including Chandigarh, have been labeled as "hotspots" for the disease [6]. A total of twenty- nine outbreaks of cholera were reported in and around Chandigarh, an inland area located in north India, from 2002 to 2015 [5]. Most of these outbreaks were due to a contaminated drinking water supply. The region has a freshwater climate, and clinical cholera cases increase annually between May and October, coinciding with hot summers and monsoons [5].

V. cholerae are autochthonous to aquatic environments. During infection, they form biofilm-like aggregates that may play an important role in pathogenesis and disease transmission. Biofilms are also important for the survival of cholera bacteria in the environment [7]. Several previous studies demonstrated that biofilms may cause delayed penetration of antimicrobial agents [8]. With fast-growing multidrug resistance among biofilm-forming *V. cholerae* isolates and a dearth of novel antibiotic research by pharmaceutical industries, there is an urgent need to discover new antibiotic alternatives. Phage therapy can be a potential alternative to antibiotics in the era of multidrug-resistant bacterial infections. The lytic phages, which disrupt bacterial metabolism and lyse the bacteria, are proposed as being useful for phage therapy [9]. Phages have many potential advantages over antibiotics. Phages are less likely to inflict "collateral damage", or the destruction of gut flora than antibiotics, since they are more host-specific [10]. Phages are common in the environment, and different phages may work together to influence the incidence and distribution of pathogenic *V. cholerae* in aquatic habitats. *Vibrio* phages may also be important in the environmental control of cholera.

We characterized phage VMJ710 isolated in 2015 from a community sewage water sample during an outbreak of cholera in Manimajra, Chandigarh. We carried out genomic characterization and tested the phage for its antibiofilm testing against the MDR *V. cholerae* O1 pathogenic strain. Using an infant mouse model, we tested the preventive effect of the phage to cause diarrhea by giving the phages before the bacterial testing. We also demonstrate the therapeutic potential of this phage.

2. Materials and Methods

2.1. Sample Collection

We collected sixty-eight sewage water samples from four surveillance sites in Chandigarh between February 2015 and November 2016. Figure S1 shows three sites of sewage sample collection that included hospital sewage, community sewage at Ramdarbar and Raipur Khurd, and three cholera outbreak sites (Manimajra, Ambala, and Ludhiana). The collected water samples were transferred to the enteric laboratory, Postgraduate Institute of Medical Education and Research (PGIMER), at room temperature and processed within 18–24 h after collection.

2.2. Bacterial Strains and Growth Conditions Used

Suspected colonies of *V. cholerae* were identified using routine biochemical tests (positive for oxidase, catalase, indole, lysine and ornithine decarboxylase, string test, nitrate reduction, fermentation of mannose and sucrose, negative for arginine dihydrolase and arabinose fermentation). The strains were confirmed via a serotyping kit (Denka Seiken Co., Ltd., Tokyo, Japan). *V. cholerae* O1 biotype El Tor serotype Ogawa VMJ1 strain isolated from the 2015 outbreak of Manimajra, Chandigarh, was used for the isolation and propagation of phage VMJ710. The phage VMJ710 was tested against 26 MDR *V. cholerae* O1 biotype El Tor serotype Ogawa strains (PGIMER culture collection) via the spot assay test on trypticase soy agar (TSA, Becton, Dickinson and Company, Franklin Lakes, NJ, USA) (Tables S1 and S2). *V. cholerae* O1 strains obtained from clinical cases of cholera and sewage water showed acquired non-susceptibility to at least one agent in three or more antimicrobial categories of the following classes of antibiotics: Third-generation cephalosporins, fluoroquinolones, aminoglycosides, tetracyclines, ampicillin, chloramphenicol, and cotrimoxazole were defined as MDR *V. cholerae* [11]. *V. cholerae* serotype O1 (strain ATCC 39315/El Tor Inaba N16961) was used to standardize the induction of cholera in mice. A streptomycin-resistant *V. cholerae* O1 (ELPGI212) strain isolated from a clinical case of cholera during the 2015 outbreak was used for testing antibiofilm activity and efficacy in mouse models.

2.3. Phage Isolation

In total, 3 mL of sewage water was mixed with 500 µL of bacterial culture (10^8 colony forming units (CFU)/mL) and 2 mL of trypticase soy broth (TSB, Becton, Dickinson and Company, Franklin Lakes, NJ, USA). The whole mixture was incubated for 24 h, and the next day, the mixture was centrifuged at $7800\times g$ for 15 min. The supernatant was collected and filtered with a 0.45 µm filter (Pall Corporation, Cortland, NY, USA). To assess phage activity (qualitatively) against the host strain, a spot assay followed by a plaque assay was used [11,12].

2.4. Phage Purification

A single plaque was picked up via a sterile 1 mL micropipette tip, and this single phage was amplified, and all steps were repeated twice [13]. Ten percent of polyethylene glycol 8000 (PEG 8000, Sigma-Aldrich Corporation, Burlington, MA, USA) was added to the phage preparation, and incubated for 20 h at 4 °C. After centrifuging, the mixture at $257,000\times g$, the pellet was resuspended in the SM buffer (10 mM $MgSO_4\cdot6H_2O$). The resuspended mixture was filtered by a 0.45 µm syringe filter. Final purification was carried out using the dialysis method with the help of the dialysis membrane (MWCO 14,000, Himedia Laboratories, Mumbai, India) [14].

2.5. Phage Stability

The following two parameters were used to determine phage stability:

2.5.1. pH Stability

For the pH stability test, 200 µL of phage lysate (10^8 CFU/mL) was added in different tubes containing sterile SM buffer with a range of pH values from 4 to 10. The tubes were incubated at 37 °C for 2 h [15]. The phage titer was determined at a 60 min interval using the double-layer agar plate technique.

2.5.2. Thermal/Temperature Stability

For thermal stability testing, 200 µL of phage lysate (10^8 plaque-forming units (PFU)/mL) in triplicates was added to the series of 6 tubes and incubated for 2 h at -20 °C, 4 °C, 25 °C, 37 °C, 50 °C, and 60 °C, respectively. The phage titer was tested every 60 min for 2 h using the double-layer agar plate technique [15]. Incubation at -20 °C and 4 °C was achieved in a laboratory freezer and refrigerator, respectively. The water bath was used to maintain the rest of the temperatures. The phage stability at different pH and temperature values was expressed as the phage stability rate (%) [11].

2.6. Phage Morphology

Plaque morphology was determined using the plaque assay [14]. The plaque diameter was calculated in millimeters (mm). A drop of the phage suspension was applied to a carbon-coated copper grid and then replaced with a 3% uranyl acetate solution [11]. The grids were examined with the transmission electron microscope, Tecnai G20, at All India Institute of Medical Education and Sciences (AIIMS), New Delhi. The ImageJ software was used to estimate the size of the phage [16].

2.7. Genomic DNA Isolation

A phage DNA isolation kit (Norgen Biotek Corporation, Thorold, ON, Canada) was used to extract phage genomic DNA [17]. The phage DNA quality was determined using 1% agarose gel electrophoresis and a Qubit 3.0 fluorometer (Life Technologies, Carlsbad, CA, USA).

2.8. Library Preparation, Sequencing, Assembly, and Annotations

A NEBNext Ultra library preparation kit (New England Bioscience, Ipswich, MA, USA) was used to prepare the genomic DNA library, and the Illumina HiSeq 2500 platform was used for the sequencing. The Cutadap tool was used to trim adaptor sequences after processing the FastQ files [18,19]. De novo assembly of the generated reads was executed using an IVA v1.0.8 assembler with default k-mer sizes [20]. GLIMMER 3 and GeneMark tools were used to predict genes from IVA-assembled contigs [21,22]. The genome annotation was performed using the Rapid Annotation Search Tool (RAST v2) [23]. The complete phage genome was scanned with ARAGORN 1.2.36, and CRISPR-CasFinder 4.2.20 to find tRNA and CRISPR-like systems in the phage genome, respectively [24,25]. The circular map of the phage genome was constructed with the CGView tool [26].

2.9. Virulence Factors and Antibiotic Resistance Genes

To determine virulence factors and antibiotic resistance genes, in the phage genome, the entire genome was scanned using the VFDB 2019 (Virulence Factor Database) and CARD (Comprehensive Antibiotic Resistance Database) tools, respectively [27,28].

2.10. Phylogenetic Tree and Comparative Genomics

The phylogenetic tree was constructed based on multiple sequence alignment using amino acid sequences of the large subunit terminase protein of phage VMJ710 and related *Vibrio* phages with the help of ClustalX and MEGA X. The ViPTree was used to compare the whole genome of the phage VMJ710 with 3234 additional sequenced phage genomes from the virus–host DB (RefSeq release 212) [29,30]. Whole-genome comparison of phage VMJ710 with four closely related *Vibrio* phage genomes, i.e., JSF 14 (KY883639.1), JSF6 (KY883635.1), ICP1_2006 (HQ641351.1, ICP1_2017 (MN419153.1), ICP1_2012) (MH310936.1), was performed using Mauve [31]. To perform core genome analysis, open reading frames (ORFs) were divided into groups based on how many of the four phage genomes were found with respect to phage VMJ710. If an ORF was found in all phages, it was considered part of the core genome; otherwise, it was regarded as part of the accessory genome. The BRIG tool was used to map core and accessory ORFs to the BRIG alignment [32].

2.11. Host Range Testing

In total, 26 *V. cholerae* strains were used to assess the phage host range. A lawn of each strain (OD_{600} ~0.6) was made on TSA plates with a sterilized cotton swab. The host range experiment was performed by spotting 20 µL of phage lysate on the bacterial lawns, followed by overnight incubation at 37 °C. A scale system approach was used to identify the zones of clearing, with negative values denoting turbid zones (no lysis) and positive values denoting clear zones [33]. The efficiency of plating (EOP), which was computed by dividing the average PFU of the target bacteria by the average PFU of the host bacteria, was used to measure relative phage killing. Based on EOP values, phages were categorized as highly virulent (0.1 < EOP < 1), moderately virulent (0.001 < EOP < 0.099), avirulent (no plaques formed), and reference (EOP = 1).

2.12. Time-Kill Assay

To measure the activity (quantitatively) of the phage, a time–kill assay was performed according to a previously described method with some modifications [15]. The bacterial suspension was combined with phage preparation to achieve initial MOIs of 0.01, 0.1, and 1. The control group contained a mixture of bacterial culture and TSB. The experiment was executed at 37 °C using a spectrophotometer (Tecan Group Limited, Mannedorf, Switzerland) with orbital shaking for 4 s. At an interval of 2 h, the OD_{600} values were measured automatically for 20 h.

2.13. Antibiofilm Activity of Phage VMJ710 against V. cholerae

The microtiter-plate-based crystal violet assay was used to measure the biofilm capacity of 30 MDR *V. cholerae* O1 isolates [34]. Strains were classified into the following categories: $OD \leq OD_C$ = non-adherent; $OD_C < OD \leq (2 \times OD_C)$ = weakly adherent; $(2 \times OD_C) < OD \leq (4 \times OD_C)$ = moderately adherent; and $(4 \times OD_C) < OD$ = strongly adherent [35,36]. To test the antibiofilm activity of the phages, 24-hour-old preformed biofilms were grown on the 13 mm polystyrene coverslips and treated at three different phage titers (10^6, 10^7, and 10^8 PFU). After 24 h of incubation, coverslips were washed, vortexed, and placed into an ultrasonic bath for 4 min at 35 kHz frequency to detach the bacterial cells. To enumerate the bacterial count, detached bacterial cells were serially diluted in normal saline (0.85% NaCl) and spread onto TSA plates. For biofilm imaging, biofilms formed on 13 mm coverslips were examined with a scanning electron microscope [11].

2.14. Phage Efficacy Testing against V. cholerae O1 Using an Infant BALB/c Mouse Model

2.14.1. Animals and Maintenance

Four-to-five-day-old BALB/c male mice were procured from the animal house, PGIMER Chandigarh, India. The animals were housed in clean polypropylene cages maintained in the animal house with a controlled temperature of (23 ± 2) °C, relative humidity at 50–60%, alternating 12/12 h light/dark cycle, and adequate ventilation. All experimental studies were approved as per guidelines by the institutional animal ethics committee (Ref. No. 93/IEAC/648). The number of animals used in each experiment is depicted in the different legends in the respective figures.

2.14.2. Mouse Cholera Infection Model

Cholera infection was established using the previously described mouse model by Yen et al., with slight modifications [37]. To standardize the minimum infective dose, mice were divided into 4 groups and inoculated with *V. cholerae* O1 El Tor N16961 using a 2 mL syringe fitted with an 18-oral gavage needle. Each of the mice in the group was then inoculated with 50 μL of culture (10^6, 10^7, 10^8 CFU), and the same volume of phosphate-buffered saline (PBS) was given to the negative control group. Mice were sacrificed with the administration of a ketamine/xylazine (K:100 mg/kg + X: 20 mg/kg) cocktail after 24 h, and the small intestines were removed which were mechanically homogenized in PBS buffer (137 mM NaCl, 2.7 mM KCl, 10 mM Na_2HPO_4, 1.8 mM KH_2PO_4, pH 7.2), and serial dilutions were plated onto Luria Bertani broth (LB broth, Himedia Laboratories, Mumbai, India) media containing 100 μg/mL of streptomycin.

To standardize the incubation period, the minimum infective dose (standardized above) was administered orogastrically. The percentage of mice positive in intestine culture and survival percentage were observed after a specific time interval for each group of mice (4, 6, 8, 10, 12, and 24 h).

2.14.3. Prophylactic and Treatment Efficacy Testing against *V. cholerae*

To estimate whether the phage VMJ710 would be able to survive in the mice intestine, a phage retention study was performed. Mice were dosed with a phage volume of 50 µL (1×10^9 PFU). This experiment was divided into four groups and sacrificed at a specific time interval after phage administration, i.e., at 2, 6, 12, and 24 h. The small intestines were removed and homogenized, as explained in the previous section. The homogenized material was further centrifuged and filtered with a 0.22 µm syringe filter. The quantification of bacteriophages present in the supernatant was performed as explained above in Section 2.3.

To study the prophylactic efficacy of phage VMJ710, mice were dosed with 50 µL of phage volume (1×10^7 PFU) at 6, 12, and 24 h (three prophylactic groups) before the bacterial challenge (Figure 1). After 48 h of bacterial inoculation, the number of viable *V. cholerae* in the small intestine of sacrificed animals was quantified as described above.

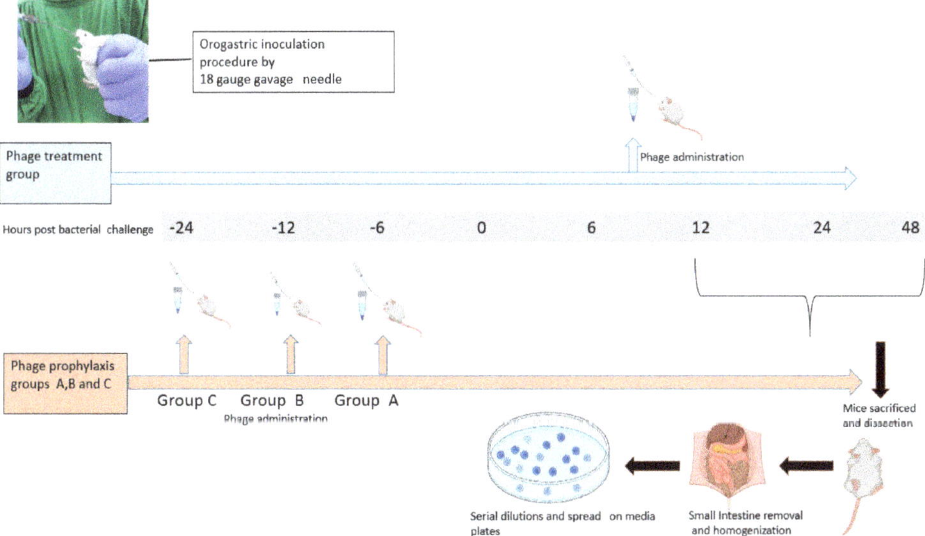

Figure 1. Overview of the procedure used to estimate the efficacy of phage VMJ710 against *V. cholerae* infection using an infant mouse model.

To test the treatment efficacy of *Vibrio* phage VMJ710 against *V. cholerae* strain ELPGI212, animals were divided into control and treatment groups (Figure 1). The treatment group received phage (1×10^9 PFU, at MOI = 1) after 8 h of bacterial challenge. After 12, 24, and 48 h, the number of viable *V. cholerae* in the small intestine of sacrificed animals was quantified as described above. Heamotoxylin and Eosin (H&E) staining was also performed to study the histopathology of the small intestines of mice treated with phage VMJ710.

2.15. Nucleotide Sequence Accession Number

The genome sequence of the phage has been submitted to the GenBank database (accession no. MN402506). The raw reads of the phage genome are available at SRA accession number SRR9686322, BioProject accession number PRJNA553871, and BioSample SAMN12253299.

2.16. Statistical Analysis

GraphPad Prism 9.0 software was used to run all statistical tests. The time–kill assay, biofilm formation, and phage stability assays were carried out in triplicates. The data were presented as mean ± SD. The Kruskal–Wallis test along with the Dunn post hoc multiple-comparison tests was used to conduct statistical analysis for the time–kill assay and quantitative biofilm assay. A *p*-value of less than 0.05 was considered statistically significant.

3. Results

3.1. Sample Collection and Phage Isolation

A total of five phages active against *V. cholerae* O1 were isolated from sixty-eight sewage samples. Three *Vibrio* phages, VMJ710, VMJ3, and LDH4, were isolated from the cholera outbreak sites. Only phage VMJ710 (obtained from the community sewage water, Manimajra) could be propagated further.

3.2. Phage Morphology

Phage VMJ710 produced a clear plaque (4–5 mm in diameter) (Figure 2a). The purified phage particles examined under transmission electron microscopy showed phage VMJ710 to have an icosahedral head of 85 ± 2.4 nm in diameter with a long contractile tail of 130 ± 5 nm, and therefore was classified under the class *Caudoviricetes* (Figure 2b).

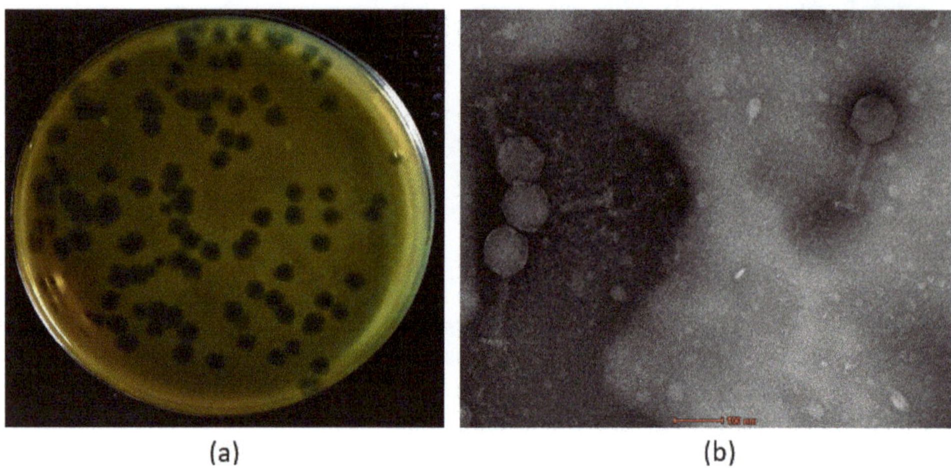

(a)   (b)

Figure 2. (a) Plaque morphology of phage VMJ710. (b) Transmission electron microscopy phage of VMJ710 with a scale bar (red) of 100 nm.

3.3. Phage Stability

Vibrio phage VMJ710 was highly stable at −20 °C and 4 °C (>95%) after 2 h of incubation. The stability rate was approximately 90% at 50 °C but decreased sharply to <40% at 60 °C (Figure 3a).

The phage was highly stable at pH 7 and pH 8 with a >95% stability rate after 120 min of incubation and was unstable at pH 10 with a stability rate of <75% after 120 min (Figure 3b).

3.4. Genome Features of Vibrio Phage VMJ710

The complete linear double-stranded DNA genome of phage VMJ710 was 121.4 kb in size, with a GC content of 37.1%. A total of 3,562,707 paired-end raw reads with 150 bp lengths were generated by the Illumina HiSeq sequencer. There were 215 ORFs predicted, and the mean ORF density was 1.76 ORFs per kb. A total of 118 ORFs were present in the direct strand, and 97 ORFs were present in the complementary strand of the phage genome. The putative functions of each ORF are summarized in Table S3. Thirty-five ORFs (17.6%) were predicted to encode functional proteins, whereas 179 ORFs (83.2%) were predicted as hypothetical proteins. Thirty-six functional proteins were classified into different functional groups (Figure 4, Table S3). Out of the thirty-six ORFs, twenty were predicted to encode for

DNA replication/metabolism-related proteins such as anaerobic nucleoside diphosphate reductase, HNH homing endonuclease, anaerobic NTP reductase small subunit, putative helicase, recombination-associated protein, putative exodeoxyribonuclease, putative thymidylate synthase, putative adenine methyl transferase, ribonucleotides diphosphate reductase, ribonucleotides diphosphate, reductase beta chain, ATP dependent protease subunit, DNA ligase, putative ribose phosphate pyrophosphokinase, putative antirepressor protein nicotinate phosphoribosyl transferase, DNA replicative helicase/primase, DNA polymerase, and ribonuclease H and PhoH family protein. The ORF 55 was predicted to encode large subunits of terminase involved in phage DNA packaging. It was predicted that seven ORFs encoded structural proteins such as the putative tail fiber protein, putative baseplate component, putative baseplate assembly protein, tail length tape measure protein, and putative major head protein. The ORF 130 was predicted to encode for the host lysis protein putative baseplate hub subunit and tail lysozyme. Furthermore, BLASTP analysis of the VMJ710 genome revealed no similarities to genes encoding for integrase, recombinase, and excisionase. Consequently, the phage VMJ710 was considered a lytic bacteriophage and was selected for further studies. In addition, genome analysis showed that the phage VMJ710 does not contain gene encoding for virulence factors, antibiotic resistance, and CRISPR-Cas.

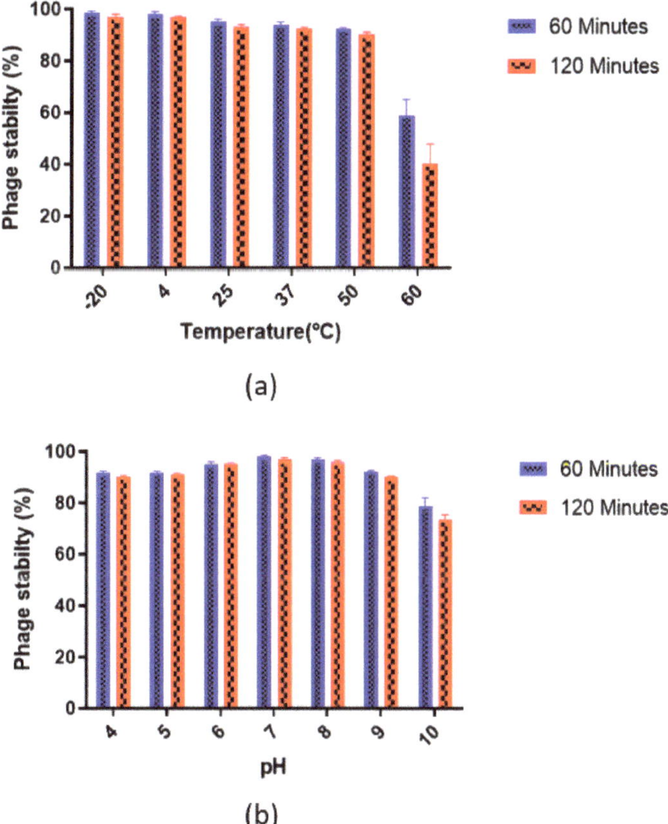

Figure 3. *Vibrio* phage VMJ710 stability rate: (a) Thermal stability; (b) pH stability. Each data point represents the mean result of experiments performed in triplicates, while the error bars represent the standard deviation.

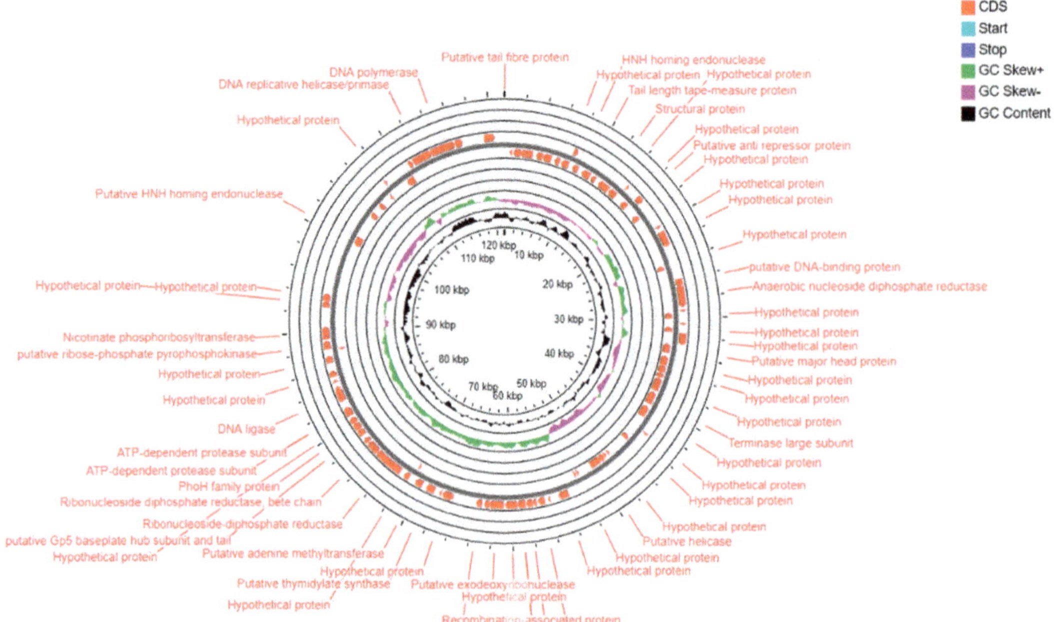

Figure 4. Circular genome map of phage VMJ710.

3.5. Phylogenetic Analysis

A phylogenetic tree generated using MEGA X (large terminase subunit based) with other related phages (with an identity of >53.2% in BLASTp) showed that the phage VMJ710 is closely related to the class *Caudoviricetes* phage ICP1_2012_A (Figure 5a). The whole genome-based phylogenetic tree of phage VMJ710 was constructed using Viptree to determine the exact taxonomic position (Figure 5b,c). Phage VMJ710 was placed next to the closely related phages with 121.4–133.6 kb genomes. Phage VMJ710 was classified under the class *Caudoviricetes* in the taxonomical branch *Duplodnaviridae* > *Heunggongvirae* > *Uroviricota* > *Caudoviricetes*.

3.6. Comparative Genomics and Core Genome Analysis

The complete genome of the phage VMJ710 was compared with four related phages (Table 1). According to the Progressive Mauve analysis, the genome of each phage stated above has three homologous local collinear blocks (LCBs), and the boundaries of colored blocks represent the breakpoints of genome organization (Figure 6). Mauve revealed a few conserved portions of the aforementioned phages that appear to be devoid of genomic changes internally. Some potential orthologous genes have also been observed and are denoted by the black vertical bars (Figure 6). We observed 92.09% (198/215) core and 7.90% (17/215) accessory ORFs among these five phages. The Brig tool analysis showed that the most variable genome regions of insertions or deletions among the above-mentioned phages encoding for accessory ORFs are shown as gaps (Figure S3). No significant areas of GC content variation were found. Despite being present in all of the genomes included in our investigation, the sequence conservation of the core-genome ORFs varied. The core ORFs were divided into three categories based on average pairwise nucleotide and amino acid sequence identity: conserved-core, synonymous-core, and divergent-core (Figure 7, Table S4).

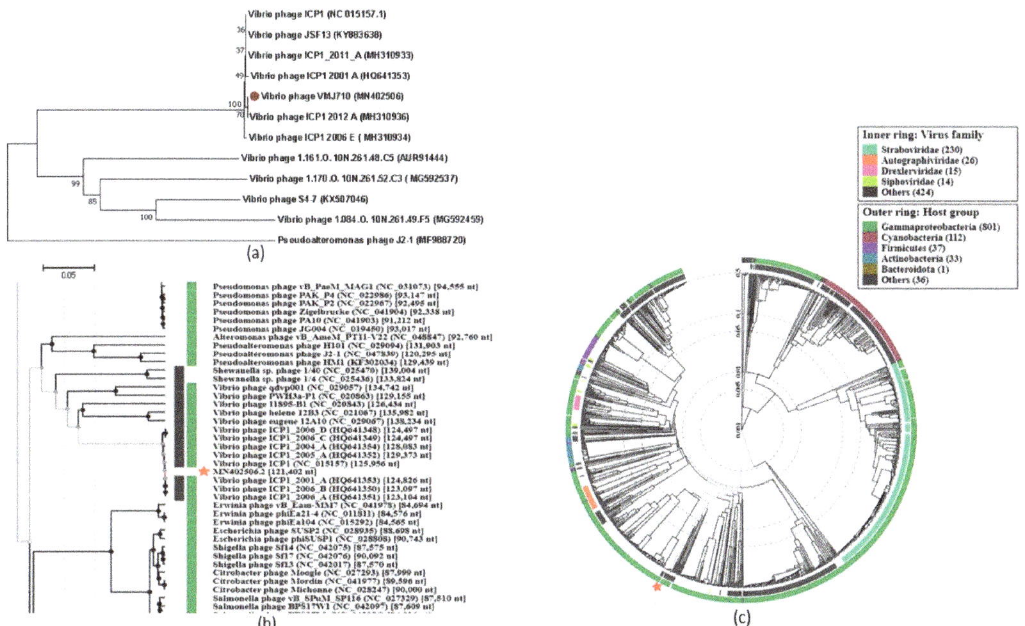

Figure 5. (a) Phylogenetic relationship between *Vibrio* phage VMJ710 (red sphere) and the other related phages. (b,c) are rectangular and circular phylogenetic trees generated using ViPTree [30]. The external and internal rings are colored according to the host bacterial group and virus family, respectively. Red star represents genome of *Vibrio* phage VMJ710.

Table 1. Properties of *Vibrio* phage VMJ710 and related four *Vibrio* phages.

Vibrio Phage Name	Taxonomic Class	Isolation Year	Isolation Source	Genomic Size (bp)	G+C Content	Genbank Accession Number
ICP1	*Caudoviricetes*	2001	Stool	125,956	37.1	HQ641347
ICP1_2012_A	*Caudoviricetes*	2012	Stool	121,418	37.2	MH310936
VMJ710	*Caudoviricetes*	2015	Water	121,402	37.2	MN402506
JSF13	*Caudoviricetes*	2017	Water	128,814	37.2	KY883638
ICP1_2011_A	*Caudoviricetes*	2011	Stool	126,861	37.1	MH310933

The conserved core was made up of 87 ORFs that had 100% nucleotide sequence and amino acid sequence identity. The synonymous core includes 27 ORFs with a small amount of nucleotide variability between genomes, but all mutations were silent, resulting in identical amino acid sequences. The remaining 84 ORFs with divergent nucleotide and amino acid pairwise identities made up the divergent core. These divergent ORFs had pairwise identity similarities ranging from 99.9 to 92.7% at the nucleotide level and from 99.95% to 94.16% for amino acid sequences (Table S4).

The accessory ORFs were distributed in three separate groups of patterns A, B, and C throughout all three genomes (Table S5). These ranged from occurrences in three genomes ($n = 2$) to singletons ($n = 2$) with an alternative pattern of accessory ORFs occurrence.

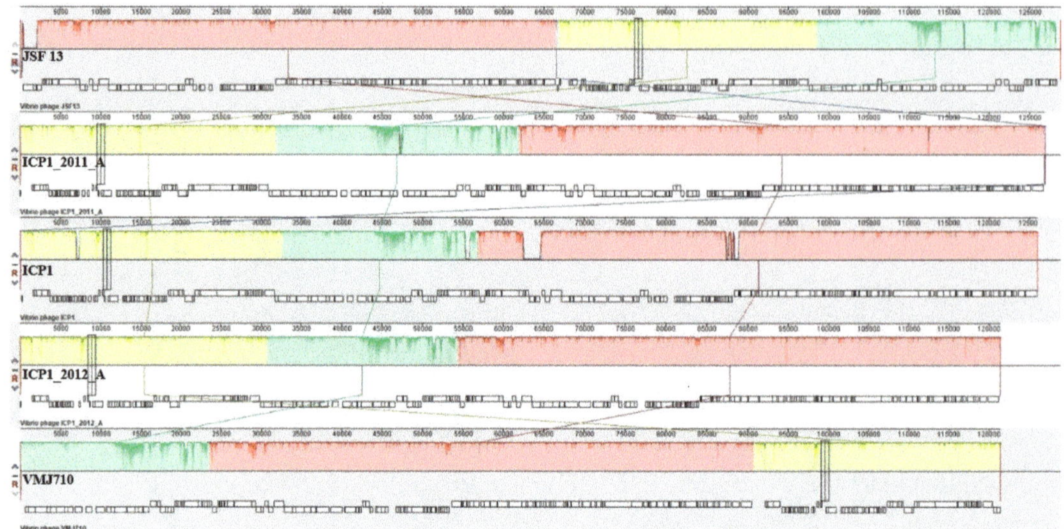

Figure 6. Whole-genome sequence comparison between the genomes of phage VMJ710 and four selected *Vibrio* phages using Mauve 2.0: Annotated CDS (genes) are depicted as white rectangles, with reverse-strand genes, relocated downward. The height of the similarity profile indicates the degree of genomic sequence similarity between the matched regions. Three different colored local collinear blocks (LCBs) illustrate the homologous areas between VMJ710 and four related phages.

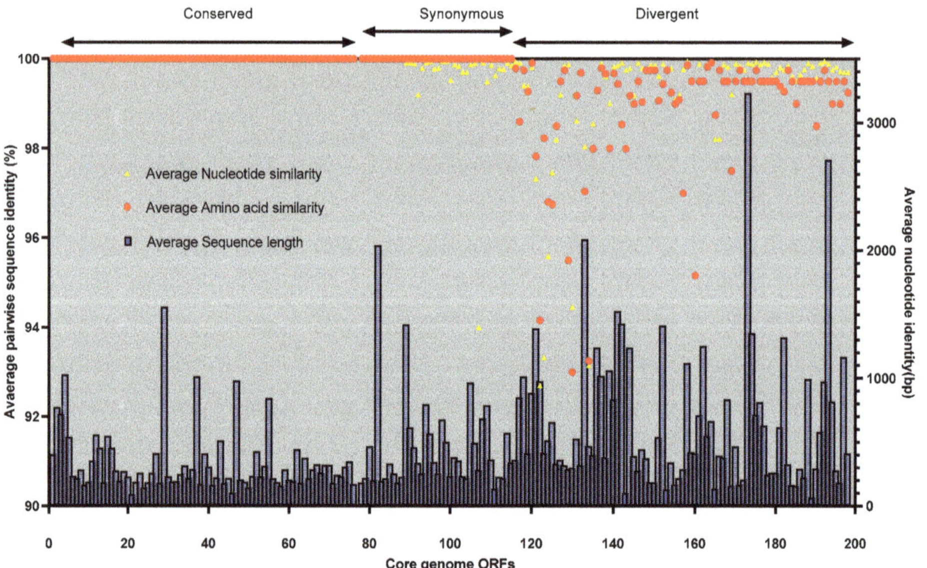

Figure 7. ORF divergence in the core genome. The average pairwise similarity of both DNA nucleotide sequence (yellow triangles and amino acid residue (red dots) alignments are used to organize all ICP1 core-genome ORFs.

3.7. Conserved Functional Domains

The majority of ORFs in the accessory and core genomes were categorized as hypothetical proteins. Only 35 of the core ORFs (17.6%), including 13 in the conserved core, 5 in the synonymous core, and 17 in the divergent core, had a projected function. The accessory core only has one ORF with a putative function (Table S5).

3.8. Host Range Testing

Phage VMJ710 was tested against 26 MDR *V. cholerae* isolates and was observed to be active against 12 (46.1%) strains (Figure S2). The highest EOP (>1) was observed against *V. cholerae* 219, 235, ELPGI212, 220, 15-238 isolates, and the lowest EOP (0.001–0.099) was against *V. cholerae* 231, 229, and 221 isolates (Table 2).

Table 2. The clinical isolate data including source, date of isolation, antibiotic sensitivity, and killing effect of the phage VMJ710 on the individual bacterial isolates.

Date of Collection, Antibiotic Sensitivity, Source of Isolation, and Phage Killing of *V. cholerae* Strains.

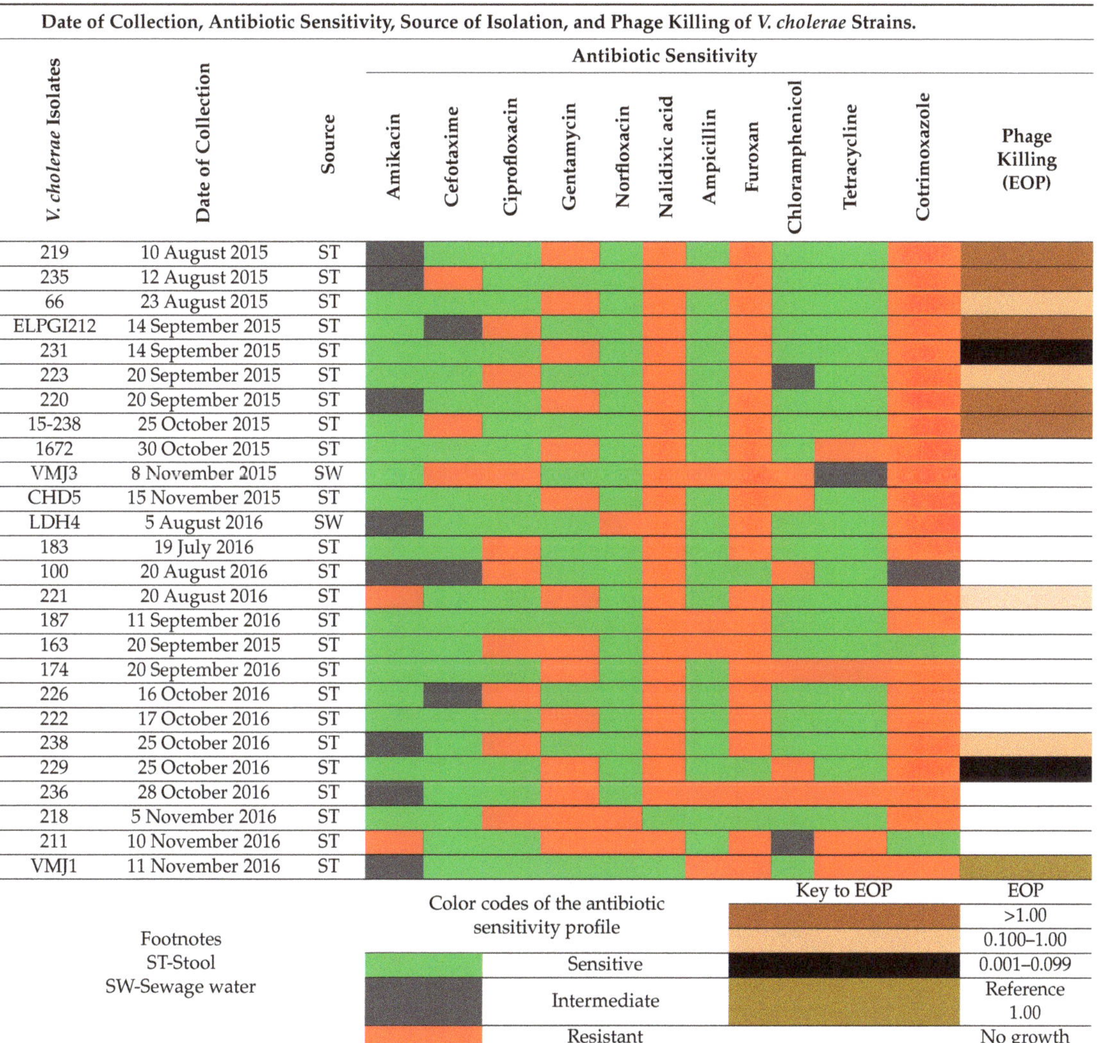

V. cholerae Isolates	Date of Collection	Source	Antibiotic Sensitivity (Amikacin, Cefotaxime, Ciprofloxacin, Gentamycin, Norfloxacin, Nalidixic acid, Ampicillin, Furoxan, Chloramphenicol, Tetracycline, Cotrimoxazole)	Phage Killing (EOP)
219	10 August 2015	ST		
235	12 August 2015	ST		
66	23 August 2015	ST		
ELPGI212	14 September 2015	ST		
231	14 September 2015	ST		
223	20 September 2015	ST		
220	20 September 2015	ST		
15-238	25 October 2015	ST		
1672	30 October 2015	ST		
VMJ3	8 November 2015	SW		
CHD5	15 November 2015	ST		
LDH4	5 August 2016	SW		
183	19 July 2016	ST		
100	20 August 2016	ST		
221	20 August 2016	ST		
187	11 September 2016	ST		
163	20 September 2015	ST		
174	20 September 2016	ST		
226	16 October 2016	ST		
222	17 October 2016	ST		
238	25 October 2016	ST		
229	25 October 2016	ST		
236	28 October 2016	ST		
218	5 November 2016	ST		
211	10 November 2016	ST		
VMJ1	11 November 2016	ST		

Footnotes
ST-Stool
SW-Sewage water

Color codes of the antibiotic sensitivity profile:
- Sensitive
- Intermediate
- Resistant

Key to EOP:

EOP	
>1.00	
0.100–1.00	
0.001–0.099	
Reference 1.00	
No growth	

The lytic ability of phage was tested against *V. cholerae* ELPGI212 strain at MOIs 0.01, 0.1, and 1.0 using the time–kill assay. No significant reduction in bacterial growth was observed at any MOI up to 2 h of incubation. When compared with the control group, bacterial growth was reduced ($p < 0.05$) after 4–6 h of treatment at all MOIs (Figure 8). Phage VMJ710 was found to be most effective at MOI 1 compared with MOI 0.01 and MOI 0.1.

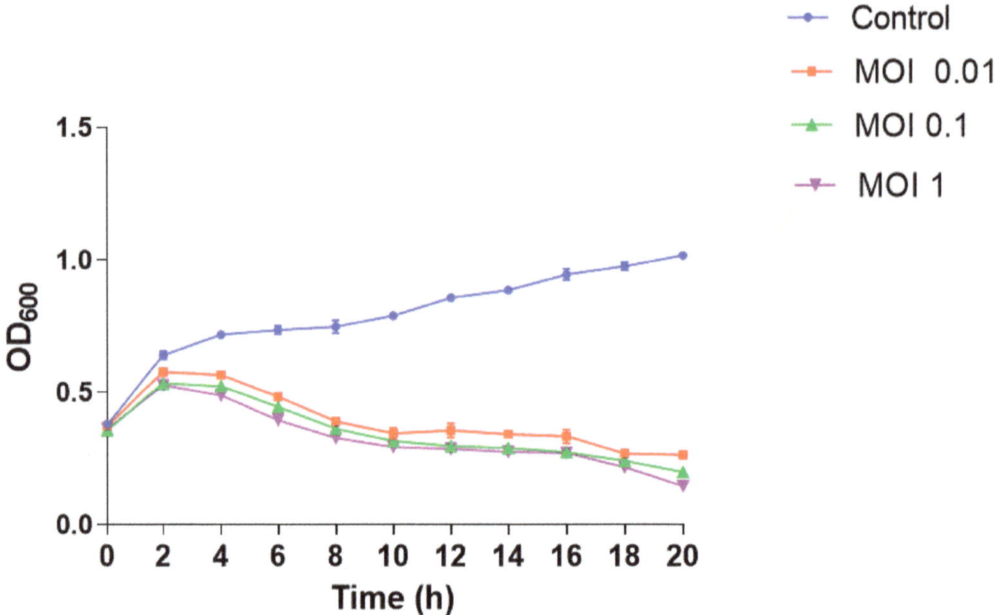

Figure 8. Time–kill assay of phage VMJ710 against MDR *V. cholerae* strains at MOI 1.0, MOI 0.1, and MOI 1. Error bars indicate the standard deviation among triplicate experiments.

3.9. Anti-Biofilm Activity of Phage VMJ710 against Multidrug-Resistant V. cholerae

The quantitative estimation of biofilm formation of 26 MDR *V. cholerae* strains by the microtiter well test method categorized the strains as follows: non–adherent (7.69%), weak (23.07%), moderate (38.4%), and strong (30.7%) (Figure S4). It demonstrates the production of aggregates and microcolonies, resulting in the formation of a compact and dense biofilm structure covering the surface of the polystyrene coverslip (Figure S5). As 24 h old biofilms were exposed to phages at three different titers (10^6, 10^7, and 10^8 PFU), the biofilm viable counts significantly decreased ($p < 0.01$) (Figure 9a). The phage titer of 10^8 PFU was found to be most effective to degrade preformed biofilms (Figure 9a, b). The structural architecture of an established 24 h biofilm at $5000\times$ is shown in Figure S5. The SEM results showed bacterial dispersal, clear disruption, and a reduction in the bacterial biofilm structure (Figure 9b).

3.10. Efficacy of Vibrio Phage VMJ710 against MDR V. cholerae ELPGI212 in Mice

A bacterial inoculum of 10^7 CFU (minimum infective dose) after 8 h was found to be the most optimum to induce cholera infection in the infant mice.

Before testing the prophylactic effect of the phage to prevent cholera infection, we tested whether phages could be retained in the mice intestine in the absence of host bacteria. When animals were dosed with a high titer (1×10^9 PFU), 1.7×10^8 PFU/g (SD $\pm$ 13,769,531) of phages could be recovered after 12 h (Figure 10a). However, at 24 h, a steep decline in the phage titer was observed. The data provided in Figure 10b showed that phage prophylaxis was maximum when mice were dosed with phage VMJ710 before 6 h of bacterial inoculation where the bacterial load was reduced by approximately three orders of magnitude in comparison to the infection control group.

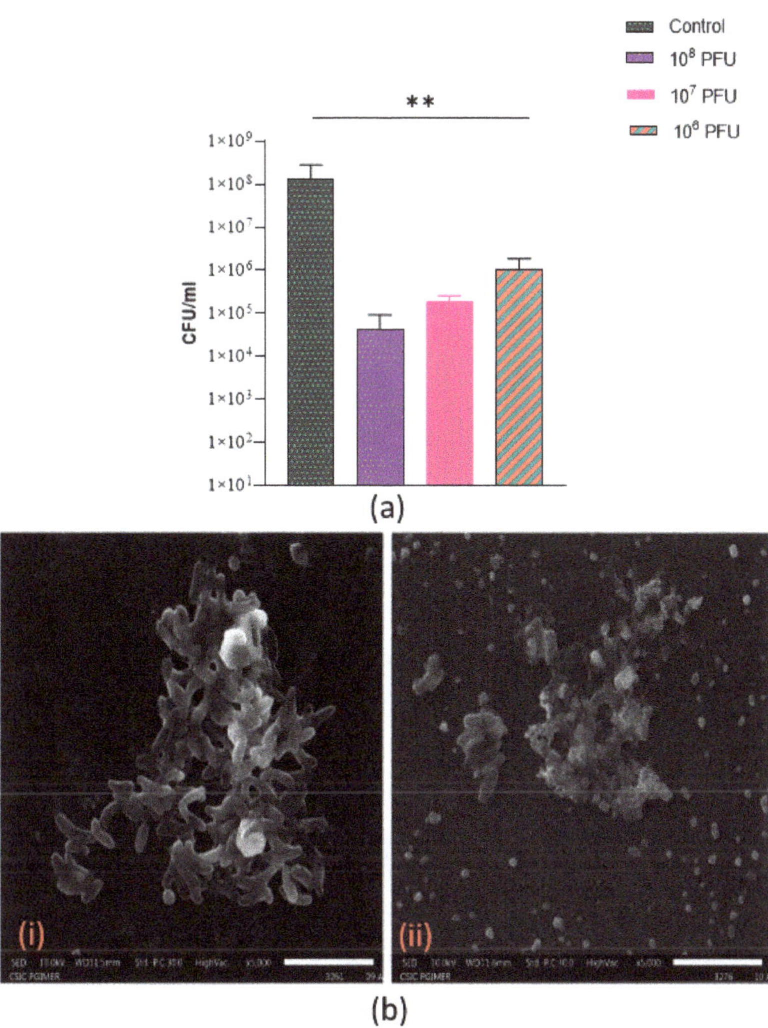

Figure 9. (**a**) Activity of phage VMJ710 at three different titers (10^6, 10^7, 10^8 PFU) on preformed biofilm structure after 24 h incubation in terms of biofilm viable counts. Error bars indicate Mean ± SD. ** indicates a statistically significant difference at the *p*-value < 0.01. (**b**) SEM images at 5000× (magnification) (i) Control group (ii) Effect of phage VMJ710 (10^8 PFU) on biofilm after 24 h of incubation. Scale bars (white)—5 μm.

In the treatment group, a reduction in the bacterial count was observed from 1.0×10^8 CFU/g (SD ± 84,660,772) to 7.6×10^4 CFU/g (SD ± 6046) after 24 h, and furthermore, no *V. cholerae* cell was detected after 48 h in comparison to the untreated group, where bacterial counts raised to 1.6×10^{10} CFU/g (SD ± 1,737,354,310) (Figure 10c). Animal survival percentage was 83% in the treatment group and 57% in the infection control group (10d). Histopathological analysis of the small intestine of mice infected with *V. cholerae* ELPGI212 showed infiltration of lymphocytes, plasma cells, and disruption of the overlying mucosa, where intestinal microvilli were normal, and no intestinal damage was observed in the phage-treated group. Phage-treated mice showed less damage to intestinal architecture and reduced inflammatory exudate (Figure 11).

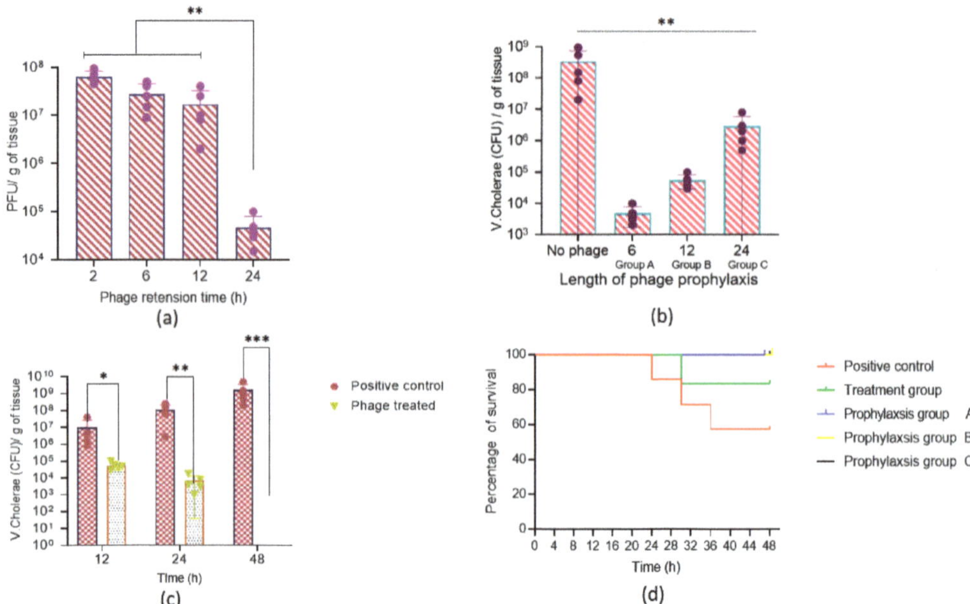

Figure 10. (**a**) *Vibrio* phage number (PFU/g of tissue) retained in the mouse intestine without host bacteria. (**b**) *V. cholerae* (CFU/g of tissue) cells recovered when phages were administered before bacterial challenge as follows: Prophylaxis group A (6 h), prophylaxis group B (12 h), and prophylaxis group C (24 h). (**c**) Bacterial load (CFU/g) recovered from mice intestine tissue after treatment with phage at 12, 24, and 48 h post-infection. (**d**) Percent survival of mice in different groups. Error bars indicate Mean $\pm$ SD. The significant difference indicated by *$p < 0.05$, ** $p < 0.01$, and *** $p < 0.001$.

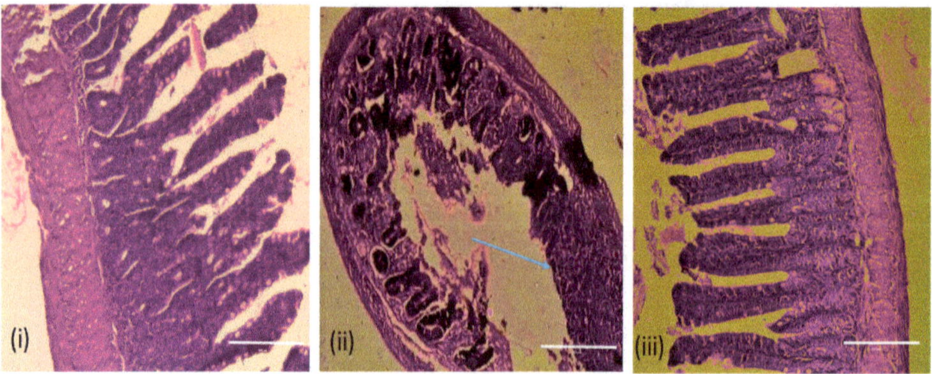

Figure 11. Histopathology of the mouse small intestine: (**i**) Negative control showing normal villus architecture. (**ii**) Infection-only control group; disruption of overlying mucosa and inflammatory cell infiltration (blue arrow). (**iii**) Intestinal architecture after phage treatment at 10^8 PFU. Scale bar (resolution)-200 µm.

4. Discussion

Bacteriophages are natural predators of bacteria and are abundantly available. Phages, like all viruses, are very species-specific in terms of their hosts, infecting only a single bacterial species or even distinct strains within a species. Phages are promising alternatives to treat multi-drug-resistant bacteria and the biocontrol of cholera. We isolated *Vibrio* phage

VMJ710 from the community sewage water sample in Manimajra, Chandigarh, in 2015 when an outbreak of cholera was occurring in this region. VMJ710 belongs to the class *Caudoviricetes* and is very similar (BLASTn identity 99.96%) to phage ICP1_2012 which was isolated from patients' stool samples that were collected in India. All-tailed bacterial and archaeal viruses with icosahedral capsids and double-stranded DNA genomes are classified into the class *Caudoviricetes* [38].

Phage VMJ710 has a long contractile tail of 130 ± 5 nm and an icosahedral capsid of 85 ± 2.4 nm length, similar to phage ICP1, which has an 86 nm long icosahedral capsid and a 106 nm long tail [39]. ICP1 phages are dominant phages widespread throughout the Bay of Bengal's coastline zones [40]. In Bangladesh, during cholera outbreaks, this group of phages is found along with *V. cholerae* in affected patients [40]. The whole-genome sequence of VMJ710 shows the most similarity (BLASTn identity 86–92.97%) to four phages (ICP1_2012_A (MH310936), ICP1 (NC_015157), ICP1_2011_A (MH310933), and JSF13 (KY883638) isolated from different geographic regions across the world in BLASTn analysis. The genome size range of previously studied ICPI (class *Caudoviricetes*) related *Vibrio* phages is 121.4–133.6 kb, similar to VMJ710. Comparative genomics of VMJ710 and related four phages using BRIG and Mauve tools revealed the conservation of nucleotide homology to a larger extent. Phage VMJ710 is closely related to ICP1-like phages of the class *Caudoviricetes*, as demonstrated in the phylogenetic study based on the terminase large sequence and ViPTree. Some of the ICP1 phages possess the CRISPR-Cas (clustered regularly interspersed short palindromic repeats-CRISPR-associated proteins) system, which is involved in bacterial defense against predators such as phages [41]. Another mechanism to resist is the prevention of phage reproduction by hosting phage-inducible chromosomal islands (PICI) [40]. CRISPR-Cas systems are typically found in bacteria and archaea; however, *V. cholerae-* specific ICP1 phages have recently been revealed to contain the CRISPR-Cas system that inactivates PICI-like elements (PLE) in *V. cholerae* [42]. In a previous study, CRISPR-Cas-related sequences were found in five (JSF5, JSF6, JSF13, JSF14, and JSF17) of the 29 sequenced *Vibrio* phages (17.2%) [41]. VMJ710 lacks a CRISPR-Cas sequence, whereas the other two closely related phages (JSF13 and ICP1_2011_A) possess a CRISPR-Cas system. It is presumed that enhanced interactions of phages with cholera bacteria during seasonal epidemics of cholera may be facilitated by the phage-encoded CRISPR-Cas system.

Like previously studied *Vibrio* ICP1 phages, the majority of the phage VMJ710 genome is related to hypothetical proteins, and more than half of ORFs (20/36) with the predicted functions are involved in replication/metabolism-related proteins. A putative Gp5 base-plate hub subunit and a tail lysozyme with N-acetylmuramidase activity were encoded by ORF 132 of phage VMJ710. These components are crucial for locally digesting the peptidoglycan layer to allow the tube to enter the periplasm [43]. ORFs 10 and 11 encode for HNH homing proteins that are involved in endonuclease activity, such as site-specific homing endonucleases [44]. HNH proteins belong to a large Pfam protein family and are associated with nuclease activity in all kingdoms of life [44]. Tailed bacteriophages use terminase enzymes to bundle their enormous double-stranded DNA genomes into a preformed protein shell known as the "prohead". The ORF 55 was predicted to encode for a large subunit terminase protein which is involved in the translocation of the viral capsid DNA during the final stage of phage assembly. The most prevalent proteins found in the majority of phages in the class *Caudoviricetes*, are terminases [45]. Endonuclease proteins with the HNH motif may interact with phage terminase proteins to facilitate the packaging and maturation of viral DNA [44].

This phage has a host range of 46 % and no bacterial resistance was observed against phage VMJ710 in comparison to a study by Yen et al., who observed bacterial resistance against the ICP1 phage after 4–6 h of treatment [37]. The host range of VMJ710 and ICP1-like phages is limited to *V. cholerae* O1, whereas the host range of ICP2 and ICP3 includes non-O1 *V. cholerae* strains. ICP2 and ICP3 can lyse the *V. cholerae* O139 strain MO10 and the non-O139 strain CR034-24, respectively [39]. Biofilm-forming cells are more resistant to antibacterial agents and may contribute to higher levels of antimicrobial resistance [46–49] Phages have been considered potential agents to control biofilms [50]. Extracellular polysaccharide material is a well-known component of stable bacterial biofilms and can protect bacteria from desiccation, predation, and bacteriophage attack [51–53]. Numerous studies have revealed that the components of the biofilm, its age, the efficacy of the phage, and the length of the treatment play a role in efficient biofilm eradication [54,55]. The findings of our study confirmed the concept that the use of phage can reduce *Vibrio* biofilms, in particular those grown on plastic surfaces. Phage treatment has significantly removed the biofilms grown on polystyrene coverslip surfaces. The SEM results were consistent with the quantitative data, which showed a considerable decrease in biofilm-forming cells and biomass. The phage was stable over a wide range of temperatures and pH when incubated without a host, suggesting that it has good thermal and pH stability. Thus, it would be easy to preserve and beneficial in different environmental conditions in phage therapy.

Our study is unique as we tested both the preventive and treatment potential of phage VMJ710 against a biofilm-forming MDR *V. cholerae* isolate in a mouse model. The infant mouse model is the most commonly used in a majority of cholera studies [56]. Adult mice are not able to colonize efficiently by *V. cholerae* without the elimination of the gut microflora, whereas infant mice are efficiently colonized [57]. Though the exact reason for the colonization of infant mice is not well studied but might be because of their immature or poor immune system. Bacterial counts were reduced significantly, and animals survived within 24–48 h post-bacterial challenge when phages were administered by orogastric inoculation. This indicated that phage preparation needs to be administered within a specified time. The histopathological examination of the intestine of the phage-treated mice revealed lesser tissue damage and almost normal intestinal walls, crypts, and overlying mucosa. This reveals that phage-treated animals had less severe infections and could withstand a deadly bacterial attack.

We also showed that the phage can prevent the establishment of *V. cholerae* infection and prevent disease when the phage was administered 6–12 h before the bacterial challenge. Our results are similar to those of a previous study by Yen et al., where a cocktail of three highly characterized virulent phages (ICP1, ICP2, and ICP3) was similarly tested in infant mice. When phages were administered orally up to 24 h before *V. cholerae* challenge, the colonization of the intestinal tract was reduced, and cholera-like illness was prevented [37]. In our study, though the most significant reduction in bacterial counts occurred at 6–12 h, mouse survival was the same in the 24 h group, suggesting that phages administered orally up to 24 h before *V. cholerae* infection could prevent cholera. This is helpful as a pre-emptive therapy in cholera-outbreak-affected areas. The phage can be given to close contacts of cholera cases, where the secondary attack rate may be as high as 50%. Due to the increasing resistance of *V. cholerae* to antibiotics, alternative therapies are needed. Biocontrol of the disease, as well as the environmental reservoir via phages, can also be a potential alternative to control cholera. Genome analysis provided no evidence of lysogenic genes (obligately lytic), genes related to potential virulence factors, or antibiotic resistance. Based on all these characteristics, phage VM710 is a suitable and promising candidate as a biocontrol and therapeutic agent. However, more such phages need to be discovered for formulating suitable cocktails. Further trials will be needed to ensure the safety of phages for human use.

5. Conclusions

We isolated a *Vibrio* phage VMJ710 from the community sewage sample during an outbreak of cholera in Chandigarh, India. WGS revealed that the phage was lytic and devoid of genes associated with lysogeny, virulence, or antibiotic resistance. The phage has anti-biofilm activity and is stable under different environmental conditions. This phage can be a suitable candidate for oral phage administration and potentially combatting cholera infections caused by pathogenic MDR *V. cholerae* strains.

Supplementary Materials: The following supporting information can be downloaded at: https://www.mdpi.com/article/10.3390/antibiotics12061046/s1, Figure S1: Water sample collection sites: 1-Ludhiana, 2-Manimajra, 3-sewage treatment plant, Ramdarbar, 4-main sewage drain, PGIMER, 5-Sewage treatment plant, Raipur Khurd, 6-Ambala. Map was created using the online version of Scribble Maps software. Figure S2: Representative picture of host range determination of *Vibrio* phage VMJ710 by spot assay. Figure S3. Whole-genome sequence comparison between the genomes of phage VMJ710 and four selected *Vibrio* phages using the *BRIG tool* v0.95: Reference genome: phage VMJ710 (Innermost black ring) genome. The phage genome sections that have less than 50% or no resemblance to VMJ710 are shown by a gap in the relevant genome ring. Figure S4: Biofilm formation capacity of 26 MDR *Vibrio cholerae* strains in crystal violet assay. Figure S5: The structural architecture of an established 24-hour-old *V. cholerae* biofilm at 5000×. Table S1: Antibiotic sensitivity pattern of 26 MDR *V. cholerae* strains. Table S2: Host range testing results of *Vibrio* phage VMJ710 against 26 *V. cholerae* strains. Table S3: Functional annotation of *Vibrio* phage VMJ710 genome. Table S4: Details of core ORFs details with an average nucleotide sequence length. Table S5: Accessory-genome ORF occurrence matrix

Author Contributions: Conceptualization, N.T. and N.C.; Methodology, N.T., N.C., H.K., V.M., V.K. and A.B., Validation, N.T. and B.M.; Analysis, N.C. and N.T.; Investigation, N.T. and N.C.; Project, N.T. and N.C.; Resources, N.T. and B.M.; Data curation; N.C. and N.T.; Writing—original draft preparation, N.T. and N.C.; Writing—review and editing, N.C. and N.T.; Supervision, N.T. and B.M.; Funding acquisition, N.T. All authors have read and agreed to the published version of the manuscript.

Funding: The University Grants Commission, New Delhi, funded this research through a contingency grant for a Ph.D. student scholarship (313690).

Institutional Review Board Statement: All animal experiments at this study was approved the Institutional Animal Ethics Committee (IAEC number Ref.no 93/IEAC/648).

Informed Consent Statement: Not applicable.

Data Availability Statement: The data presented in this study are available in this article and data used to support the findings of this study are available from the corresponding author upon request.

Acknowledgments: We are grateful to the Municipal corporation, Chandigarh for permitting us to collect water samples from different sewage treatment plants.

Conflicts of Interest: The authors declare no conflict of interest.

References

1. Ramamurthy, T.; Mutreja, A.; Weill, F.; Das, B.; Ghosh, A.; Nair, G.B. Revisiting the Global Epidemiology of Cholera in Conjunction With the Genomics of *Vibrio Cholerae*. *Front. Public Health* **2019**, *7*, 203. [CrossRef] [PubMed]
2. Faruque, S.M.; Kamruzzaman, M.; Meraj, I.M.; Chowdhury, N.; Nair, G.B.; Sack, R.B.; Colwell, R.R.; Sack, D.A. Pathogenic Potential of Environmental *Vibrio Cholerae* Strains Carrying Genetic Variants of the Toxin-Coregulated Pilus Pathogenicity Island. *Infect. Immun.* **2003**, *71*, 1020–1025. [CrossRef] [PubMed]
3. Thapa Shrestha, U.; Adhikari, N.; Maharjan, R.; Banjara, M.R.; Rijal, K.R.; Basnyat, S.R.; Agrawal, V.P. Multidrug Resistant *Vibrio Cholerae* O1 from Clinical and Environmental Samples in Kathmandu City. *BMC Infect. Dis.* **2015**, *15*, 104. [CrossRef]
4. Nelson, E.J.; Nelson, D.S.; Salam, M.A.; Sack, D.A. Antibiotics for Both Moderate and Severe Cholera. *N. Engl. J. Med.* **2011**, *364*, 5–7. [CrossRef] [PubMed]
5. Taneja, N.; Mishra, A.; Batra, N.; Gupta, P.; Mahindroo, J.; Mohan, B. Inland Cholera in Freshwater Environs of North India. *Vaccine* **2020**, *38*, A63–A72. [CrossRef] [PubMed]
6. Ali, M.; Gupta, S.S.; Arora, N.; Khasnobis, P.; Venkatesh, S.; Sur, D.; Nair, G.B.; Sack, D.A.; Ganguly, N.K. Identification of Burden Hotspots and Risk Factors for Cholera in India: An Observational Study. *PLoS ONE* **2017**, *12*, e0183100. [CrossRef]
7. Naser, I.B.; Hoque, M.M.; Abdullah, A.; Bari, S.M.N.; Ghosh, A.N.; Faruque, S.M. Environmental Bacteriophages Active on Biofilms and Planktonic Forms of Toxigenic *Vibrio Cholerae*: Potential Relevance in Cholera Epidemiology. *PLoS ONE* **2017**, *12*, e0180838. [CrossRef]
8. Khatoon, Z.; McTiernan, C.D.; Suuronen, E.J.; Mah, T.F.; Alarcon, E.I. Bacterial Biofilm Formation on Implantable Devices and Approaches to Its Treatment and Prevention. *Heliyon* **2018**, *4*, e01067. [CrossRef]
9. Lin, D.M.; Koskella, B.; Lin, H.C. Phage Therapy: An Alternative to Antibiotics in the Age of Multi-Drug Resistance. *World J. Gastrointest. Pharmacol. Ther.* **2017**, *8*, 162–173. [CrossRef]
10. Chan, B.K.; Abedon, S.T.; Loc-carrillo, C. Phage Cocktails and the Future of Phage Therapy. *Futur. Microbiol.* **2013**, *8*, 769–783. [CrossRef]

11. Chaudhary, N.; Mohan, B.; Mavuduru, R.S.; Kumar, Y.; Taneja, N. Characterization, Genome Analysis and in Vitro Activity of a Novel Phage VB_EcoA_RDN8.1 Active against Multi-Drug Resistant and Extensively Drug-Resistant Biofilm-Forming Uropathogenic Escherichia Coli Isolates, India. *J. Appl. Microbiol.* **2022**, *132*, 3387–3404. [CrossRef] [PubMed]

12. Chaudhary, N.; Narayan, C.; Mohan, B.; Taneja, N. Indian Journal of Medical Microbiology Characterization and in Vitro Activity of a Lytic Phage RDN37 Isolated from Community Sewage Water Active against MDR Uropathogenic *E. Coli*. *Indian J. Med. Microbiol.* **2021**, *39*, 343–348. [CrossRef] [PubMed]

13. Chaudhary, N.; Singh, D.; Maurya, R.K.; Mohan, B.; Tanejaa, N. Complete Genome Sequence of Salmonella Phage VB_SenA_SM5, Active against Multidrug-Resistant *Salmonella Enterica* Serovar Typhi Isolates. *Microbiol. Resour. Announc.* **2022**, *44*, e00309-22. [CrossRef] [PubMed]

14. Chaudhary, N.; Singh, D.; Maurya, R.K.; Mohan, B.; Mavuduru, R.S.; Taneja, N. Whole Genome Sequencing and in Vitro Activity Data of Escherichia Phage NTEC3 against Multidrug-Resistant Uropathogenic and Extensively Drug-Resistant Uropathogenic *E. coli* Isolates. *Data Br.* **2022**, *43*, 108479. [CrossRef]

15. Gu, Y.; Xu, Y.; Xu, J.; Yu, X.; Huang, X.; Liu, G.; Liu, X. Identification of Novel Bacteriophage vB _ EcoP-EG1 with Lytic Activity against Planktonic and Biofilm Forms of Uropathogenic *Escherichia coli*. *Appl. Microbiol. Biotechnol.* **2018**, *103*, 315–326. [CrossRef]

16. Surekhamol, I.S.; Deepa, G.D.; Somnath Pai, S.; Sreelakshmi, B.; Varghese, S.; Bright Singh, I.S. Isolation and Characterization of Broad Spectrum Bacteriophages Lytic to *Vibrio Harveyi* from Shrimp Farms of Kerala, India. *Lett. Appl. Microbiol.* **2014**, *58*, 197–204. [CrossRef]

17. Chaudhary, N.; Singh, D.; Narayan, C.; Samui, B.; Mohan, B.; Mavuduru, R.S.; Taneja, N. Complete Genome Sequence of *Escherichia* Phage 590B, Active against an Extensively Drug-Resistant Uropathogenic *Escherichia coli* Isolate. *Microbiol. Resour. Announc.* **2021**, *10*, 9–10. [CrossRef]

18. Martin, M. Cutadapt Removes Adapter Sequences from High-Throughput Sequencing Reads. *EMBnet. J.* **2011**, *17*, 10–12. [CrossRef]

19. Chaudhary, N.; Mohan, B.; Taneja, N. Draft Genome Sequence of *Escherichia* Phage PGN829.1, Active against Highly Drug-Resistant Uropathogenic *Escherichia coli*. *Microbiol. Resour. Announc.* **2018**, *7*, 20. [CrossRef]

20. Hunt, M.; Gall, A.; Ong, S.H.; Brener, J.; Ferns, B.; Goulder, P.; Nastouli, E.; Keane, J.A.; Kellam, P.; Otto, T.D. IVA: Accurate de Novo Assembly of RNA Virus Genomes. *Bioinformatics* **2015**, *31*, 2374–2376. [CrossRef]

21. Besemer, J.; Borodovsky, M. GeneMark: Web Software for Gene Finding in Prokaryotes, Eukaryotes and Viruses. *Nucleic Acids Res.* **2005**, *33*, 451–454. [CrossRef] [PubMed]

22. Delcher, A.L.; Bratke, K.A.; Powers, E.C.; Salzberg, S.L. Identifying Bacterial Genes and Endosymbiont DNA with Glimmer. *Bioinformatics* **2007**, *23*, 673–679. [CrossRef] [PubMed]

23. Aziz, R.K.; Bartels, D.; Best, A.; DeJongh, M.; Disz, T.; Edwards, R.A.; Formsma, K.; Gerdes, S.; Glass, E.M.; Kubal, M.; et al. The RAST Server: Rapid Annotations Using Subsystems Technology. *BMC Genom.* **2008**, *9*, 75. [CrossRef] [PubMed]

24. Laslett, D.; Canback, B. ARAGORN, a Program to Detect TRNA Genes and TmRNA Genes in Nucleotide Sequences. *Nucleic Acids Res.* **2004**, *32*, 11–16. [CrossRef]

25. Couvin, D.; Bernheim, A.; Toffano-Nioche, C.; Touchon, M.; Michalik, J.; Néron, B.; Rocha, E.P.C.; Vergnaud, G.; Gautheret, D.; Pourcel, C. CRISPRCasFinder, an Update of CRISRFinder, Includes a Portable Version, Enhanced Performance and Integrates Search for Cas Proteins. *Nucleic Acids Res.* **2018**, *46*, W246–W251. [CrossRef]

26. Grant, J.R.; Stothard, P. The CGView Server: A Comparative Genomics Tool for Circular Genomes. *Nucleic Acids Res.* **2008**, *36*, W181–W184. [CrossRef]

27. Liu, B.; Zheng, D.; Jin, Q.; Chen, L.; Yang, J. VFDB 2019: A Comparative Pathogenomic Platform with an Interactive Web Interface. *Nucleic Acids Res.* **2019**, *47*, D687–D692. [CrossRef]

28. Alcock, B.P.; Raphenya, A.R.; Lau, T.T.Y.; Tsang, K.K.; Bouchard, M.; Edalatmand, A.; Huynh, W.; Nguyen, A.L.V.; Cheng, A.A.; Liu, S.; et al. CARD 2020: Antibiotic Resistome Surveillance with the Comprehensive Antibiotic Resistance Database. *Nucleic Acids Res.* **2020**, *48*, D517–D525. [CrossRef]

29. Chaudhary, N.; Maurya, R.K.; Singh, D.; Mohan, B.; Taneja, N. Genome Analysis and Antibiofilm Activity of Phage 590B. *Pathogen* **2022**, *11*, 1448. [CrossRef]

30. Nishimura, Y.; Yoshida, T.; Kuronishi, M.; Uehara, H.; Ogata, H.; Goto, S. ViPTree: The Viral Proteomic Tree Server. *Bioinformatics* **2017**, *33*, 2379–2380. [CrossRef]

31. Darling, A.C.E.; Mau, B.; Blattner, F.R.; Perna, N.T. Mauve: Multiple Alignment of Conserved Genomic Sequence with Rearrangements. *Genom. Res.* **2004**, *14*, 1394–1403. [CrossRef] [PubMed]

32. Alikhan, N.F.; Petty, N.K.; Ben Zakour, N.L.; Beatson, S.A. BLAST Ring Image Generator (BRIG): Simple Prokaryote Genome Comparisons. *BMC Genom.* **2011**, *12*, 402. [CrossRef] [PubMed]

33. Kutter, E. Phage Host Range and Efficiency of Plating. In *Methods in Molecular Biology*; Humana Press: Totowa, NJ, USA, 2009; Volume 501, pp. 141–149, ISBN 9781588296825.

34. Kaur, H.; Kalia, M.; Chaudhary, N.; Singh, V.; Yadav, V.K.; Modgil, V.; Kant, V.; Mohan, B.; Bhatia, A.; Taneja, N. Repurposing of FDA Approved Drugs against Uropathogenic *Escherichia coli*: In Silico, in Vitro, and in Vivo Analysis. *Microb. Pathog.* **2022**, *169*, 105665. [CrossRef]

35. Dua, P.; Karmakar, A.; Ghosh, C. Virulence Gene Profiles, Biofilm Formation, and Antimicrobial Resistance of *Vibrio Cholerae* Non-O1/Non-O139 Bacteria Isolated from West Bengal, India. *Heliyon* **2018**, *4*, e01040. [CrossRef] [PubMed]

36. Kaur, S.; Sharma, P.; Kalia, N.; Singh, J.; Kaur, S. Anti-Biofilm Properties of the Fecal Probiotic *Lactobacilli* against *Vibrio* Spp. *Front. Cell. Infect. Microbiol.* **2018**, *8*, 120. [CrossRef] [PubMed]
37. Yen, M.; Cairns, L.S.; Camilli, A. A Cocktail of Three Virulent Bacteriophages Prevents *Vibrio Cholerae* Infection in Animal Models. *Nat. Commun.* **2017**, *8*, 14187. [CrossRef]
38. Turner, D.; Shkoporov, A.N.; Lood, C.; Millard, A.D.; Dutilh, B.E.; Alfenas-Zerbini, P.; van Zyl, L.J.; Aziz, R.K.; Oksanen, H.M.; Poranen, M.M.; et al. Abolishment of Morphology-Based Taxa and Change to Binomial Species Names: 2022 Taxonomy Update of the ICTV Bacterial Viruses Subcommittee. *Arch. Virol.* **2023**, *168*, 74. [CrossRef]
39. Seed, K.D.; Bodi, K.L.; Kropinski, A.M.; Ackermann, H.W.; Calderwood, S.B.; Qadri, F.; Camilli, A. Evidence of a Dominant Lineage of *Vibrio Cholerae*-Specific Lytic Bacteriophages Shed by Cholera Patients over a 10-Year Period in Dhaka, Bangladesh. *MBio* **2011**, *2*, e00334-10. [CrossRef]
40. Angermeyer, A.; Das, M.M.; Singh, D.V.; Seed, K.D. Analysis of 19 Highly Conserved *Vibrio Cholerae* Bacteriophages Isolated from Environmental and Patient Sources over a Twelve-Year Period. *Viruses* **2018**, *10*, 299. [CrossRef]
41. Naser, I.B.; Hoque, M.M.; Nahid, M.A.; Tareq, T.M.; Rocky, M.K.; Faruque, S.M. Analysis of the CRISPR-Cas System in Bacteriophages Active on Epidemic Strains of *Vibrio cholerae* in Bangladesh. *Sci. Rep.* **2017**, *7*, 14880. [CrossRef]
42. Seed, K.D.; Lazinski, D.W.; Calderwood, S.B.; Camilli, A. A Bacteriophage Encodes Its Own CRISPR/Cas Adaptive Response to Evade Host Innate Immunity. *Nature* **2013**, *494*, 489–491. [CrossRef]
43. Nakagawa, H.; Arisaka, F.; Ishii, S. Isolation and Characterization of the Bacteriophage T4 Tail-Associated Lysozyme. *J. Virol.* **1985**, *54*, 460–466. [CrossRef]
44. Kala, S.; Cumby, N.; Sadowski, P.D.; Hyder, B.Z.; Kanelis, V.; Davidson, A.R.; Maxwell, K.L. HNH Proteins Are a Widespread Component of Phage DNA Packaging Machines. *Proc. Natl. Acad. Sci. USA* **2014**, *111*, 6022–6027. [CrossRef] [PubMed]
45. Casjens, S.R.; Gilcrease, E.B. Determining DNA Packaging Strategy by Analysis of the Termini of the Chromosomes in Tailed-Bacteriophage Virions. *Methods Mol. Biol.* **2009**, *502*, 91–111. [CrossRef] [PubMed]
46. Gilbert, P.; Das, J.; Foley, I. Biofilm Susceptibility to Antimicrobials. *Adv. Dent. Res.* **1997**, *11*, 160–167. [CrossRef]
47. Costerton, J.W.; Stewart, P.S.; Greenberg, E.P. Bacterial Biofilms: A Common Cause of Persistent Infections. *Science* **1999**, *284*, 1318–1322. [CrossRef]
48. Davies, D. Understanding Biofilm Resistance to Antibacterial Agents. *Nat. Rev. Drug Discov.* **2003**, *11*, 53–62. [CrossRef] [PubMed]
49. Grenier, D.; Grignon, L.; Gottschalk, M. Characterisation of Biofilm Formation by a *Streptococcus Suis Meningitis* Isolate. *Vet. J.* **2009**, *179*, 292–295. [CrossRef]
50. Chan, B.; Abedon, S. Bacteriophages and Their Enzymes in Biofilm Control. *Curr. Pharm. Des.* **2014**, *21*, 85–99. [CrossRef] [PubMed]
51. Landini, P. Cross-Talk Mechanisms in Biofilm Formation and Responses to Environmental and Physiological Stress in *Escherichia coli*. *Res. Microbiol.* **2009**, *160*, 259–266. [CrossRef] [PubMed]
52. Curtin, J.J.; Donlan, R.M. Using Bacteriophages to Reduce Formation of Catheter-Associated Biofilms by *Staphylococcus epidermidis*. *Antimicrob. Agents Chemother.* **2006**, *50*, 1268–1275. [CrossRef] [PubMed]
53. Donlan, R.M.; Costerton, J.W. Biofilms: Survival Mechanisms of Clinically Relevant Microorganisms. *Clin. Microbiol. Rev.* **2002**, *15*, 167–193. [CrossRef] [PubMed]
54. Sharma, M.; Ryu, J.H.; Beuchat, L.R. Inactivation of Escherichia Coli O157:H7 in Biofilm on Stainless Steel by Treatment with an Alkaline Cleaner and a Bacteriophage. *J. Appl. Microbiol.* **2005**, *99*, 449–459. [CrossRef] [PubMed]
55. Drilling, A.; Hons, B.; Morales, S.; Boase, S.; Jervis-bardy, J. Safety and Efficacy of Topical Bacteriophage and Ethylenediaminetetraacetic Acid Treatment Of. *Int. Forum Allergy Rhinol.* **2014**, *4*, 176–186. [CrossRef] [PubMed]
56. Abel, S.; Waldor, M.K. Infant Rabbit Model for Diarrheal Diseases. *Curr. Protoc. Microbiol.* **2015**, *38*, 6. [CrossRef]
57. Matson, J.S. Infant mouse model of *Vibrio cholerae* infection and colonization. In *Vibrio Cholerae: Methods and Protocols*; Humana Press: Totowa, NJ, USA, 2018; pp. 147–152.

antibiotics

Article

An Edible Biopolymeric Microcapsular Wrapping Integrating Lytic Bacteriophage Particles for *Salmonella enterica*: Potential for Integration into Poultry Feed

Arthur O. Pereira [1], Nicole M. A. Barros [1], Bruna R. Guerrero [1], Stephen C. Emencheta [1,2], Denicezar Â. Baldo [3], José M. Oliveira Jr. [3], Marta M. D. C. Vila [1] and Victor M. Balcão [1,4,*]

[1] PhageLab—Laboratory of Biofilms and Bacteriophages, University of Sorocaba, Sorocaba 18023-000, SP, Brazil; artopereira@gmail.com (A.O.P.); a.alencastronicole@gmail.com (N.M.A.B.); riberabruna@gmail.com (B.R.G.); stephen.emencheta@unn.edu.ng (S.C.E.); marta.vila@prof.uniso.br (M.M.D.C.V.)

[2] Department of Pharmaceutical Microbiology and Biotechnology, University of Nigeria, Nsukka 410001, Enugu, Nigeria

[3] LaFiNAU—Laboratory of Applied Nuclear Physics, University of Sorocaba, Sorocaba 18023-000, SP, Brazil; denicezar.baldo@prof.uniso.br (D.Â.B.); jose.oliveira@prof.uniso.br (J.M.O.J.)

[4] Department of Biology and CESAM, University of Aveiro, Campus Universitário de Santiago, P-3810-193 Aveiro, Portugal

* Correspondence: victor.balcao@prof.uniso.br; Tel.: +55-15-21017029

Abstract: This research work aimed at developing an edible biopolymeric microcapsular wrapping (EBMW) integrating lytic bacteriophage particles for *Salmonella enterica*, with potential application in poultry feed for biocontrol of that pathogen. This pathogen is known as one of the main microorganisms responsible for contamination in the food industry and in foodstuff. The current techniques for decontamination and pathogen control in the food industry can be very expensive, not very selective, and even outdated, such as the use of broad-spectrum antibiotics that end up selecting resistant bacteria. Hence, there is a need for new technologies for pathogen biocontrol. In this context, bacteriophage-based biocontrol appears as a potential alternative. As a cocktail, both phages were able to significantly reduce the bacterial load after 12 h of treatment, at either multiplicity of infection (MOI) 1 and 10, by 84.3% and 87.6%, respectively. Entrapment of the phage virions within the EBMW matrix did not exert any deleterious effect upon their lytic activity. The results obtained showed high promise for integration in poultry feed aiming at controlling *Salmonella enterica*, since the edible biopolymeric microcapsular wrapping integrating lytic bacteriophage particles developed was successful in maintaining lytic phage viability while fully stabilizing the phage particles.

Keywords: bacteriophage particles; *Salmonella enterica*; edible biopolymeric wrapping; foodborne illness; phage cocktail; antibacterial control

check for **updates**

Citation: Pereira, A.O.; Barros, N.M.A.; Guerrero, B.R.; Emencheta, S.C.; Baldo, D.Â.; Oliveira Jr., J.M.; Vila, M.M.D.C.; Balcão, V.M. An Edible Biopolymeric Microcapsular Wrapping Integrating Lytic Bacteriophage Particles for *Salmonella enterica*: Potential for Integration into Poultry Feed. *Antibiotics* **2023**, *12*, 988. https://doi.org/10.3390/antibiotics12060988

Academic Editors: Juhee Ahn, Aneta Skaradzińska and Anna Nowaczek

Received: 28 April 2023
Revised: 28 May 2023
Accepted: 29 May 2023
Published: 31 May 2023

1. Introduction

Foodborne diseases are a major cause of morbidity and mortality worldwide. According to the World Health Organization (WHO), it is estimated that diarrheal diseases alone (most of which are caused by foodstuff contaminated by pathogenic microorganisms) kill 1.9 million children per year [1]. Foodborne diseases are a global public health concern and, according to the WHO, it is estimated that one out of ten cases can be fatal, especially in children under five years of age, causing about 420 thousand deaths in the American continent [2]. In Brazil, most foodborne diseases are caused by the pathogens *Salmonella enterica*, *Escherichia coli*, and *Staphylococcus aureus* [3,4], causing diarrhea, abdominal pain, vomiting, and/or nausea [4].

Discovered and described in 1885 by Daniel Salmon, a veterinary bacteriologist, the genus *Salmonella* is considered a member of the Gram-negative *Enterobacteriaceae* family [5]. It is commonly divided into two species, *Salmonella enterica* and *Salmonella bongori*, and has over 2500 known serotypes that differ in their wide host range and ability to cause disease, with over 50% of the identified serotypes belonging to the *Salmonella enterica* species, which is responsible for the vast majority of *Salmonella* infections in humans [6]. *Salmonella* is part of the microbiota of birds, and there is a diverse amount of *Salmonella* serovars. While the serovars *Salmonella pullorum* and *Salmonella gallinarum* are important pathogens of birds, the serovars *Salmonella enteritidis* and *Salmonella typhimurium* have a broad host spectrum [7]. Among other foodborne pathogens, *Salmonella* alone is responsible for several tens of millions of incidents of salmonellosis worldwide, on a yearly basis, associated with more than 150,000 deaths [4,5].

From the second half of the twentieth century, the emergence of bacterial strains resistant to multiple drugs (viz. multi-drug resistant) has been a reality arising from broad, indiscriminate utilization of chemical antibiotics in areas as diverse as (but not limited to) human medicine, animal medicine, food industry, and agriculture [6,8]. In addition, such multi-drug resistant bacteria may be transmitted from direct contact between farmers and animals and the environment [9]. Antibiotics that promote animal growth have been used since the last three quarters of a century with the aim of improving both animal health and performance. Despite this, and due to the appearance of resistance to antibiotics in bacteria associated with public health hazards, routine supplementation of antibiotics in animal production has been drastically reduced and even banned in some countries [10].

Poultry production is one of the world's sectors that most uses antibiotics, and reducing the use of antibiotics is one of the biggest challenges for this industry globally [9]. In this context, alternative approaches have become necessary, with the application of lytic bacteriophages (or phages) being a potential alternative to combat bacterial diseases in the agricultural industry [11], with enormous potential in the fight to reduce the burden of infectious diseases [2,12].

Bacteriophages are viruses devoid of metabolic machinery of their own that exclusively infect susceptible bacterial cells, hence being obligate intracellular parasites that require a bacterial host cell to replicate [13,14]. The use of bacteriophages to biocontrol bacteria has unique advantages, including that these viral particles are natural, self-multiplying, and highly specific antibacterial agents [13,14]. In addition, bacteriophages specifically target their bacterial host cells while not affecting the local microbiota, are self-replicating and self-limiting nano-entities while there are still viable target host cells, can adapt to the major defense mechanisms of the target host cells, display virtually nil toxicity, are easy and economical to isolate, and can tolerate various conditions prevailing in food matrices [13]. In this way, researchers have sought to use them to treat various types of bacterial infections in humans and animals as well as in environmental applications [5,13,15]. According to several researchers, the concept of combating pathogenic bacteria in food by using phage particles can be addressed at all stages of the entire food chain, specifically in preventing or reducing colonization and disease in livestock via phage therapy [16].

In the poultry industry, bacteriophages have also been used in a wide variety of applications, such as treating live birds, adding to poultry products, and disinfecting processing equipment. It was shown that treatment with phages administered to chickens via aerosol or oral gavage was able to control bacterial infections and decrease mortality [17]. Recent studies have reported success in reducing *Salmonella* spp. by the application of bacteriophages in chickens and products derived from them. Additionally, bacteriophage supplementation has been shown to improve feed efficiency, reduce pathogens in broilers, and improve production and egg quality in laying hens [11].

For the success of antibacterial therapy with phages, several obstacles still have to be overcome, with one of the problems of phage therapy for birds (especially in large commercial aviaries) being the form of administration. In this sense, the incorporation of phage particles into poultry feeds could be an interesting alternative. For this, the phage must be able to survive in the feed and the gastrointestinal tract of the birds. The acidic environment of the stomach can promote the deactivation of phage particles [18]. Abiotic factors such as pH, temperature, and light radiation are parameters known to affect the stability of bacteriophages and their infectious ability [19]. Luminal pH in the gastrointestinal tract of birds ranges from highly acidic in the proventriculus (pH 2.0–5.0) to slightly basic in the small intestine (pH 5.0–7.0). All feeds fed to chickens are thus subjected to gastric pH in the range of 2.0 to 5.0 [20]. In the case of bacteriophages, for their efficient delivery into the gastrointestinal environment and a safe passage through the acidic environment, the protection of phage particles is a very important factor to achieve the desired antibacterial therapeutic effect of bacteriophages. From this perspective, preventive strategies are necessary to protect the phage particles. One such strategy is the microencapsulation of phages [18,21] aiming at their potential incorporation into poultry feed, which was the strategy followed in the research work entertained herein. Essentially, phage encapsulation is a process whereby the phage particles are coated with appropriate biopolymeric materials to segregate them from the surrounding environment, thus protecting the bacteriophages from the aggressive environment of the bird's gastrointestinal tract, which could reduce their viability or render them inactive [22] before they could exert their antibacterial action.

With all the aforementioned facts in mind, the major goal of the research work entertained herein was to isolate and characterize lytic bacteriophage particles for *Salmonella enterica* and promote their structural and functional stabilization within biopolymeric microcapsular wrappings aiming at potential applications in poultry feed for the biocontrol of *Salmonella enterica* in live poultry and the associated foodstuff thereof (eggs and carcasses).

2. Materials and Methods

2.1. Biological Material

Bacterial host for phage isolation: The collection *Salmonella enterica* CCCD-S004 strain utilized in this work as host for phage isolation was acquired from CEFAR Diagnóstica (São Paulo, SP, Brazil). **Bacteriophages**: The two phages utilized in this study (ph001L and ph001T) were previously isolated from samples of lake water (ph001L) and soil with hen faeces (ph001T) collected near the Veterinary Hospital at UNISO (geographic coordinates: 23°29′58.7″ S; 47°23′45.2″ W), Sorocaba/SP (Brazil). **Collection strains for host-range assays**: The bacterial strains utilized in the extended host-range assays were obtained either from ATCC (American Type Culture Collection, Gaithersburg, MD, USA) (viz. *Aeromonas hydrophyla* ATCC 7966, *Salmonella thyphimurium* ATCC 13311, *Escherichia coli* ATCC 8739, *Enterococcus faecalis* ATCC-29212, *Klebsiella pneumoniae* ATCC-13883, *Salmonella enterica* subsp. Enteritidis ATCC 13076, *Proteus mirabilis* ATCC 25933, *Pseudomonas aeruginosa* ATCC 27853, *Pseudomonas aeruginosa* ATCC 9027, *Staphylococcus aureus* ATCC 25923, *Staphylococcus aureus* ATCC 6538, *Salmonella enterica* subsp. *enterica* serovar Typhimurium ATCC 14028, *Bacillus cereus* ATCC 14579, and *Escherichia coli* ATCC 25922), CEFAR (São Paulo, SP, Brazil) (viz. *Pseudomonas aeruginosa* CCCD-P004, *Salmonella enterica* CCCD-S004, and *Proteus mirabilis* CCCD-P001), IBSBF (Phytobacteria Culture Collection of Instituto Biológico Campinas, SP, Brazil) (viz. *Pseudomonas syringae* pv. *garcae* IBSBF-158), and NCTC (National Collection of Type Cultures, UK Health Security Agency (UKHSA), Salisbury, UK) (viz. *Klebsiella pneumoniae* NCTC-13439). All manipulation of materials, bacteria, bacteriophages, culture media, and evaluation of microbiological activity was carried out in a Filterflux® Class II, type B2, biological safety cabinet model SP-SBIIB2-126 from SP-LABOR (Presidente Prudente, SP, Brazil).

2.2. Chemicals

The chemicals utilized in this study were purchased from Dinâmica Química Contemporânea Ltd.a (Diadema, SP, Brazil). Tryptic Soy Agar (TSA) and Tryptic Soy Broth (TSB) culture media were purchased from Sigma-Aldrich Brazil (Cotia, SP, Brazil), and bacteriologic solid agar was purchased from Gibco Diagnostics (Madison, WI, USA). Sterilizing filtration systems Stericup™-GP (with 0.22 μm pore diameter polyethersulphate membrane) were acquired from Merck-Millipore (Darmstadt, Germany). Tap water was ultra-purified to a final resistivity of 18.18 MΩ·cm and conductivity of 0.05 μS·cm^{-1} in a Master System All MS2000 (Gehaka, São Paulo, SP, Brazil).

2.3. Preparation of a Salmonella enterica CCCD-S004 Growth Curve

The host bacteria (in lyophilized form, from CEFAR collection) was hydrated in TSB liquid medium, plated on solid TSA, and incubated at 37 °C for 12 h. A single CFU was then withdrawn with a sterile loop, inoculated in 250 mL sterile TSB, and incubated at 37 °C for 12 h. At predetermined time intervals, the optical density of the culture was evaluated spectrophotometrically at 610 nm.

2.4. Phage Enrichment, Isolation, Propagation, and Enumeration

Phage enrichment from samples of lake water and soil with hen faeces collected near the Veterinary Hospital at UNISO was performed according to the procedure described elsewhere [23–25], with small modifications, using *Salmonella enterica* CCCD-S004 (in exponential growth).

Isolation of phage plaques was carried out using the conventional double-layer agar method described in previous works [23–26], and titres (PFU/mL, plaque-forming units/mL) of the phage suspensions produced thereof were determined.

A *Salmonella enterica* overnight culture (100 µL) was mixed with 5 mL of molten top agar–TSB (30 g/L TSB, 6 g/L agar, 0.05 g/L $CaCl_2$, 0.12 g/L $MgSO_4$, pH 7.4) in test tubes, tapped gently, and poured onto TSA plates which were gently swirled and allowed to dry out for 1–2 min, followed by overnight incubation at 37 °C. Sterile paper strips were wetted in the phage enrichment suspension and dragged several times on a Petri plate containing a bacterial lawn. The plate was incubated at 37 °C for 24 h and then the morphology of the phage plaques was observed. Different phage plaques were pierced with sterile toothpicks, which were then stuck several times (in a line) in Petri plates with bacterial lawn. Sterile paper strips were then used to streak the phages as described above. More successive single-plaque isolation cycles were performed to obtain pure phage isolates. The plates with different plaque morphologies were then incubated overnight at 37 °C, and the last two steps were repeated until all phage plaques were uniform. The plates were stored at 4 °C until needed. To each plate used, 5 mL of SM buffer was added, and the plates were then further incubated with shaking (70 rpm) at 4 °C for 18 h. After incubation, the SM buffer with phages was collected and added to chloroform up to a final ratio of 10% (v/v). The phage suspensions were centrifuged ($9000\times g$, 4 °C, 10 min) to remove intact bacteria or bacterial debris. The aqueous phases were collected into a single sterile flask and stored at 4 °C as phage stocks.

Bacteriophage enumeration was carried out via the double agar overlay technique, as follows. Serial dilutions of the concentrated stock bacteriophage suspensions produced were prepared sequentially, starting by adding 50 µL of each concentrated stock bacteriophage suspension to 450 µL SM buffer. An amount of 5 µL-droplets of each bacteriophage dilution were plated in triplicate in a lawn of the bacterial host and the plates were allowed to dry out for 10 min, after which they were incubated overnight at 37 °C. Following incubation, the bacteriophage plaques formed in each serial dilution were counted, considering only those dilutions with 3–30 bacteriophage plaques. The bacteriophage titre (PFU/mL) of the concentrated stock bacteriophage suspensions was then calculated as *number of phage plaques formed* $\times \left(1/_{dilution}\right) \times \left(1/V_{bacteriophage\ inoculum\ (mL)}\right)$.

2.5. Phage–PEG Precipitation

Phage suspensions (10^{10}–10^{11} PFU/mL) were added to a sterile mixture of polyethylene glycol (PEG) 8000 (Sigma-Aldrich, St. Louis, MO, USA) (10%, w/w) and NaCl (1 M) (Sigma-Aldrich, St. Louis, MO, USA), in a volumetric proportion of 2:1, respectively. The resulting suspensions were incubated overnight at 4 °C and then centrifuged at 11,000 rpm (4 °C, 45 min). The supernatant was then discarded, and the pellet was resuspended and homogenized in a 5 mM $MgSO_4$ aqueous solution (Sigma-Aldrich, St. Louis, MO, USA).

2.6. UV-Vis Spectral Scans for Determination of Phage Particle Extinction Coefficient

Determination of the phage particle molar extinction coefficient was based on the procedure described elsewhere [23–27], using the wavelengths producing the maximum absorption of phage particles (viz. 252 nm (phage ph001L) or 251 nm (phage ph001T)) and 320 nm (wavelength where phage chromophores produce little light absorption). All spectrophotometric readings were performed in a UV-Vis spectrophotometer from Agilent (model Cary 60 UV-Vis, Santa Clara, CA, USA).

2.7. Transmission Electron Microscopy (TEM) Analyses

Phage particles of PEG-concentrated suspensions were centrifuged (4 °C, 150 min, 45,000 rpm, 124,740× g) in a benchtop Beckman-Coulter ultracentrifuge (model Optima TLX micro-ultracentrifuge) with a TLA-55 Fixed-Angle Rotor (Indianapolis, IN, USA) and underwent negative staining with uranyl acetate (Sigma-Aldrich, St. Louis, MO, USA) at 2% (w/v) and pH 7.0, following the procedure described in previous works [23,24], prior to analysis via transmission electron microscopy in a transmission electron microscope from JEOL (model JEM 2100, Tokyo, Japan), encompassing an LaB_6 filament, operating at 200 kV and with resolution of 0.23 nm; a high-resolution CCD camera from GATAN Inc. (model ORIUS™ 832.J4850 SC1000B, Pleasanton, CA, USA) with a resolution of 11 Mp (4.0 × 2.7 k·pixels/9 × 9 μm^2) was utilized for the acquisition of digital images, via the software Gatan Microscopy Suite (DigitalMicrograph from Gatan Inc., version 2.11.1404.0, Pleasanton, CA, USA).

To determine virion capsid and tail dimensions, 7 phage particles were measured for each phage using the public domain ImageJ software (version 1.52a) from the National Institute of Health (Bethesda, MD, USA).

2.8. Host Range of Isolated Phage Particles: Spot Test and Efficiency of Plating (EOP)

Phage host-range was determined by spot-testing using the bacterial strains listed in Table 4, according to the procedure described elsewhere [25,28,29]. For those bacterial strains that produced positive spot tests, the EOP was calculated by comparison with the efficacy for *Salmonella enterica* CCCD-S004 (isolation host, EOP = 100%), as EOP $= \left(Average\ PFU_{target\ bacteria}\ /\ Average\ PFU_{host\ bacteria} \right) \times 100$ [23–25,28,30,31]. All EOP data displayed in Table 4 represent averages of three separate experiments, and were scored as high ($\geq$50%), moderate (10–0.1%), low (0.1–0.001%), or inefficient ($\leq$0.001%), relative to the isolation host (100%) [31].

2.9. Phage One-Step Growth (OSG) Analyses

Growth parameters for the two phages were extracted from the one-step growth curves using *Salmonella enterica* CCCD-S004 (1 × 10^8 CFU/mL) and ph001L (1 × 10^5 PFU/mL) and ph001T (1 × 10^5 PFU/mL) at MOI $\leq$ 0.001 [23,25], with three independent experiments. Adjusting a typical sigmoidal model such as a four-parameter logistic (4-PL) regression equation, viz. $Log\ (P_t) = P_\infty + \left\{ (P_0 - P_\infty) / \left(1 + \left(\frac{t}{\psi} \right)^{\zeta} \right) \right\}$, to the experimental phage growth data was a natural sequence. This allowed one to determine phage growth characteristics such as eclipse, latent and intracellular accumulation periods, and virion progeny yield [23–25,29]. In the model just described, P_t, P_0, and P_∞ are phage concentrations (PFU/mL) at times t, 0, and ∞, respectively, ψ is the curve inflection point, ζ is the curve steepness (Hill's slope), and t is the incubation time (min). The model was fitted to the experimental phage growth data via nonlinear regression analysis using the function "Solver" within Microsoft Excel (Microsoft, Redmond, WA, USA).

2.10. Phage Adsorption Analyses

A *Salmonella enterica* CCCD-S004 suspension in exponential growth (OD$_{610nm}$ $\approx$ 0.5, $\approx$ 5 × 10^8 CFU/mL) was added to bacteriophage particles at 8 × 10^5 CFU/mL in order to produce MOI 0.001 [32], so that phage particles had a bacterial cell to adsorb onto, and the determinations proceeded in three independent assays as previously described [23–25]. Phage particle adsorption onto the host cells was expressed as normalized phage concentration in the supernatant along incubation time. If one assumes that phage virions have the ability to adsorb onto susceptible cells and establish a reversible complex involving both bacterial cells and adsorbed virions leading either to an infected bacterium or to a non-infected counterpart, according to the mechanistic representation deployed by several researchers [33–35],

$$\text{viz. Free virion (P)} + \text{bacterium } (X_0) \underset{\phi}{\overset{\delta \cdot X_0}{\rightleftharpoons}} \text{Reversible complex } \{\text{virion} - \text{bacterium}\} \ (\Delta), \text{resulting}$$

in postulation of the mathematical model $\frac{P_t}{P_0} = \frac{\phi + \left\{ \delta \times X_0 \times e^{-(\delta \times X_0 + \phi) \times t} \right\}}{\delta \times X_0 + \phi}$, where P_t and P_0 are phage concentrations (PFU·mL^{-1}) at times t and 0, respectively, δ is the (first order) phage virion adsorption rate onto susceptible bacteria (CFU^{-1}·mL·min^{-1}), ϕ is the (first order) phage virion desorption rate from reversible virion–bacteria complexes (mL·min^{-1}), X_0 is the initial concentration of uninfected (but susceptible) bacteria (CFU/mL), and t is infection time (h). The model was then fitted to the experimental phage adsorption data via nonlinear regression analysis using the function "Solver" within Microsoft Excel (Microsoft, Redmond, WA, USA), allowing determination of the phage adsorption rate.

2.11. Bacteria Inactivation Experiments In Vitro by the Two Phages

Inactivation of planktonic host cells (10^5 CFU/mL, exponential growth) by the phages was studied at MOI values 0.01, 0.1, 1, 10, 100, and 1000. For each MOI experiment (performed in triplicate in three independent assays), a bacterial control (BC) was also added, comprising only planktonic host cells. Both BC and treatments (BP-B, bacteria, and phage–bacterial concentration) were incubated with the same time/temperature parameters, viz. 37 °C and 12 h. Two mL-aliquots of BC and treatment samples (BP-B) were withdrawn at predetermined intervals of time up to a total treatment timeframe of 12 h, viz. 0, 15, 30, 45, 60, 90, 120, 150, 180, 210, 240, 270, 300, 330, 360, 420, 480, 540, 600, 660, and 720 min, and their absorbance duly measured at a wavelength of 610 nm.

2.12. Assessment of the Outcome of Abiotic Factors upon Phage Viability

The aftermath of T, pH, and solar radiation on phage (ph001L and ph001T, 10^{7-8} PFU/mL) viability was studied in phosphate-buffered saline (PBS, 10 mL). For the pH and T experiments, aliquots were withdrawn every 2 h during the first 12 h and then every 12 h up to 72 h of incubation. For the solar radiation experiments, aliquots were withdrawn every 1 h up to 7 h of exposure to direct sunlight. Phage concentrations were evaluated in triplicate in three independent experiments through double-layer agar plating followed by overnight incubation at 37 °C.

2.12.1. pH Studies

The aftermath of pH on phages ph001L and ph001T viability was studied via addition of phage suspensions to sterile PBS with different pH values (viz. 3.0, 6.5, 8.0, 9.0, 10.0, and 12.0), under constant temperature (25 °C).

2.12.2. Temperature Studies

The aftermath of T on phages ph001L and ph001T viability was studied via addition of phage suspensions to sterile PBS (pH 7.0), followed by incubation at a constant temperature (25, 41, and 50 °C).

2.12.3. Solar Radiation Studies

The aftermath of solar radiation on phages ph001L and ph001T viability was studied via addition of phage suspensions to sterile PBS (pH 7.0), followed by exposure to natural sunlight (these were the test samples, S). Control samples (SR-C) were not exposed to solar radiation. These studies were carried out on a shiny day with ambient T varying from 29 °C (09h00) to 35 °C (end of the experiment), with a solar irradiance of ca. 4.561 kWh/m^2 (data obtained from https://globalsolaratlas.info/map?c=-23.499789,-47.400936,11&s=-23.499789,-47.400936&m=site (accessed on 3 April 2023)). Solar radiation (specifically, UV irradiation) is the most important factor for the loss of phage infectivity in the environment [28]. Short radiation wavelengths (UV-B, 290–320 nm) impart irreversible damages to the phage virion genome and result in modification of viral proteins and formation of (lethal) photoproducts [28]. In the experiments performed herein, one used small transparent flasks made of ordinary (non-mineral) glass, so that UV-A and UV-B radiation from sunlight could pass through and hit the phage suspensions. At the same time, the opening of the flasks were maintained oriented towards the sun, in such a position that sunlight directly hit the surface of the phage suspensions.

2.13. Formulation of the Edible Biopolymeric Microcapsular Wrapping (EBMW) Integrating the Bacteriophage Cocktail

Preparation of the EBMW formulation integrating (or not) the phage cocktail followed the internal gelation procedure described elsewhere [36–38], with modifications, aimed at structurally and functionally stabilizing the phage particles. As the source of calcium ions, one used calcium chloride. The phage MOIs used in the formulation of the EBMWs were defined considering an initial bacterial contamination "load" of 1.0×10^7 CFU·mL^{-1}. Hence, the process (Table 1) was initiated via the preparation of sodium alginate dispersions in ca. 80% (w/w) of the total mass of ultrapure water containing the phage cocktail. Occasionally, these dispersions were stirred and allowed to stand by for at least 1 h, to allow complete hydration of the sodium alginate. For the EBMW formulation, 5.0 mL gelatin solution at a concentration of 10 mg/mL and 5 mL of the phage cocktail at a given MOI were thoroughly mixed together in a beaker and, afterwards, the resulting suspension was added to 20 mL of sodium alginate at 2% (w/w). The suspension thus produced (dispersion A, Table 1) was then dripped on 10 mL of a 2.65 mol·dm^{-3} calcium chloride solution containing chitosan at 0.3% (w/w) (dispersion B, Table 1). After the formation of the EBMWs, the pH was measured and adjusted to 5.5, and the EBMW particles were stored at 4 °C.

Table 1. Final (qualitative and quantitative) compositions of all edible biopolymeric microcapsular wrapping (EBMW) formulations encompassing calcium alginate biopolymeric matrices with entrapped phage particles.

Component		EBMW Formulation				
		EBMW 1: No Phages	EBMW 2: MOI 1	EBMW 3: MOI 10	EBMW 4: MOI 100	EBMW 5: MOI 1000
Phage ph001L (amount of virions)		-	8.63×10^7	5.18×10^8	5.18×10^9	9.94×10^9
Phage ph001T (amount of virions)		-	5.70×10^7	5.70×10^8	5.70×10^9	1.09×10^{10}
Dispersion A	Phage buffer (mL)	5	5	5	5	5
	Gelatin (mg)	50	50	50	50	50
	Phage cocktail (virions)	-	1.43×10^8	1.09×10^9	1.09×10^{10}	2.09×10^{10}
	Sodium alginate at 2% (w/w) (mL)	20	20	20	20	20
Dispersion B	Calcium chloride 2.65 mol dm^{-3} with chitosan at 0.3% (w/w) (mL)	10	10	10	10	10

2.14. Assessment of the Lytic Viability of Entrapped Bacteriophage Particles within the EBMW Formulations

To check the preservation of lytic viability of the bacteriophage particles entrapped within EBMW, a sample of the formulation was placed on a bacterial lawn of the host bacteria on a Petri plate followed by incubation at 37 °C for 24 h. After this time period, the presence (or not) of clear lysis zones surrounding the EBMW sample was observed.

2.15. Bacteria Inactivation Experiments In Vitro by the EBMW Particles with Entrapped Phage Cocktail

Inactivation of planktonic *(Salmonella enterica)* host cells (OD$_{610nm} \approx 0.5$, exponential growth) by the EBMW particles with entrapped phage cocktail at MOI values 100 and 1000 was studied. For each EBMW formulation (performed in triplicate in three independent assays), a bacterial control (BC) was also prepared, comprising only planktonic host cells. Both BC and treatments (EBMW-BP-B, bacteria, and EBMW formulation—bacterial concentration) were incubated with the same time/temperature parameters, viz. 37 °C and 12 h. A total of 10 microparticles of a given formulation were placed in a 0.5 mL conical-bottom bioreactor containing a Teflon-coated conical stirring bar. An amount of 500 microlitres of bacterial suspension (OD$_{610nm} \approx 0.5$, exponential growth) was then added, magnetic stirring (75 rpm) was initiated, and the bioreactor was placed in an incubation chamber set at 37 °C. Five µL-aliquots of BC and treatment samples (EBMW-BP-B) were withdrawn at predetermined intervals of time up to a total treatment timeframe of 12 h, viz. 0, 15, 30, 45, 60, 120, 240, 360, 480, and 720 min, and serial diluted in 45 µL SM buffer. The bacterial concentration in both BC and EBMW-BP-B dilutions was determined in triplicate in solid TSA medium via the drop (5 µL)-plate method after an incubation period of 12 h at 37 °C.

2.16. Phage Release Experiments from the EBMW Particles with Entrapped Phage Cocktail

To verify the preservation of lytic activity of the entrapped phage particles and to assess their release from the EBMW particles, a simple experiment was performed. A total of 10 EBMW microparticles with entrapped phages at MOI 1000 were placed in a 0.5 mL conical-bottom bioreactor containing a Teflon-coated conical stirring bar. An amount of 500 microlitres of SM buffer was then added, magnetic stirring (75 rpm) was initiated, and the bioreactor was left at room temperature (ca. 25 °C). Five µL-aliquots of the supernatant were withdrawn at predetermined intervals of time up to a total timeframe of 3 h, viz. 0, 5, 10, 15, 20, 25, 30, 45, 60, 75, 90, and 120 min, and serial diluted in 45 µL SM buffer. The phage titre in all dilutions was determined in triplicate by the double agar-layer method [23–25], using 5 µL droplets plated in triplicate in *Salmonella enterica* bacterial lawn (exponential growth phase, $OD_{610nm} \approx 0.5$), after an incubation period of 12 h at 37 °C.

2.17. Physicochemical Characterization of EBMW Formulation

The physicochemical characterization of the EBMW formulations (Table 1) involved a wide array of analyses (Fourier transform infrared spectrophotometry (FTIR), differential scanning calorimetry (DSC), energy-dispersive X-ray fluorescence (EDXRF), X-ray tomography (XRT), and scanning electron microscopy (SEM)), which will be detailed next.

2.17.1. FTIR Tests

FTIR spectra of EBMW formulations 1 and 5 were obtained in a FTIR spectrophotometer from Thermo Scientific (model Nicolet 6700, Madison, WI, USA) coupled with an ATR module (germanium crystal) (Smart Omni Sampler), from 4000 cm^{-1} to 675 cm^{-1} (resolution: 4 cm^{-1}, 1024 scans), with Happ–Genzel apodization.

2.17.2. Thermal Analyses via DSC Tests

Thermal analyses (DSC) of EBMW formulations (16.6180 mg of plain EBMW, and 16.0130 mg of EBMW with phage cocktail at MOI 1000) were performed in a microcalorimeter from METTLER TOLEDO (model DSC-1, Schwerzenbach, Switzerland), according to Rocha et al. [39], using high-pressure aluminum pans sealed by pressure (with the lid punctured with a small hole) containing the samples and a reference aluminum pan with plain air sealed inside. Samples were heated from ca. 25 °C up to 250 °C, at 10 °C·min^{-1}, under a constant N_2 flow of 50 mL·min^{-1}, with the heat absorbed by the samples being recorded at a sampling rate of 0.2 s per data point.

2.17.3. EDXRF Tests

The elemental makeup of EBMW formulations (plain and loaded with bacteriophage cocktail at MOI 1000) was determined using an EDXRF spectrometer (model Epsilon 1, Malvern Panalytical, Cambridge, UK) equipped with a 5 W, 10–50 kV, Ag anode X-ray tube, with energy resolution of 125 eV, filters of Ag, Cu, Ti, and Al for the X-ray beams, and a high-resolution 25 mm^2 silicon drift detector (SDD) operating at P_{atm}. All tests were performed with a measuring timeframe of 300 s using atmospheric air, and the spectra were obtained sequentially from 0 keV to 30 keV (resolution of 0.02 keV).

2.17.4. XRT Tests

Tomographic images of the EBMW formulation entrapping the phage cocktail at MOI 1000 were obtained in an X-ray transmission tomograph [40] from Bruker microCT (model SkyScan 1174, Kontich, Belgium). The sample was placed on top of a metallic support coated with adhesive tape which was then placed inside the tomograph chamber. Image slices of the sample were then collected at an operating voltage of 31 kV and electric current of 661 µA. A high number of radiographs (image slices) of the sample were collected via measurement of the X-ray intensities transmitted through the sample at different angular positions (rotation of 180° with angular increments of 0.7° originating 217 radiographs per image (exposure time per radiograph of 2500 ms), each of which holding 1024×1304 (width $\times$ height) pixels with a spatial resolution of 6.70 µm), so that a tomographic image could be produced. Mathematical algorithms were then utilized to reconstruct the three-dimensional (3D) tomographic images ($652 \times 652 \times 652$ pixels) of the EBMW particle, via composition of the bi-dimensional (2D) images acquired. With all the radiographs collected at each angular position, the software NRecon™ (Bruker, version 1.6.9.4, Kontich, Belgium) (using the Feldkamp et al. [41] algorithm to reconstruct the tomographic images), CTVox™ (Bruker, version 2.6.0 r908-64bit), CTan™ (Bruker, version 1.13.5.1-64bit), and CTvol™ (Bruker, version 2.2.3.0-64bit) were utilized for processing all the digital radiographs (image slices).

2.17.5. SEM Tests

The surface and morphology of an EBMW particle were analyzed in a SEM (JEOL, model JSM-IT200, Tokyo, Japan) at high-vacuum. The samples were sputter-coated with a 92 Å-thick Au film via cathodic pulverization, in a metalizing device (JEOL, Sputter Coater model DII-29010SCTR Smart Coater, Tokyo, Japan). Photomicrographs were collected via random scanning using electron beams at acceleration speeds of 10.0 keV.

2.18. Statistical Tests

Statistical tests of lack of fit of the mathematical model for phage adsorption (i.e., the expectation function) to the experimental phage adsorption data were undertaken, aiming at testing the goodness of the nonlinear fittings. These statistical tests are based on the fact that the subspace containing the experimental data replications is orthogonal to the subspace containing both the experimental data averages and the expectation function [23,24,28,29,42–44]. For this, the F-ratio (lack of fit mean square ($SS_{\text{lack of fit}}/NDF_{\text{lack of fit}}$) over the replications mean square ($SS_{\text{replications}}/NDF_{\text{replications}}$)) was compared with the statistical F-value ($F(\nu_{\text{NDF, lack of fit}}, \nu_{\text{NDF, replications}}; \alpha = 5\%)$). SS = sum of squares, NDF = number of degrees of freedom.

The data gathered in the in vitro phage–bacteria inactivation assays was statistically analyzed with GraphPad Prism 7.04 (GraphPad Software, San Diego, CA, USA). While the normal distribution of the experimental inactivation data was verified by a Kolmogorov–Smirnov test, the homoscedasticity was verified by the Levene's test. The significance of the differences recorded for bacterial concentration was evaluated via comparison between the results of treatment samples for each MOI (BP-B, bacteria and phage—bacterial concentration in the treatment) with the corresponding bacterial concentrations in the control (BC, bacterial concentration in the control) for the different inactivation times, using two-way ANOVA and Bonferroni post hoc tests. A value of $p < 0.05$ was considered to be statistically significant.

3. Results

In the present research work, the formulation of a bioactive edible biopolymeric microcapsular wrapping (EBMW) integrating a structurally and functionally stabilized cocktail of two newly isolated lytic phages for *Salmonella enterica* has been proposed, aiming at potential applications in poultry feed for the biocontrol of this pathogen. Two different virulent phages were selected based on their ability to form clear plaques of lysis, which were amplified in a *Salmonella enterica* CCCD-S004 bacterial strain and, to assess their infectious potential for biotechnological applications such as biocontrol of the aforementioned pathogen in live poultry, physicochemical and biological characterization was undertaken, together with the characterization of the EBMW formulation integrating the lytic cocktail.

3.1. Bacterial Growth Curve

For isolating lytic bacteriophage particles, one decided to use a collection bacterium (viz. *Salmonella enterica* CCCD-S004), and therefore the production of a growth curve (Figure 1) was mandatory to observe the growth characteristics of the bacterial cells. For up to 5 h of growth, approximately, the bacterium remained in a latency period and, after this short timeframe, its exponential growth phase began and extended up to 9 h of growth (data not shown). Then, the onset of the stationary phase period could be observed. These results were very important to the subsequent research work, since an active bacterial culture in the exponential phase is necessary for all phases of phage infection, isolation, and amplification. A nonlinear fitting of the Gompertz function was performed on the experimental bacterial growth data, allowing one to estimate the maximum biomass concentration at $t = \infty$ as 1.013×10^9 CFU/mL (corresponding to a maximum absorbance of 1.1049), and of the lag period as 345 min (ca. 5.75 h).

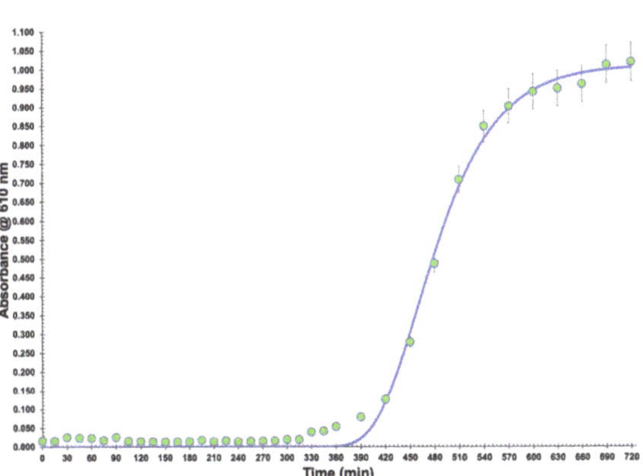

Figure 1. Growth curve of the host bacteria (*Salmonella enterica* CCCD-S004). The nonlinear fitting performed (Gompertz function) allowed estimation of the maximum biomass concentration at $t = \infty$ of 1.013×10^9 CFU/mL (corresponding to a maximum absorbance of 1.1049), and of the lag period (345 min, ca. 5.75 h). Values represent the mean of three experiments; error bars represent the standard deviation.

3.2. Phage Virion Morphology via Transmission Electron Microscopy (TEM) Analyses

Phages ph001L and ph001T, isolated from lake water and soil containing hen faeces in the surroundings of the Veterinary Hospital located within the UNISO Campus (Sorocaba, SP, Brazil), produced clear plaques with very small dimensions on the host (*Salmonella enterica* CCCD-S004) lawn (Figure 2), with absence of secondary halo surrounding them, indicating that these phages probably do not produce depolymerase enzymes [45,46]. TEM photomicrographs of phages ph001L and ph001T can be observed in Figure 2.

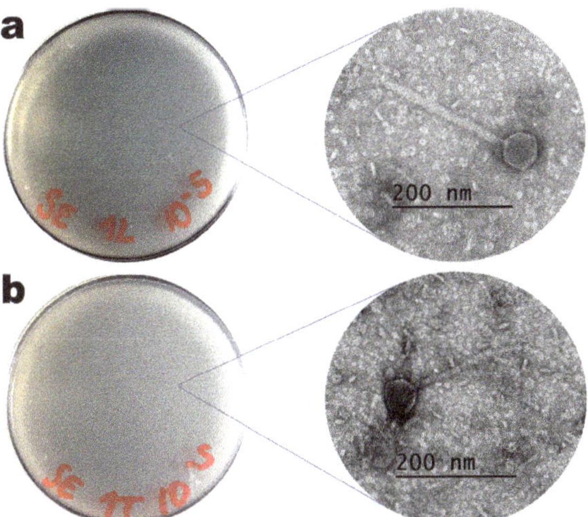

Figure 2. Images of lysis plaques and negative-staining TEM photomicrographs of phages ph001L (**a**) and ph001T (**b**).

Based on the morphological analysis entailed by TEM (Figure 2), both phages displayed siphovirus morphotypes and were putatively identified as belonging to the class Caudoviricetes. Phages ph001L and ph001T displayed perfect icosahedral heads and long, flexible, non-contractile tails; their approximate dimensions are displayed in Table 2.

Table 2. Approximate dimensions of the two newly isolated bacteriophages.

Dimension	Phage ph001L	Phage ph001T
Head length (nm)	56.1 ± 1.7	52.6 ± 1.7
Head width (nm)	56.9 ± 2.6	57.1 ± 0.0
Tail length (nm)	217.9 ± 1.8	185.2 ± 2.7
Tail thickness (nm)	16.8 ± 1.4	12.8 ± 1.9

3.3. Phage Particle Extinction Coefficients

UV-Vis spectral scans of PEG-concentrated phages ph001L and ph001T (Figure 3a) were used to obtain the wavelength producing the maximum absorption of radiation of both phages, giving rise to the data displayed in Table 3 that was used to determine the phage particle extinction coefficient (Figure 3b).

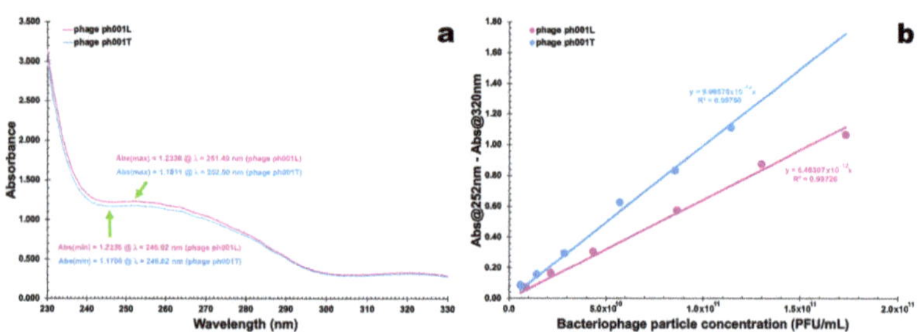

Figure 3. UV-Vis spectral scans of PEG-concentrated phages ph001L and ph001T (**a**) and linear relationships between corrected absorbance and phage particle concentration (**b**).

Table 3. Data used to determine the (whole) phage particle extinction coefficients.

Phage Suspension Volume (μL)	Dilution Volume (μL)	Virion	Number of Phage Virions Withdrawn from Suspension	Virion Particle Concentration (PFU/mL)	Absorbance at 251 (or 252) nm	Absorbance at 320 nm	Absorbance at 251 (or 252) nm—Absorbance at 320 nm
10	2000	ph001L	1.73×10^{10}	8.65×10^{9}	0.106	0.032	0.075
		ph001T	1.14×10^{10}	5.70×10^{9}	0.135	0.045	0.090
25	2000	ph001L	4.33×10^{10}	2.16×10^{10}	0.250	0.082	0.168
		ph001T	2.85×10^{10}	1.43×10^{10}	0.226	0.065	0.161
50	2000	ph001L	8.65×10^{10}	4.33×10^{10}	0.451	0.141	0.310
		ph001T	5.70×10^{10}	2.85×10^{10}	0.429	0.130	0.299
100	2000	ph001L	1.73×10^{11}	8.65×10^{10}	0.841	0.263	0.577
		ph001T	1.14×10^{11}	5.70×10^{10}	0.905	0.277	0.628
150	2000	ph001L	2.60×10^{11}	1.30×10^{11}	1.272	0.394	0.878
		ph001T	1.71×10^{11}	8.55×10^{10}	1.210	0.374	0.836
200	2000	ph001L	3.46×10^{11}	1.73×10^{11}	1.544	0.476	1.068
		ph001T	2.28×10^{11}	1.14×10^{11}	1.610	0.495	1.115

For the two phages, maximum absorption was observed around 251 nm (phage ph001L) and 252 nm (phage ph001T), and the minimum absorption that was observed around 245 nm was an indication that bacterial cell debris were virtually absent from the preparation with the concomitant presence of a high concentration of phage virions.

Fitting the Beer–Lambert linear relationship to the experimental data displayed in Table 3 ($\text{Abs}_{251(2)\,\text{nm}} - \text{Abs}_{320\,\text{nm}} = f$ (phage particle concentration, PFU/mL)), allowed obtaining the molar extinction coefficient of the newly isolated phages as $\varepsilon_{\text{phage ph001L}} = 6.463 \times 10^{-12}$ (PFU/mL)$^{-1} \cdot$cm^{-1} and $\varepsilon_{\text{phage ph001T}} = 9.986 \times 10^{-12}$ (PFU/mL)$^{-1} \cdot$cm^{-1}. By subtracting $\text{Abs}_{320\text{nm}}$, a wavelength where there is little absorption of radiation from phage chromophores, a raw correction for light scattering from phage particles and non-phage particulate contaminants was carried out [23–28].

Phage virions are essentially made of structural proteins such as capsid, tail, baseplate, and spike, whose chromophores (essentially the side chains of TRP, TYR, and PHE and disulfide bonds of CYS moieties) exhibit a maximum absorption of radiation around 278–280 nm in the ultraviolet wavelength region. Hence, quenching of radiation owing to protein chromophore absorption and scattering by whole virion particles corrected by the quenching of radiation owing solely to protein chromophores, leads to the hallmark absorbance for a particular virion at a particular concentration [47]. Structural protein chromophores of phage virions have a nearly zero absorption of radiation (completely due to scattering [48]) at 320 nm, which one uses to correct for radiation scattering from virions and other contaminating particulates. Therefore, the structural proteins of the two phages isolated in the research work described herein contributed substantially to their absorption spectrum and were responsible for the wide plateau between 250 and 280 nm in the spectra, with a shallow maximum at 251 nm (phage ph001L) or 252 nm (phage ph001T).

According to previous studies [23–26,29], the molar extinction coefficient is yet another parameter that allows one to differentiate between isolated phages and, as can be observed in Figure 3b, the two phages are indeed different despite exhibiting the same siphovirus morphotype.

3.4. Phage Host Range and Efficiency of Plating (EOP)

Spot testing indicated that phages ph001L and ph001T could form completely cleared zones on 7 of the 19 strains tested (Table 4). Beyond the host isolation strain, phages ph001L and ph001T infected *Salmonella enterica* subsp. Enteritidis ATCC 13076, *Pseudomonas syringae* pv. *Garcae* IBSBF-158, *Escherichia coli* ATCC 25922, *Escherichia coli* ATCC 8739, *Klebsiella pneumoniae* ATCC-13883, and *Klebsiella pneumoniae* NCTC-13439, with moderate efficacies (Table 4).

3.5. Phage One-Step Growth (OSG) Analyses

Fitting the experimental one-step phage growth data to the 4-PL model via nonlinear regression enabled one to determine the virion growth features for phages ph001L and ph001T (Figure 4).

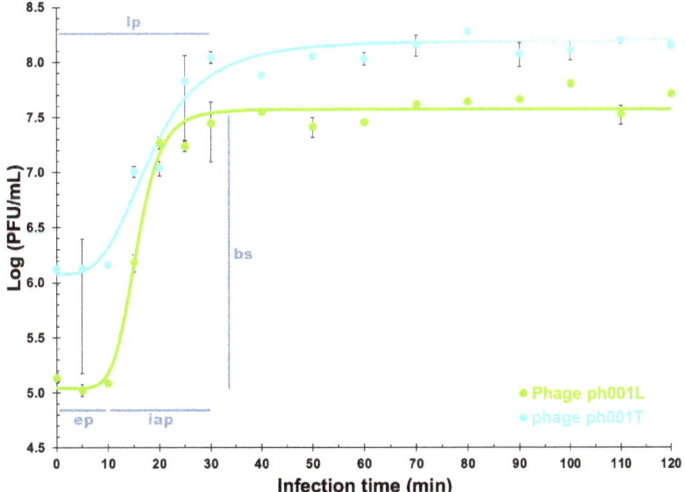

Figure 4. Analysis of the growth curves in a single synchronous cycle of phages (10^5 PFU/mL) ph001L and ph001T on a late exponential phage culture of the host (*Salmonella enterica* CCCD-S004, 10^8 CFU/mL) (MOI $\leq$ 0.001). The fitted 4-PL model to the experimental phage virion growth data, represented by blue and green lines, enabled estimation of phage virion growth features such as eclipse (ep), latent (lp), and host intracellular accumulation (iap) periods, and burst size (bs), as illustrated in Figure 4. All experimental data values represent means of triplicate determinations in three independent experiments. Error bars represent asymmetric standard deviations.

Table 4. Host range of phages ph001L and ph001T, evaluated on 19 collection strains. (+): clear zone of lysis; (−): absence of lysis.

Bacterial Strains	Source	Phage ph001L				Phage ph001T			
		Spot Test	Titre in Target Bacteria (PFU/mL)	EOP (%)	Score	Spot Test	Titre in Target Bacteria (PFU/mL)	EOP (%)	Score
Salmonella enterica CCCD-S004	Collection, CEFAR	+	1.42×10^{12}	100 (host)	High	+	1.82×10^{12}	100 (host)	High
Salmonella enterica subsp. Enteritidis ATCC 13076	Collection, ATCC	+	6.87×10^{11}	48.4	Moderate	+	6.06×10^{11}	33.3	Moderate
Pseudomonas syringae pv. *Garcae* IBSBF-158	Collection, IBSBF	+	2.10×10^{11}	14.8	Moderate	+	3.26×10^{11}	17.9	Moderate
Escherichia coli ATCC 25922	Collection, ATCC	+	2.18×10^{11}	15.4	Moderate	+	1.78×10^{11}	9.8	Moderate
Escherichia coli ATCC 8739	Collection, ATCC	+	1.89×10^{11}	13.3	Moderate	+	2.25×10^{11}	12.4	Moderate
Klebsiella pneumoniae ATCC-13883	Collection, ATCC	+	1.04×10^{11}	7.3	Moderate	+	4.12×10^{11}	22.7	Moderate
Klebsiella pneumoniae NCTC-13439	Collection, NCTC	+	9.41×10^{10}	6.6	Moderate	+	3.29×10^{11}	18.1	Moderate
Aeromonas hydrophyla ATCC-7966	Collection, ATCC	−	−	−	−	−	−	−	−
Enterococcus faecalis ATCC-29212	Collection, ATCC	−	−	−	−	−	−	−	−
Proteus mirabilis ATCC 25933	Collection, ATCC	−	−	−	−	−	−	−	−
Pseudomonas aeruginosa ATCC 27853	Collection, ATCC	−	−	−	−	−	−	−	−
Pseudomonas aeruginosa ATCC 9027	Collection, ATCC	−	−	−	−	−	−	−	−
Staphylococcus aureus ATCC 25923	Collection, ATCC	−	−	−	−	−	−	−	−
Staphylococcus aureus ATCC 6538	Collection, ATCC	−	−	−	−	−	−	−	−
Salmonella enterica subsp. Enterica serovar Typhimurium ATCC 14028	Collection, ATCC	−	−	−	−	−	−	−	−
Pseudomonas aeruginosa CCCD-P004	Collection, CEFAR	−	−	−	−	−	−	−	−
Bacillus cereus ATCC 14579	Collection, CEFAR	−	−	−	−	−	−	−	−
Proteus mirabilis CCCD-P001	Collection, CEFAR	−	−	−	−	−	−	−	−
Salmonella thyphimurium ATCC 13311	Collection, ATCC	−	−	−	−	−	−	−	−

The phage growth parameters were determined from the nonlinear fittings performed to the experimental phage growth data. Phage ph001L presented an eclipse period (ep) of 10 min, a latent period (lp) of 25 min, and an intracellular accumulation period (iap) of 15 min, with a virion morphogenesis yield (bs) of 466 virions/host cell, whereas phage ph001T presented an eclipse period (ep) of 10 min, a latent period (lp) of 40 min, and an intracellular accumulation period (iap) of 30 min, with a virion morphogenesis yield (bs) of 132 virions/host cell.

3.6. Phage Adsorption Analyses

Phages ph001L and ph001T adsorption assays showed that approximately 90% of the phage particles adsorb to *Salmonella enterica* CCCD-S004 cells after 30 min and 100% adsorbed after 60 min (Figure 5).

Fitting the experimental phage virion adsorption data to the adsorption decay model via nonlinear regression enabled estimation of the phage virion adsorption rates onto their host cells (δ) and desorption rates from virion–bacteria complexes (φ): $\delta_{ph001L} = 8.000 \times 10^{-10}$ CFU^{-1}·mL·min^{-1} and $\varphi_{ph001L} = 1.900 \times 10^{-3}$ mL·min^{-1} ($X_0 = 1.0 \times 10^8$ CFU·mL^{-1}; r^2 (coefficient of determination) = 0.96384); $\delta_{ph001T} = 7.000 \times 10^{-10}$ CFU^{-1}·mL·min^{-1} and $\varphi_{ph001T} = 4.000 \times 10^{-4}$ mL·min^{-1} ($X_0 = 1.0 \times 10^8$ CFU·mL^{-1}; r^2 (coefficient of determination) = 0.96791).

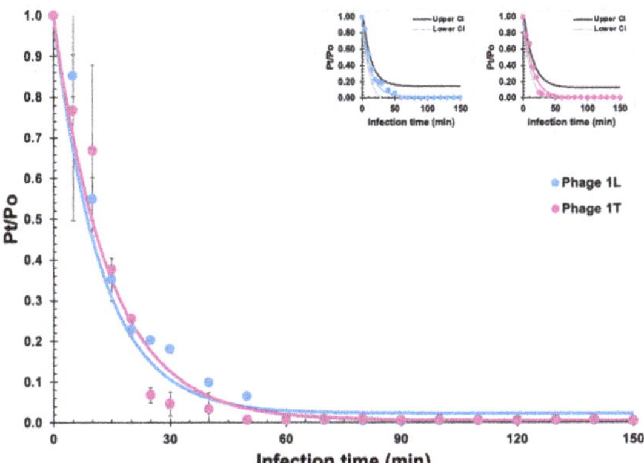

Figure 5. Adsorption curves of phages ph001L and ph001T particles onto their host cells. The fitted adsorption decay model to the experimental phage virion adsorption data, represented by blue and magenta lines, enabled estimation of phage virion adsorption features such as adsorption and desorption rates. All experimental data values represent means of triplicate determinations in three independent experiments. Error bars represent standard deviations.

A statistical test of lack of fit of the adsorption decay model depicted in Section 3.9 was made due to the not-so-small standard deviations of the experimental data points during the first 20 min of the assay, indicating no lack of fit of the mathematical model at a significance level of 0.05 (95% confidence), (phage ph001L: calculated $F_{ratio} = 0.0868$, standard $F_{ratio} = 2.0147$, p-value = 0.9999; phage ph001T: calculated $F_{ratio} = 0.0772$, standard $F_{ratio} = 2.0147$, p-value = 0.9999). Because "lack of fit" arises from the oscillation of experimental data points around the model fitted, a p-value > 0.10 (lack of fit statistically not significant), allows one to conclude that the mathematical model fits (i.e., predicts) the actual response data. Such a conclusion was illustrated via inclusion of small plots in Figure 5, containing upper and lower 95% confidence intervals of the nonlinear fittings performed to the phage adsorption data.

3.7. In Vitro Phage–Bacteria Inactivation Assays

The bacterial concentration in the control (BC) increased substantially (ANOVA, $p < 0.05$) during the 12 h of incubation (Figure 6(a1,b1,c1)). When applying the two phages in an independent fashion, phage ph001L did show the worst performance in terms of bacterial reduction (ANOVA, $p > 0.05$) (Figure 6(a1,a2)). At MOI 0.01, phage ph001L managed to reduce the bacterial load by 32.5% after 12 h of incubation with its host. This number increased to 42.5% at MOI 0.1 after 12 h of incubation, and to 56.2% at MOI 1 after 12 h of incubation (Figure 6(a1,a2)). At MOI 1, higher bacterial inactivation was attained after 8 h, viz. 74.6% (ANOVA, $p < 0.05$), with phage ph001L. At MOI 10, the performance of this phage was similar to at MOI 1 (ANOVA, $p > 0.05$). However, at MOI 100, phage ph001L reduced the bacterial load by 58.0% after 6 h of incubation with the host, which was reduced to only 26.6% after 12 h, due to bacterial regrowth (Figure 6(a1,a2)). At MOI values 100 and 1000, the performance of phage ph001L was similar (ANOVA, $p > 0.05$) (Figure 6(a1,a2)). Increasing the MOI from 0.01 → 0.1 → 1 → 100 did significantly increase the inactivation factor after 6 h of incubation (ANOVA, $p < 0.05$), for independently applied phage ph001L (Figure 6), but after 12 h of incubation MOI 100 was not very effective in maintaining bacterial reduction, with the increase in MOI from 0.01 → 0.1 → 1 succeeding in maintaining a significant bacterial reduction (Figure 6(a1,a2)). This scenario was completely changed when phage ph001T was used. After 6 h of incubation with its host, phage ph001T managed to reduce the bacterial load by ca. 93.0% at MOI 10 or 99.2% at MOI 1000, reductions that were maintained at both MOI up to 9 h of incubation (Figure 6(b1,b2)). These bacterial reductions were, however, reduced to not-so-high values after 12 h of incubation, viz. 77.7% (MOI 1), 79.2% (MOI 10), 62.3% (MOI 100), or 67.2% (MOI 1000) (Figure 6(b1,b2)). For phage ph001T, MOI 0.01 was the worst in terms of bacterial reduction performance (Figure 6(b1,b2)), attaining only 24.3% (ANOVA, $p > 0.05$) of bacterial load reduction after 12 h of incubation with the host. Increasing the MOI from $0.01 \rightarrow 0.1 \rightarrow 1 \rightarrow 10 \rightarrow 100 \rightarrow 1000$ did significantly increase the inactivation factor after 6 h of incubation (ANOVA, $p < 0.05$), for independently applied phage ph001T (Figure 6(b1,b2)). This same trend could be noticed at 7 h, 8 h, and 9 h of incubation of this phage and its host (Figure 6(b1,b2)). As a cocktail, both phages were able to significantly reduce the bacterial load at both MOI 1 and MOI 10, by 84.3% and 87.6%, respectively, after 12 h of incubation with the bacterial host (Figure 6(c1,c2)). At MOI values 1 and 10, the phage cocktail was able to effectively control the bacteria after only 6 h of incubation, maintaining the bacterial reduction at high levels of 84.3% and 87.6%, respectively, up to 12 h of incubation (Figure 6(c1,c2)). The phage cocktail at MOI 10 proved to be the most effective, by significantly (ANOVA, $p < 0.05$) reducing the bacterial load between 6 h and 12 h of incubation with the host (Figure 6(c1,c2)), although MOI 1000 was highly effective between 6 h and 9 h of incubation (Figure 6(c1,c2)). No statistical difference was found for the results produced by the phage cocktail at MOI 1 and MOI 10 after 12 h of Incubation (ANOVA, $p > 0.05$). When phage ph001L was used, a significant bacterial regrowth was observed after 8 h of incubation (ANOVA, $p < 0.05$) for MOI 1, 10, 100, and 1000 (Figure 6(a1,a2)). Regarding phage ph001T, after 8 h of incubation, a slight bacterial regrowth at MOI 1, 10, 100, and 1000 could be observed, until the end of the treatment (Figure 6(b1,b2)). When the two-phage cocktail was used, only a slight (and similar) bacterial regrowth could be observed at MOI 100 and 1000 after 9 h of incubation with the host (ANOVA, $p > 0.05$). Despite this, by the end of the incubation timeframe, the bacterial densities in the different treatments using either independent phages ph001L and ph001T or a cocktail of both phages were significantly lower than that observed for the bacterial control (BC, Figure 6).

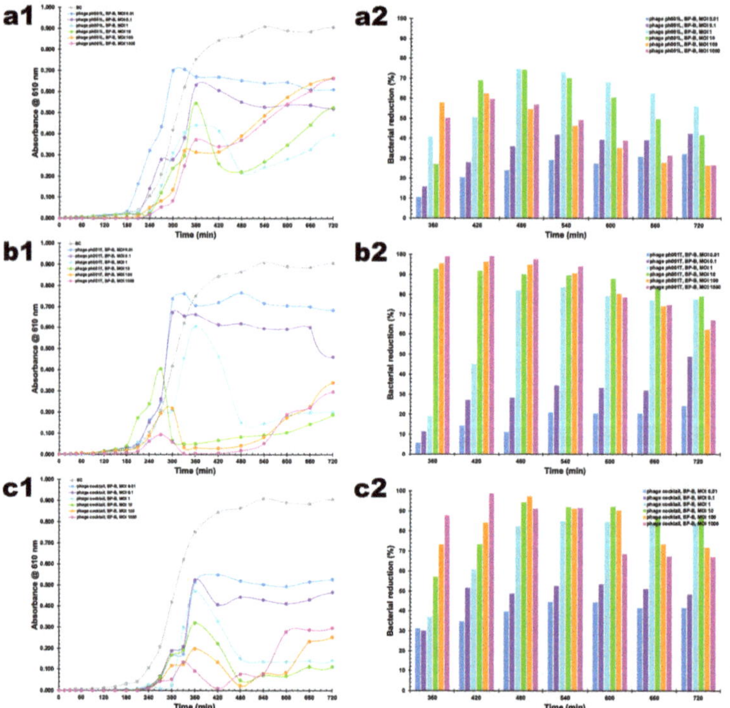

Figure 6. In vitro inactivation of *Salmonella enterica* CCCD-S004 by (**a1**) independent phage ph001L, (**b1**) independent phage ph001T, and by their cocktail (**c1**), at a multiplicity of infection (MOI) of 0.01, 0.1, 1, 10, 100, and 1000, and bacterial reductions (%) produced at all MOI for phages ph001L (**a2**), ph001T (**b2**), and their cocktail (**c2**), during a 12 h treatment timeframe. Bacterial concentration: BC, bacterial control; BP-B, bacteria with phage. Values represent the mean of three independent assays and error bars represent the standard deviation.

3.8. Assessment of the Outcome of Abiotic Factors upon Phage Viability

3.8.1. pH Studies

When different pH values (3.0, 6.5, 8.0, 9.0, 10.0, 12.0) were tested, it was observed that phage ph001L concentration decreased with the decrease in pH; however, the differences among pH values 6.5 and 8.0 were not statistically significant (Figure 7(a1), ANOVA, $p > 0.05$) up to 12 h of incubation, after which phage ph001L viability decreased more at pH 8.0 than at pH 6.5 (Figure 7(a1), ANOVA, $p < 0.05$). However, at pH 3.0, phage ph001L endured during the first 2 h (Figure 7(a1), ANOVA, $p > 0.05$) but lost all its lytic viability (Figure 7(a1), ANOVA, $p < 0.05$) after 4 h. At pH values 6.5 and 8.0, phage ph001L persisted as viable for at least 72 h at 25 °C (Figure 7(a1)). Regarding phage ph001T, 2 h of incubation at pH 3.0 were sufficient to completely inactivate it (Figure 7(b1), ANOVA, $p < 0.05$). After 72 h of incubation at pH 6.5 the abundance of phage ph001T decreased by about two orders of magnitude (Figure 7(b1), ANOVA, $p < 0.05$). Nevertheless, the reduction of lytic viability was more significant at pH 6.5 for this phage than at pH 8.0, after 24 h of incubation (Figure 7(b1), ANOVA, $p < 0.05$). At pH 9.0, phage ph001T lost 1.07 log·PFU/mL after 2 h, 2.35 log·PFU/mL after 6 h, and completely lost its lytic viability after 12 h (Figure 7(b1)). On the other hand, phage ph001L was relatively stable at pH 9.0 up to 12 h of incubation at that pH (Figure 7(a1)), but progressively lost 1.28 log·PFU/mL after 36 h of incubation, 1.59 log·PFU/mL after 48 h, and 1.73 log·PFU/mL after 72 h. At pH values 10 and 12, both phages lost completely their lytic activity immediately after contacting the buffer at those high pH values (Figure 7(a1,b1)).

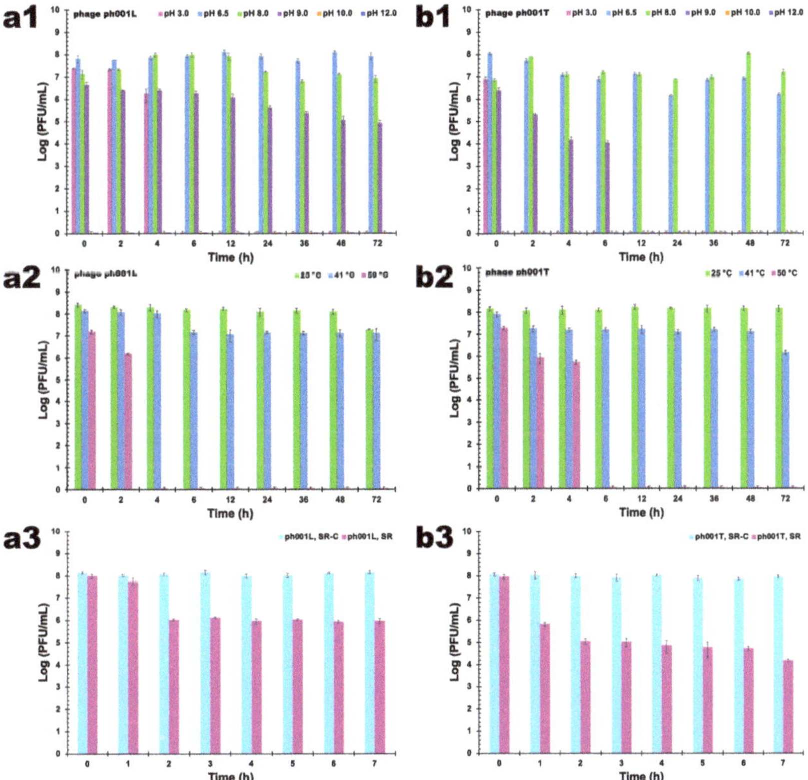

Figure 7. Survival of phages ph001L and ph001T following exposure to different pH values ((**a1**): ph001L; (**b1**): ph001T), different temperature values ((**a2**): ph001L; (**b2**): ph001T), and solar radiation ((**a3**): ph001L; (**b3**): ph001T). All experimental data values represent means of triplicate determinations in three independent experiments. Error bars represent the standard deviation. SR: phage exposed to direct sunlight; SR-C: phage not exposed to sunlight (control).

3.8.2. Temperature Studies

The reduction in the concentration of viable phage ph001L particles was much higher at 50 °C than at 25 °C and 41 °C (Figure 6(a2), ANOVA, $p < 0.05$). A maximum decrease of ca. 1 log·PFU/mL was observed after 72 h when the phage ph001L samples were kept at a temperature of 25 °C, a trend that was also observed at 41 °C (Figure 7(a2)). However, at 50 °C, phage ph001L viability decreased 1 log·PFU/mL after only 2 h of incubation, after which it completely lost its lytic viability (Figure 7(a2), ANOVA, $p < 0.05$). Regarding phage ph001T, it was completely stable at 25 °C (Figure 7(b2), ANOVA, $p > 0.05$), but at 41 °C lost 0.8 log·PFU/mL after 48 h of incubation and ca. 1.8 log·PFU/mL after 72 h (Figure 7(b2), ANOVA, $p < 0.05$). At 50 °C, phage ph001T lost ca. 1.6 log·PFU/mL after only 4 h of incubation, but after this timeframe it ceased to be viable (Figure 7(b2), ANOVA, $p < 0.05$).

3.8.3. Solar radiation Studies

Exposure of phages ph001L and ph001T to direct sunlight for 7 h promoted a decrease of 2.2 log·PFU/mL (phage ph001L, Figure 7(a3), ANOVA, $p < 0.05$) and 3.8 log·PFU/mL (phage ph001T, Figure 7(b3), ANOVA, $p < 0.05$) in the abundance of viable phage virion particles, when compared to the controls (SR-C).

3.9. Preparation and Characterization of the Edible Biopolymeric Microcapsular Wrapping (EBMW) Integrating the Bacteriophage Cocktail

The edible biopolymeric microcapsular wrapping was prepared with sodium alginate by inotropic gelling. The technique allowed the preparation of a microcapsular wrapping with translucent and uniform characteristics, with average diameters of 2 μm (Figure 8). Sodium alginate was chosen for the production of the edible microcapsular wrappings for its interesting characteristics linked to biocompatibility, biodegradability, non-toxicity, and gelling capacity [49]. Sodium alginate has been successfully applied in edible films and coatings, aiming at food protection and also as carriers of some food preserving agents (antioxidants and antimicrobials) [50]. Furthermore, other research works have indicated that alginate-based matrices are suitable for phage incorporation and protection [36–38,51–53].

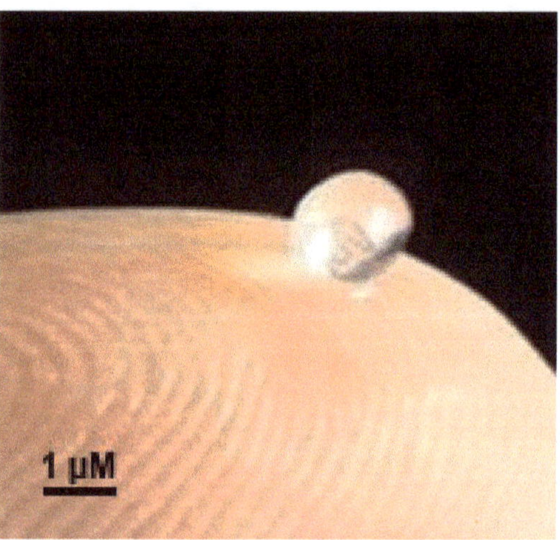

Figure 8. Image of a calcium alginate microcapsular wrapping obtained by ionotropic gelling and integrating a cocktail of two lytic bacteriophages for *Salmonella enterica* CCCD-S004.

3.10. Assessment of the Lytic Viability of Entrapped Bacteriophage Particles within the EBMW Formulations

Entrapment of the phage virion particles in the chitosan-coated calcium alginate biopolymeric matrix of the EBMW particles promoted structural and functional stabilization of said virions, with maintenance of their lytic viability (Figure 9). Maintenance of the lytic activity of the phage virion particles within the microcapsular wrapping was evaluated since immobilization on different matrices can affect both their viability and availability. The process of obtaining films, coatings, and hydrogels integrating phage particles ends up exposing them to stressful conditions such as mixing, stirring, or drying [54]. Figure 8 displays images of Petri plates containing a lawn of *Salmonella enterica* CCCD-S004 and, on top of it, the microcapsular wrappings (integrating or not phage particles). Lysis zones can be seen in the lawn, surrounding the microcapsular wrappings integrating the phage cocktail at MOI 1 (Figure 9b, inserted arrow), 10 (Figure 9c, inserted arrow), 100 (Figure 9d, inserted arrow), and 1000 (Figure 9e, inserted arrow), indicating maintenance of the lytic activity of the phage particles on the host bacteria upon immobilization within the microcapsular wrapping matrices. No lysis zone could be observed for the control EBMW (Figure 9a).

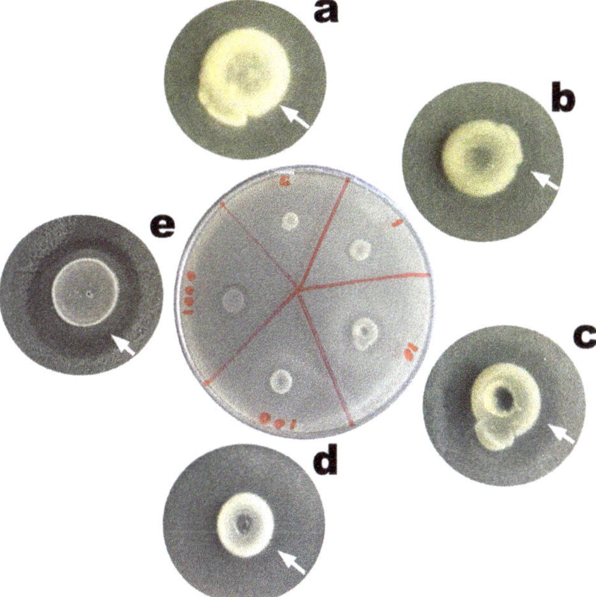

Figure 9. Results from assessment of the lytic viability of entrapped bacteriophage particles within the EBMW formulations. (**a**) EBMW matrix devoid of phage particles, and bioreactive EBMW matrices integrating the phage cocktail at (**b**) MOI 1, (**c**) MOI 10, (**d**) MOI 100, and (**e**) MOI 1000.

No lysis zone could be observed in the bacterial lawn surrounding the control microcapsular wrapping matrix (Figure 9a). On the contrary, clear zones of lysis surrounding the EBMW matrices integrating the phage cocktail was most evident for all MOI values tested (Figure 9), with the lysis area increasing in general with increasing MOI, leading to the conclusion that integration of the phage particles within the EBMW matrix formulation did not interfere with the lytic activity of the entrapped phage particles.

To try to explain the lysis promoted by the entrapped phage particle cocktail when in contact with a lawn of the host (*Salmonella enterica* CCCD-S004), a putative mechanism was put forward (Figure 10).

Figure 10 displays an illustration for the putative interactions between the Ca alginate matrix and the chitosan coating at different pH values. Chitosan is electrostatically bound to the surface of the Ca alginate matrix at a lower pH (top-agar surface, pH equal to ca. 6). At a higher pH (*Salmonella enterica* lawn surface, pH equal to ca. 9), chitosan becomes deprotonated and acquires a net negative charge, and the repulsion forces acting on the (also negatively charged) Ca alginate matrix prevents surface rebinding. The increase in pH promotes a disentanglement of the two polymers, destructuring the particle and promoting release of the phage virions into the outer medium, where the phage virions can contact and infect the bacterial host cells, promoting their lysis, as can be observed in Figure 9.

The EBMW matrices integrating the phage cocktail at MOI 100 and MOI 1000 were also used in microscale bacterial inactivation assays, and the results obtained are displayed in Figure 11.

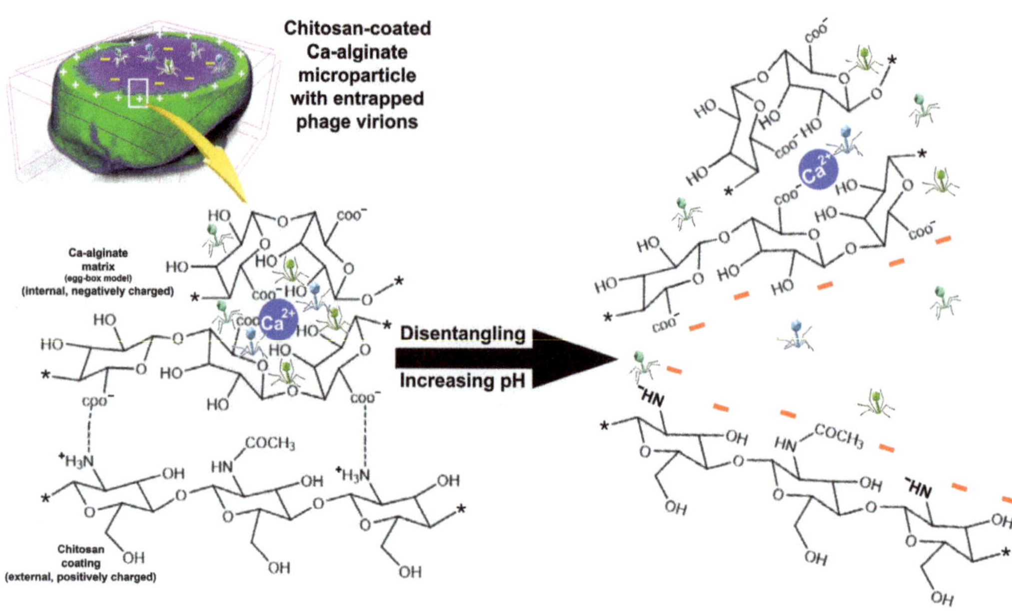

Figure 10. Putative mechanism for the release of entrapped phage virions during incubation with a bacterial lawn of the host, showing the interactions (dotted lines representing hydrogen bonding) between calcium alginate and chitosan. ⋆ denotes repeating polymer monomers.

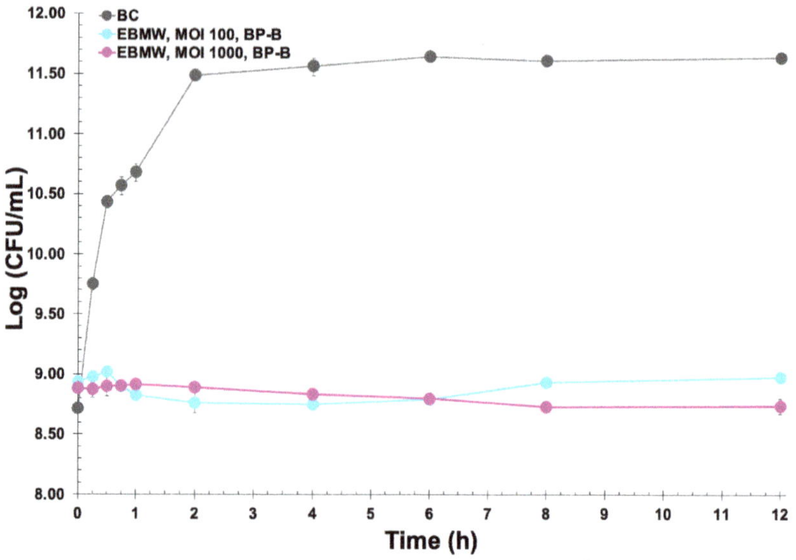

Figure 11. In vitro inactivation of *Salmonella enterica* CCCD-S004 by the entrapped phage cocktail (EBMW) at MOI 100 and MOI 1000, during a 12 h treatment timeframe. Bacterial concentration: BC, bacterial control; BP-B, bacteria with EBMW particles. Values represent the mean of three independent assays and error bars represent the standard deviation.

As can be observed from inspection of the data in Figure 11, the bioactive (lytic) EBMW formulation containing the cocktail of bacteriophages at MOI 1000 was able to promote a reduction in the bacterial load, albeit slight, due most probably to the non-disintegration of the particles within the bacterial suspension and concomitant non-release of all bacteriophage particles. The EBMW with the entrapped phage cocktail at MOI 100 was able to promote a slight decrease of 2.81 log·CFU/mL after 4 h of treatment, which decreased to only 2.68 log·CFU/mL after 8 h and 2.65 log·CFU/mL after 12 h of treatment. On the other hand, the EBMW with the entrapped phage cocktail at MOI 1000 was able to promote a nearly identical decrease in bacterial load after 4 h of treatment, which increased to 2.88 log·CFU/mL after 8 h and endured up to the end of treatment, with a decrease of 2.90 log·CFU/mL after 12 h.

From the simple experiment that was designed and implemented aiming at evaluating the process of release of the phage virions from the EBMW formulation with entrapped phage particles at MOI 1000, it is clear that the EBMW particles released the virions progressively with time and that they retained their lytic activity (Figure 12).

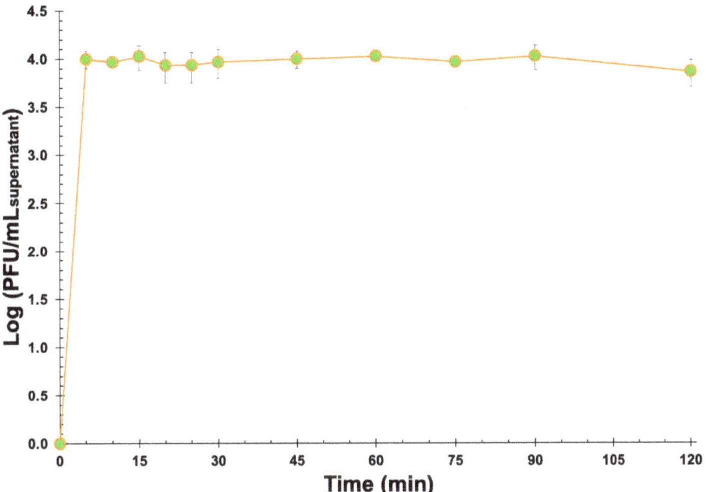

Figure 12. In vitro phage virion release profile from the EBMW particles integrating the phage cocktail at MOI 1000, into plain SM buffer. Values represent the mean of three independent assays and error bars represent the standard deviation.

3.11. Fourier Transform InfraRed Spectrometry (FTIR) Analyses

Figure 13 shows the FTIR spectra of the chitosan-coated EBMW integrating the phage cocktail at MOI 1000 and of the chitosan-coated EBMW devoid of phage particles (control sample).

The FTIR spectra of the chitosan-coated EBMW integrating or not the phage cocktail are very similar (Figure 13), allowing one to conclude that the phage particle did not engage in any type of chemical reaction with the EBMW biopolymeric matrix. The broad peak between 3000 and 3650 cm^{-1} corresponds to the elongation of the OH- groups present in both the alginate polymer chain and residual water molecules [55,56]. The peaks at 1420.05/1419.78 cm^{-1} and 1635.87/1635.66 cm^{-1} may be attributed, respectively, to asymmetrical and symmetrical axial distortions of -COO- groupings, indicative of the existence of carboxylic acid residues in the calcium alginate matrix [55]. Stretching of C=C were encountered at ca. 1636 cm^{-1} for the calcium alginate matrix coated with chitosan, arising most likely from isolated alkenes. The existence of N-acetyl moieties originating from chitosan was ascertained by the peaks appearing at ca. 1635 cm^{-1} (primary amide C=O stretching) and 1295 cm^{-1} (tertiary amide C-N stretching). No characteristic N-H bondings from secondary amides were found in the spectra of the formulations tested. The peaks at ca. 1420 cm^{-1} were attributed to bonding of CH_2 groups. The small peak at ca. 1144 cm^{-1} was assigned to asymmetrical stretching of -C-O-C- groups whereas the peaks at 1078 cm^{-1} and 1028/1029 cm^{-1} were assigned to stretching vibrations of the -C-O-C- bond of the ether groups from the chitosan coating [57–61]. The peak found at ca. 902 cm^{-1} was attributed to -C-H- groups bonding out of the plane of the sugar rings in chitosan moieties [62], whereas the peak at 1078 cm^{-1} was most likely due to stretching of -C-N- bonds from aliphatic amines [63]. Abnormally, the absorption peak that appeared at 2360 cm^{-1} was most likely due to antisymmetric stretching of CO_2 molecules from air entrapped inadvertently in the EBMW matrices during their formation [64]. As can be noticed from inspection of Figure 13, the same peaks can be observed (with only minor variations in peak intensity) in the FTIR spectra of plain EBMW microparticles and EBMW microparticles integrating the phage virion cocktail at MOI 1000, strongly suggesting that the chemical features of phage virions were conserved during entrapment within the microcapsular wrappings.

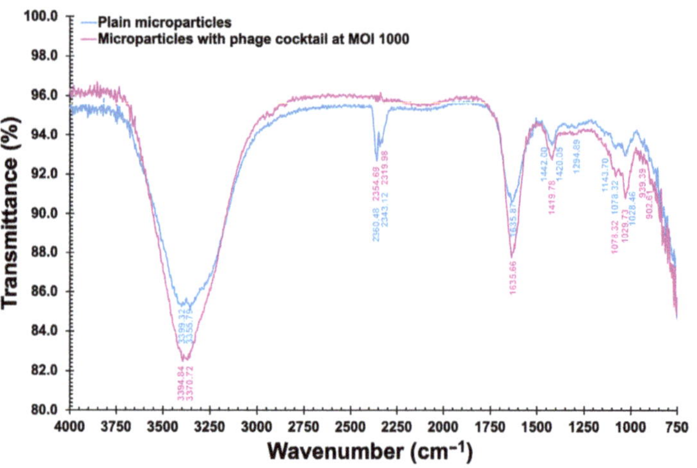

Figure 13. Fourier transform infrared (FTIR) spectra of the chitosan-coated EBMW integrating the phage cocktail at MOI 1000 (pink line) and of the chitosan-coated EBMW devoid of phage particles (blue line).

3.12. Thermal Characterization of the EBMW Formulations via DSC

DSC thermograms of a plain EBMW formulation and of an EBMW formulation integrating the phage virion cocktail at MOI 1000 are displayed in Figure 14.

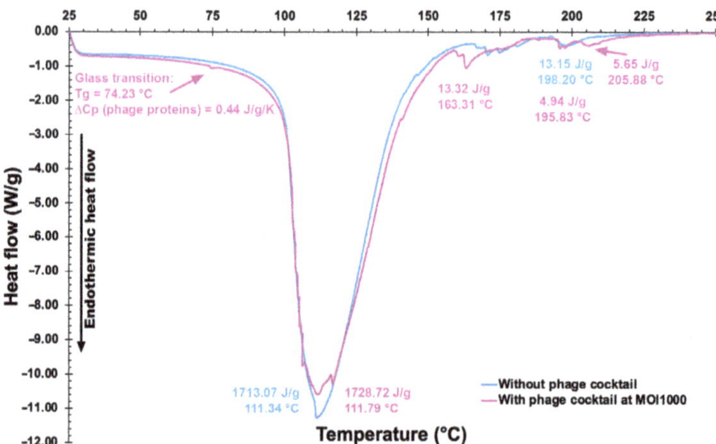

Figure 14. Differential scanning calorimetry thermograms of the EBMW formulation devoid of phage particles (blue line) and of the EBMW formulation integrating the cocktail of phage particles (magenta line).

Very similar thermal events can be perceived for both EBMW particles, with the sample containing the phage virion cocktail displaying a slightly higher melting enthalpy at virtually the same temperature, viz. 111 °C. The peak temperature of heat absorption of the two particle formulations were very close to one another and virtually equal to the mid-point of the calcium alginate melting range, viz. ca. 111.5 °C [65]. The prepared microcapsular wrappings were basically made of chitosan-coated calcium alginate. Thus, one can observe thermal events similar to each other. The first endothermic events (sample and control) are probably related to coating dehydration [66]. The second endothermic events at 163.31 °C (sample) may be due to a depolymerization process. The process of thermal disintegration of (bio)polymers encompass sequential steps of dehydration, depolymerization, and disruption of -C-O- and -C-C- bonds with concomitant production of CO, CO_2, and H_2 [67]. The EBMW sample also showed two small endothermic events at higher temperatures, viz. 195.83 °C and 205.88 °C, which might be due to the influence of components in the buffer solution (where the phages are diluted) that increase the conformational stability by electrostatic interactions of the present components [68]. Phages encapsulated in a glassy matrix having a low moisture content, such as the EBMW, may result in better storage stability at low and ambient storage temperatures [69]. In addition, the glass transition temperature (Tg) detected at Tg = 74.23 °C in the EBMW sample containing the phage cocktail might be due to the phage proteins, a value slightly higher than the Tg reported by other researchers [70] for microencapsulated phage against *Salmonella*. According to several researchers [71], protein moieties can be retained dried within a vitreous sugar matrix at temperatures (T) lower by at least 50 °C than Tg, primarily for the reason that at (Tg − T) > 50 °C, protein moieties are sufficiently stagnant with decreased reactivity. Perhaps such vitreous stabilization rationale may also be applied to protein-based entities such as phage virions.

3.13. Elemental Profile of the EBMW Formulations Obtained by EDXRF

The elemental profiles of EBMW formulations, with and without phage particles, are displayed in Figure 15. Relatively high concentrations of magnesium (Mg), chlorine (Cl), and calcium (Ca) were found for both formulations, originating probably from the bacteriophage suspension and calcium chloride utilized to prepare the formulations. Al (most likely originating from the sodium alginate itself or chitosan, being probably a contaminant) was detected at ca. 0.4%, and other elements such as phosphorus, sulfur, iron, and silver were detected in very small amounts.

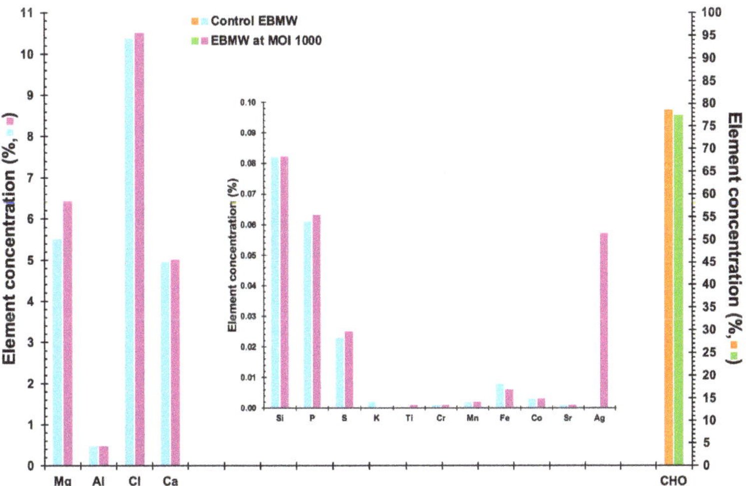

Figure 15. Elemental profiles of EBMW formulations, with and without phage particles.

The most common substances were, as expected, carbon, hydrogen, and oxygen, accounting for ca. 79% and 77% of EBMW formulations without and with the phage cocktail, respectively (Figure 15). Calcium alginate, the basis of the edible microcapsular wrapping, is made almost entirely of alginate extracted from seaweed [72], and thus, CHO, calcium, and chlorine, in greater proportions, are derived, probably from alginate. The element calcium (Ca) was also found in higher concentration due to its addition during the inotropic gelation process. The elements phosphorus (P) and magnesium (Mg) also stood out, probably coming from the phage suspensions utilized. However, these elements are not at all considered to be toxic.

3.14. XRT Analysis of the EBMW Integrating the Cocktail of Phage Particles

The optimized EBMW developed may be considered a natural polymer composite exhibiting a very special porous microstructure which enables the imprisonment of the phage particles. From the tomographic analyses via X-ray transmission performed to an EBMW particle loaded with the phage particle cocktail (Figure 16), a homogeneous surface can be observed.

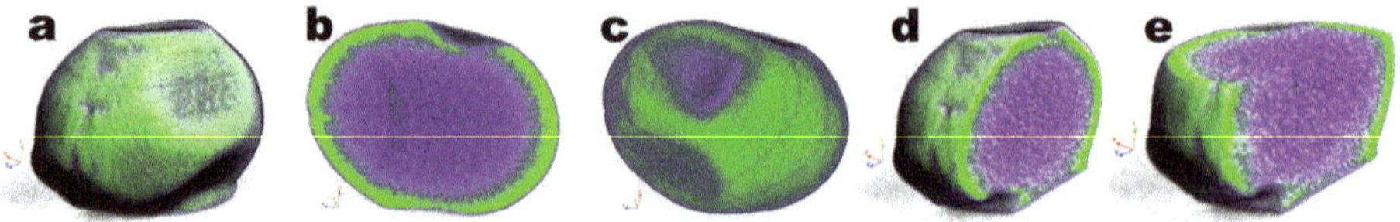

Figure 16. Images obtained by tomographic analyses via X-ray transmission of the EBMW particle loaded with phage particles, being (**a**) front view of a EBMW, (**b**) vertical cut of a EBMW, (**c**) top view of a EBMW, (**d**) frontal cut, and (**e**) EBMW particle with front and top cuts. Three-dimensional image slices were gathered using an operating voltage set at 31 kV and electric current with 661 µA.

The chitosan layer coating the calcium alginate matrix is in greater evidence (in green in Figure 16) since, due to its stronger atomic density, it absorbs radiation to a greater extent. On the other hand, the void spaces show up pinpointed in light green within the polymeric network matrix (in purple in Figure 16) in the reconstructed three-dimensional image (Figure 16b,d,e). This closely compares with the information gathered in the FTIR tests (Figure 13), viz. that the virion particles likely did not establish any covalent bondings with the calcium alginate matrix. This realization is clearly significant, meaning that by not establishing permanent bonding with the calcium alginate matrix, the phage particles become readily available and maintain their lytic bioactivity, as was demonstrated before (Figures 9, 11 and 12). A comparative porosity analysis of the EBMW formulation integrating the phage particles can be found in Table 5, resulting from 2D and 3D morphological analyses.

Table 5. Bi- and three-dimensional morphological parameters of the EBMW containing the cocktail of phage virions at MOI 1000.

Parameter	EBMW Entrapping the Cocktail of Virion Particles at MOI 1000	
	Bi-Dimensional (2D) Morphological Analysis	**Three-Dimensional (3D) Morphological Analysis**
Number of layers	-	101.0
Pixel size (µm)	-	6.70
Total VOI (volume of interest), TV (μm^3)	6.45×10^9	6.45×10^9
Object volume, Obj.V (μm^3)	8.67×10^8	8.66×10^8
Percent object volume, Obj.V/TV (%)	13.44	13.42
Total VOI surface, TS (μm^2)	2.74×10^7	2.74×10^7
Object surface, Obj.S (μm^2)	6.13×10^6	5.41×10^6
Total intersection surface, i.S (μm^2)	0	1.92×10^6
Object surface/volume ratio, Obj.S/Obj.V (μm^{-1})	7.07×10^{-3}	6.25×10^{-3}
Mean number of objects per slice, Obj.N	1.15	-
Average object area per slice, Av.Obj.Ar (μm^2)	1.20×10^6	-

Table 5. *Cont.*

Parameter	EBMW Entrapping the Cocktail of Virion Particles at MOI 1000	
	Bi-Dimensional (2D) Morphological Analysis	Three-Dimensional (3D) Morphological Analysis
Average moment of inertia (x), Av.MMI (x) (μm^4)	1.79×10^{11}	1.51×10^{14}
Average moment of inertia (y), Av.MMI (y) (μm^4)	1.00×10^{11}	9.83×10^{13}
Average moment of inertia (z), Av.MMI (z) (μm^4)	-	1.90×10^{14}
Mean eccentricity, Ecc	0.69	-
Cross-sectional thickness, Cs.Th (μm)	573.70	-
Object surface density, Obj.S/TV (μm^{-1})	-	8.38×10^{-4}
Mean surface convexity index, SCv.I (μm^{-1})	2.17×10^{-4}	3.14×10^{-3}
Degree of anisotropy, DA	-	2.75 (0.64)
Eigenvalue 1	-	2.86×10^{-2}
Eigenvalue 2	-	4.32×10^{-2}
Eigenvalue 3	-	7.88×10^{-2}
Number of closed pores, Po.N (cl)	-	0
Volume of closed pores, Po.V (cl) (μm^3)	-	0
Surface of closed pores, Po.S (cl) (μm^2)	-	0
Closed porosity (percent), Po (cl) (%)	9.47×10^{-4}	0
Mean fractal dimension, FD	1.02	2.04
Volume of open pore space, Po.V (op) (μm^3)	-	5.58×10^9
Open porosity (percent), Po (op) (%)	-	86.58
Total volume of pore space, Po.V (tot) (μm^3)	-	5.58×10^9
Total porosity (percent), Po(tot) (%)	-	86.58
Euler number, Eu.N	-	1
Connectivity, Conn	-	3
Connectivity density, Conn.Dn (μm^{-3})	-	0

When the properties of a substance, both mechanical and/or physical, differ when determined along a Cartesian coordinate system, it means that such properties are directionally dependent, i.e., are anisotropic. The degree of anisotropy (DA) $\{1 - (\text{Eigenvalue}_{min}/\text{Eigenvalue}_{max})\}$ can assume any value in the range 0 (total isotropy)–1 (total anisotropy). The EBMW loaded with phage virions exhibited a DA of 0.63632 (Table 5), and thus can be considered more anisotropic than isotropic. The results obtained for the total porosity of the EBMW (86.58%, Table 5) allow one to conclude that the particles produced were mostly porous in their structure, displaying an open porosity exactly equal to the total porosity (Table 5). Additionally, the mean fractal dimension in the 3D analysis is 2.04, giving a measure of how "intricate" a self-similar figure is and measuring roughly "how many points" lie in a given set. The mean fractal dimension obtained (2.04) has an interesting property in the sense that, as it fills the space of an area, it acts as if it is filling the space of a volume [73].

3.15. Morphological Analyses of the EBMW via SEM

Analyses of a EBMW particle via SEM allowed one to observe a homogeneous surface without any fissures or crevices (Figure 17).

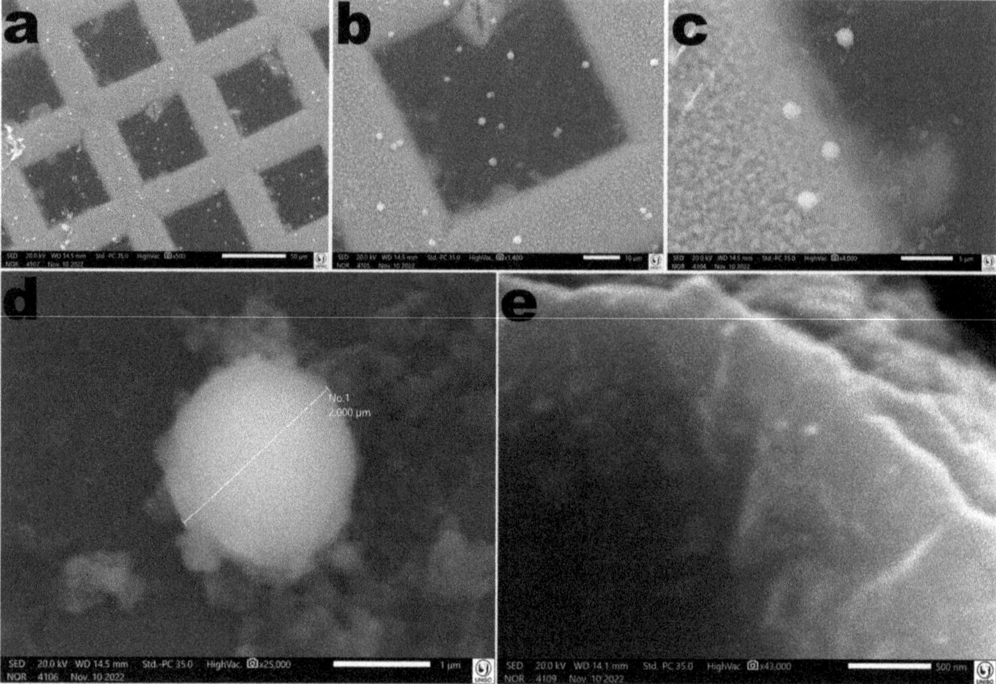

Figure 17. Photomicrographs of the vaginal egg surface at several magnifications ((**a**): ×500; (**b**): ×1400; (**c**): ×4000; (**d**): ×25,000; (**e**): ×43,000). Images obtained by scanning electron microscopy (SEM) confirmed the formation of microcapsular wrappings with homogeneous characteristics. In photomicrograph (**d**) it is possible to observe the mean diameter of the particles produced.

4. Discussion

Salmonella enterica is known as one of the main microorganisms responsible for poultry contamination. Developing new alternatives to the conventional antibiotic-based antimicrobial control for preventing and/or controlling infections by this pathogen have been quite challenging and a long-time goal within the scientific community, aiming at reducing the development of multi-drug resistant bacteria. In the research effort described herein, the structural and functional stabilization of two newly isolated lytic phages for *S. enterica* (viz. phages ph001L and ph001T, isolated from environmental samples at the Campus of UNISO in Sorocaba, SP, Brazil) within edible biopolymeric microcapsular wrappings (EBMW) has been proposed, aiming at a potential integration in poultry feed as a means to control the aforementioned pathogen. The results obtained in this study provide clear evidence that the use of the two newly isolated phages can reduce the population of pathogenic *S. enterica* cells. The two newly isolated phages produced translucent and tiny plaques on a lawn of the bacterial host, exhibiting diameters of approximately 0.1 mm (Figure 2); were identified as members of the class Caudoviricetes and displayed siphovirus morphotypes (Figure 2) with similar capsid dimensions but with different tail lengths (Table 2); and displayed distinct extinction coefficients (Figure 3b) yet of the same order of magnitude [24,25]. The phage plaques produced by both phages were clear and tiny and did not exhibit a secondary halo in the frontier of the lysis plaque of phage (Figure 2), which is a likely indication that these phages do not produce depolymerase enzymes [74].

In the present study, the host range of the two newly isolated phages was assessed by determining if they were able to form clear plaques of lysis on particular bacteria (meaning that the phages were able to productively infect the bacteria and yield progeny). According to Hyman [75], newly isolated phage particles may also infect different bacterial cells displaying similar receptors on their surface, beyond the species used in their isolation. Besides the isolation strain, phages ph001L and ph001T were able to bind to *Salmonella enterica* subsp. Enteritidis ATCC 13076, *Pseudomonas syringae* pv. *Garcae* IBSBF-158, *Escherichia coli* ATCC 25922, *Escherichia coli* ATCC 8739, *Klebsiella pneumoniae*

ATCC-13883, and *Klebsiella pneumoniae* NCTC-13439, and kill them with moderate efficacies (Table 4), yielding progeny virions at relatively high numbers and producing EOP values not so low, as was verified for those bacterial strains, representing 37% of all bacteria tested. According to several authors, while some phages can only infect one or a few bacterial strains, other phages can infect many species or even bacteria from different genera [76–79], evidence that supports the results obtained in the present research effort for the EOP determinations in the bacterial strains that produced positive spot tests with both phages. Hence, a well-known singularity of bacterial lysis prompted from the inside by phage-derived holins and lysins, a process commonly known as "lysis from within", can be speculated as a credible reasoning for these bacterial strains exhibiting specific surface receptors recognized by both phages that led to their infection and concomitant killing. Notwithstanding this realization, if practical applications are sought, new (different) lytic phages (isolated from environmental sources) need to be integrated in the cocktail in order to attain a broader lytic spectrum against more strains of *S. enterica*.

A one step of growth was clearly observed for both phages (Figure 4) during the first 10–30 min, which levelled off after this growth. The two phages produced quite large virion progenies (i.e., burst sizes), viz. 466 and 132 virions/host cell, respectively, for phages ph001L and ph001T, suggesting that both phages replicate well in the host with small latencies (25 min for phage ph001L and 40 min for phage ph001T). A number of studies that appear in the specialty literature revealed that using phages producing large virion progenies (morphogenesis yields) within short lytic cycles enhance the efficiency of bacterial control [80–82], however large morphogenesis yields are generally followed by considerably longer latencies [83]. The morphogenesis yield of phage ph001L was ca. 4 times larger than that of phage ph001T, but this did not imply a better performance of phage ph001L; on the contrary, inactivation of planktonic host cells in vitro was in general much higher with phage ph001T.

Adsorption of a free phage virion onto a bacterial host cell is the apotheosis of its existence, with the free energy reserve imparted to the virion three-dimensional conformation during its morphogenesis coming into play, with the bacterial surface receptor-specific adsorption of free phage virions dictating their host range [84]. Hence, knowing the dynamics of virion adsorption onto the bacterial host and its concomitant inactivation in in vitro experiments is of utmost importance if use of phage virions is intended to control pathogenic bacteria.

Both phages revealed virtually equal adsorption rates onto the host cells, viz. 8.0×10^{-10} CFU^{-1}·mL·min^{-1} and 7.0×10^{-10} CFU^{-1}·mL·min^{-1}, for phages ph001L and ph001T, respectively (Figure 5). These results are of the same order of magnitude as the results reported by [85] (lytic phage fSPB adsorption rate on *Salmonella* serovar Paratyphi B, 4.7×10^{-10}) and [86] (adsorption rate of phage $1 = 2.2 \times 10^{-10}$ mL·min^{-1} and adsorption rate of phage $2 = 1.8 \times 10^{-10}$ mL·min^{-1}, onto *Salmonella typhi*), one order of magnitude lower than results reported by [87] (phage PVP-SE1 adsorption rate on *Salmonella enterica* serovar *Enteritidis* strain S1400, 1.00×10^{-9} mL·CFU^{-1}·PFU^{-1}·h^{-1}) and [88] (phage SHWT1 adsorption rate on *Salmonella pullorum* SP01, $(8.8 \pm 0.5) \times 10^{-9}$ mL·min^{-1}), and two orders of magnitude lower than the results reported by [89] (phage phi1 adsorption rate onto *Salmonella enterica* $= 1.6 \times 10^{-8}$ mL·min^{-1}). According to [84], the adsorption constants in some phages are close to the maximally possible values, viz. ca. 1×10^{-8} mL·min^{-1}, but our results (as the vast majority of the adsorption rates for phages onto their bacterial hosts) were two orders of magnitude lower than such a maximum. Nevertheless, the desorption rate of phage ph001L $(1.900 \times 10^{-3}$ mL·min$^{-1})$ was much larger than that of phage ph001T $(4.000 \times 10^{-4}$ mL·min$^{-1})$, implying that fewer phage ph001L virions endured adsorbed to the host cells. Because adsorption of phage virions onto specific receptors on the host cell followed by the virion genome translocation into the host cytoplasm is required for its effective infection and concomitant virion morphogenesis [23,24,90], the much larger desorption rate for phage ph001L may have been accountable for the smaller bacterial inactivation rates promoted by this phage at all MOI studied (Figure 6). The adsorption profile showed that after 30 min ca. 90% of phages ph001L and ph001T particles were adsorbed onto the host cells (Figure 5), whereas after 60 min ca. 100% of the phage particles were adsorbed onto the host cells. As a consequence of the much higher desorption rate, phage ph001L was not able to promote a significant decrease in bacterial concentration at all MOI studied, compared with the non-treated BC (Figure 6(a1,a2)). During the first 6 h of incubation of phage ph001L in the presence of its host, the bacterial concentration was only slightly reduced at all MOI, compared with that of the bacterial control (Figure 6(a1)). On the other hand, phage ph001T, by having a much lower desorption rate, succeeded in promoting significant bacterial reductions, especially at MOI 1, 10, 100, and 1000 (Figure 6(b1,b2)). When incubating the phage cocktail integrating both phages ph001L and ph001T with its host, significant bacterial reductions were observed especially at MOI 1, 10, 100, and 1000 (Figure 6(c1,c2)). A number of studies found in the specialty literature revealed that the decrease in bacterial cell numbers is either stronger or faster at higher MOI [25,91–93]. In this work, for either phage, increasing MOI from 0.01 → 0.1 did not significantly increase the efficacy of phage-based treatment (Figure 6(a1,b1,c1)) but increasing MOI from 1 → 10 was much more effective than when increasing MOI from 10 → 100 or from 100 → 1000.

When tested against abiotic factors such as pH, temperature, and solar radiation, phage ph001L concentration decreased with decreasing pH; however, the differences among pH values 6.5 and 8.0 were not statistically significant up to 12 h of incubation, after which phage ph001L viability decreased more at pH 8.0 than at pH 6.5. However, at pH 3.0, phage ph001L endured during the first 2 h but lost all its lytic viability after 4 h. At pH values 6.5 and 8.0, phage ph001L persisted as viable for at least 72 h at 25 °C (Figure 7(a1)). Regarding phage ph001T, 2 h of incubation at pH 3.0 were sufficient to completely inactivate it (Figure 7(b1)). After 72 h of incubation at pH 6.5 the abundance of phage ph001T decreased about two orders of magnitude (Figure 7(b1)). Nevertheless, the reduction of lytic viability was more significant at pH 6.5 for this phage than at pH 8.0, after 24 h of incubation (Figure 7(b1)). At pH 9.0, phage ph001T was not very stable, fully losing its lytic viability after 12 h (Figure 7(b1)). On the other hand, phage ph001L was relatively stable at pH 9.0 up to 12 h of incubation at that pH (Figure 7(a1)), but progressively lost up to 1.73 log·PFU/mL after 72 h of incubation at that pH. At pH values 10 and 12, both phages lost completely their lytic activity immediately after contacting the buffer at those high pH values (Figure 7(a1,b1)). The reduction in the concentration of viable phage ph001L particles was much higher at 50 °C than at 25 °C and 41 °C (Figure 7(a2)). A maximum decrease of ca. 1 log·PFU/mL was observed after 72 h when the phage ph001L samples were kept at a temperature of 25 °C, a trend that was also observed at 41 °C (Figure 7(a2)). However, at 50 °C, phage ph001L viability decreased by 1 log·PFU/mL after only 2 h of incubation, after which it completely lost its lytic viability (Figure 7(a2)). Regarding phage ph001T, it was completely stable at 25 °C (Figure 7(b2)), but at 41 °C lost 0.8 log·PFU/mL after 48 h of incubation and ca. 1.8 log·PFU/mL after 72 h (Figure 7(b2)). At 50 °C, phage ph001T lost ca. 1.6 log·PFU/mL after only 4 h of incubation, but after this timeframe it ceased to be viable (Figure 7(b2)). When phage ph001L was exposed to solar radiation, the abundance of phage particles decreased by 2.2 log·PFU/mL (Figure 7(a3)) after 7 h of exposure, when compared with the phage control (SR-C). The decrease in phage abundance was 2.0 log·PFU/mL (Figure 7(a3)). When phage ph001T was exposed to solar radiation, the abundance of phage particles decreased by 3.8 log·PFU/mL (Figure 7(b3)) after 7 h of exposure, when compared with the phage control (SR-C). The decrease in phage abundance was 3.8 log·PFU/mL (Figure 7(b3)).

Salmonella enterica was not effectively inactivated by the cocktail of both isolated phages at MOI 0.01 and 0.1 (Figure 6(c1)), and all remaining MOI tested failed to fully prevent bacterial regrowth (Figure 6(c1)). A number of studies found in the specialty literature state that, because of the gargantuan variability in bacterial cell surface receptors recognized by phage virions, regrowth of bacteria after treatment with phages can be virtually circumvented by using a cocktail composed of a variety of lytic phages displaying different adsorption processes [28,29,80,94–97].

One of the many challenges faced when performing bacterial biocontrol studies using phage virions resides in proving its feasibility in real-world situations [24,28,29] and, therefore, integration of the phage virions as a cocktail within the chitosan-coated calcium alginate matrix formulation (EBMW) was performed aiming at proving its suitability for inclusion in poultry feed. The process of entrapment of the phage particles via ionotropic gelation did not impact negatively in the lytic viability of the imprisoned phage virions (Figure 9), a conclusion backed by the formation of clear zones of lysis surrounding the EBMW formulations containing phages at different MOI on the *S. enterica* lawn (Figure 9b–e). This was a clear indication that the phage virions imprisoned within the calcium alginate matrix retained their lytic viability [26,27,38,98], with the biopolymeric matrix providing suitable diffusion of the phage particles towards their host cells.

Hence, new in vitro phage–bacteria inactivation assays were performed in microscale, using the EBMW particles containing entrapped phage virion cocktails at MOI 100 and MOI 1000. These experiments, together with the putative mechanism developed (Figure 10), indeed gave a better understanding of how EBMWs with entrapped phages will work in the avian digestive tract. In bacterial suspension in TSB (pH ≈ 7), the chitosan-coated alginate microparticles remain cohesive and firm since the cationic (protonated) chitosan coating stays firmly electrostatically attracted to the Ca alginate matrix (negatively charged). This impairs the full release of phage particles into the suspension, with concomitant low levels of bacterial inactivation, as can be observed in Figure 11. In the avian digestive tract, where pH changes from 2.5–3.5 in the gizzard to 5–6 in the duodenum, 6.5–7.0 in the jejunum, 7.0–7.5 in the ileum, and 8.0 in the cecum/colon, this effect is anticipated to be much more pronounced since separation of the chitosan layer due to its deprotonation at higher pH values leads to disintegration of the EBMW particles. Hence, according to the putative phage release mechanism deployed (Figure 10), release of phage virions from the EBMW formulations is expected to occur easily.

The FTIR spectra of the chitosan-coated EBMW integrating or not the phage cocktail were very similar (Figure 13), allowing one to conclude that the phage virions did not engage in any types of chemical reactions with the EBMW biopolymeric matrix, which suggests that the chemical aspect of the phage virions was fully preserved during incorporation into the biopolysaccharide microcapsular wrapping, a very important conclusion, because if the phages were involved in any type of bonding with the biopolymeric matrix, that would prevent their release.

The prepared microcapsular wrappings were basically made of chitosan-coated calcium alginate, hence similar (major) thermal events were observed for both the EBMW formulation devoid of phage particles and the EBMW formulation integrating the cocktail of phage particles (Figure 14), with melting enthalpies of the same order of magnitude, with the EBMW formulation integrating the phage cocktail absorbing a slightly higher amount of energy (viz. 1728.72 J/g) at virtually the same melting point (Figure 10), viz. 111.79 °C, than the EBMW formulation devoid of phage particles (1713.07 J/g at 111.34 °C). Integration of the phage cocktail in the EBMW formulation promoted a slight increase in the melting enthalpy and displaced slightly the peak temperature of the major thermal event (Figure 14).

The results from the thermal characterization of the EBMW particles are linked to the stability of the particles, a crucial parameter for their successful incorporation into poultry feeding. The thermal stability of a biopolymer can be defined as its ability to withstand the action of heat while maintaining its properties (such as toughness or elasticity) at a given temperature and as can be observed from inspection of Figure 14, the melting temperature peak was around 111 °C. Since both phages retained most of their lytic activity at 41 °C (Figure 7), the thermal analysis confirmed the stability of the EBMW microparticles at temperatures up to 41 °C at least, since the biopolymeric matrix melts down at a much higher temperature (Figure 14).

As expected, the EBMW formulations were composed almost entirely of carbon, hydrogen, and oxygen (Figure 15), arising from the calcium alginate, chitosan, and also from the virions integrated into the antibacterial EBMW formulation.

Analysis of EBMW containing the cocktail of lytic virions at MOI 1000 by XRT (Figure 16) closely compared with the FTIR analysis (Figure 13) and denoted a lack of establishment of covalent bonding between the phage virions and the calcium alginate matrix, thus making the phage virions readily available with maintenance of their lytic bioactivity, as was demonstrated before (Figures 9, 11 and 12). The EBMW formulation produced is essentially anisotropic, with a highly porous (viz. 86.58%) structure and a high mean fractal dimension in the 3D analysis (Figure 16).

Additionally, morphology analysis of a EBMW particle via SEM allowed one to observe a homogeneous rugged surface (photomicrographs c, d, and e in Figure 17), devoid of any fissures or crevices whatsoever. From observation of the photomicrographs in Figure 17, a highly uniform and compact matrix structure can be clearly observed, in clear agreement with results from X-ray tomography (Figure 16). These observations are very significant, since the phage particles were uniformly dispersed within the biopolymeric matrix, and the three-dimensionally reconstructed digital slices in Figure 16 allow the clear observation of the compactness of the formulation, with plenty of hydrophilic pores without air pockets (hydrophobic in nature, which could negatively impact phage viability).

The results described herein clearly suggest that the cocktail produced with the two newly isolated lytic phages, ph001L and ph001T, have the potential to be an effective surrogate to antibiotics in controlling *Salmonella enterica*. Yet, both phages could not fully restrain bacterial regrowth in vitro, neither separately nor as a cocktail. Hence, selecting more lytic phages for phage–bacteria inactivation assays should consider not only their efficacy but also their potential for developing phage-resistant bacterial mutants, which should circumvent and virtually obliterate the downside of bacterial-acquired resistance to phages.

5. Conclusions

The results of the present work suggest that phage treatment using phages ph001L and ph001T can be an effective alternative to control *Salmonella enterica*. However, both phages were not able to fully prevent bacterial regrowth, although the bacterial reduction levels were quite high when both phages were used as a cocktail, especially at MOI 1 and 10. It was also demonstrated that microencapsulation within a biopolymeric formulation is a viable method for fully stabilizing the virion particles, both from structural and functional points of view, if integration within poultry feed is sought as a means of controlling *Salmonella enterica* in poultry.

Author Contributions: All authors participated in the conception and design of the experiments and analyzed the resulting data; A.O.P., N.M.A.B. and B.R.G. performed the experiments; S.C.E. helped with the abiotic assays; D.Â.B. and J.M.O.J. contributed with the XRF, DSC, FTIR, and SEM analyses; V.M.B. analyzed the data and wrote the paper. V.M.B. and M.M.D.C.V. supervised the work, revised the paper, and contributed with reagents and analysis tools. M.M.D.C.V. contributed with a critical reading of the manuscript. All authors have read and agreed to the published version of the manuscript.

Funding: This research was funded by São Paulo Research Foundation (FAPESP), grants 2016/08884-3 (Project PneumoPhageColor), 2016/12234-4 (Project TransAppIL), and 2022/10775-9 (Project PsgPhageKill). Funding was provided by FCT/MCTES to CESAM (UID/AMB/50017/2019). V.M.B. was visiting researcher fellow from FAPESP (2018/05522-9, Project PsaPhageKill). V.M.B. received research fellowship awards from the National Council for Scientific and Technological Development (CNPq) (grants 306113/2014-7, 308208/2017-0, 301978/2022-0). Funding for Arthur Oliveira Pereira through a full M.Sc. scholarship from Coordination for the Improvement of Higher Education Personnel (Prosup/CAPES) is hereby gratefully acknowledged.

Institutional Review Board Statement: Not applicable.

Informed Consent Statement: Not applicable.

Data Availability Statement: Data will be made available upon request.

Conflicts of Interest: The authors declare no conflict of interest whatsoever. The funders had no role in the design of the study; in the collection, analyses, or interpretation of data; in the writing of the manuscript; or in the decision to publish the results.

References

1. Ravi, Y.; Pooja, M.K.; Krushna Yadav, D.K. Review- bacteriophages in food preservation. *Int. J. Pure App. Biosci.* **2017**, *5*, 197–205. [CrossRef]
2. Anany, H.; Brovko, L.Y.; El-Arabi, T.; Griffiths, M.W. 4—Bacteriophages as antimicrobials in food products: History, biology and application. In *Handbook of Natural Antimicrobials for Food Safety and Quality*; Taylor, T.M., Ed.; Woodhead Publishing Limited: Cambridge, UK, 2015; pp. 69–87.
3. Finger, J.A.F.F.; Baroni, W.S.G.V.; Maffei, D.F.; Bastos, D.H.M.; Pinto, U.M. Overview of Foodborne Disease Outbreaks in Brazil from 2000 to 2018. *Foods* **2019**, *8*, 434. [CrossRef] [PubMed]
4. Feltes, M.M.C.; Arisseto-Bragotto, A.P.; Block, J.M. Food quality, food-borne diseases, and food safety in the Brazilian food industry. *Food Qual. Saf.* **2017**, *1*, 13–27. [CrossRef]
5. Eng, S.-K.; Pusparajah, P.; Ab Mutalib, N.-S.; Ser, H.-L.; Chan, K.-G.; Lee, L.-H. Salmonella: A review on pathogenesis, epidemiology and antibiotic resistance. *Front. Life Sci.* **2015**, *8*, 284–293. [CrossRef]
6. Pulido-Landínez, M. Food safety—Salmonella update in broilers. *Anim. Feed Sci. Technol.* **2019**, *250*, 53–58. [CrossRef]
7. Penha Filho, R.A.C.; Ferreira, J.C.; Kanashiro, A.M.I.; Darini, A.L.C.; Berchieri Junior, A. Antimicrobial susceptibility of Salmonella Gallinarum and Salmonella Pullorum isolated from ill poultry in Brazil. *Cienc. Rural* **2016**, *46*, 513–518. [CrossRef]
8. Doss, J.; Culbertson, K.; Hahn, D.; Camacho, J.; Barekzi, N. A review of phage therapy against bacterial pathogens of aquatic and terrestrial organisms. *Viruses* **2017**, *9*, 50. [CrossRef]
9. Żbikowska, K.; Michalczuk, M.; Dolka, B. The Use of Bacteriophages in the Poultry Industry. *Animals* **2020**, *10*, 872. [CrossRef]
10. Ricke, S.C.; Lee, S.I.; Kim, S.A.; Park, S.H.; Shi, Z. Prebiotics and the poultry gastrointestinal tract microbiome. *Poult. Sci.* **2020**, *99*, 670–677. [CrossRef]
11. Upadhaya, S.D.; Ahn, J.M.; Cho, J.H.; Kim, J.Y.; Kang, D.K.; Kim, S.W.; Kim, H.B.; Kim, I.H. Bacteriophage cocktail supplementation improves growth performance, gut microbiome and production traits in broiler chickens. *J. Anim. Sci. Biotechnol.* **2021**, *12*, 49. [CrossRef]
12. Kazi, M.; Annapure, U.S. Bacteriophage biocontrol of foodborne pathogens. *J. Food Sci. Technol.* **2016**, *53*, 1355–1362. [CrossRef] [PubMed]
13. Harada, L.K.; Silva, E.C.; Campos, W.F.; Del Fiol, F.S.; Vila, M.M.D.C.; Dąbrowska, K.; Krylov, V.N.; Balcão, V.M. Biotechnological applications of bacteriophages: State of the art. *Microbiol. Res.* **2018**, *212–213*, 38–58. [CrossRef] [PubMed]
14. Rios, A.C.; Moutinho, C.G.; Pinto, F.C.; Del Fiol, F.S.; Jozala, A.F.; Chaud, M.V.; Vila, M.M.D.C.; Teixeira, J.A.; Balcão, V.M. Alternatives to overcoming bacterial resistances: State-of-the-art. *Microbiol. Res.* **2016**, *191*, 51–80. [CrossRef] [PubMed]
15. Rakhuba, D.V.; Kolomiets, E.I.; Szwajcer Dey, E.; Novik, G.I. Bacteriophage receptors, mechanisms of phage adsorption and penetration into host cell. *Pol. J. Microbiol. (Pol. Tow. Mikrobiol.)* **2010**, *59*, 145–155. [CrossRef]

16. García, P.; Martínez, B.; Obeso, J.M.; Rodríguez, A. Bacteriophages and their application in food safety. *Lett. Appl. Microbiol.* **2008**, *47*, 479–485. [CrossRef]

17. Kimminau, E.A.; Russo, K.N.; Karnezos, T.P.; Oh, H.G.; Lee, J.J.; Tate, C.C.; Baxter, J.A.; Berghaus, R.D.; Hofacre, C.L. Bacteriophage in-feed application: A novel approach to preventing Salmonella Enteritidis colonization in chicks fed experimentally contaminated feed. *J. Appl. Poult. Res.* **2020**, *29*, 930–936. [CrossRef]

18. Qadir, M.I.; Mobeen, T.; Masood, A. Phage therapy: Progress in pharmacokinetics. *Braz. J. Pharm. Sci.* **2018**, *54*, e17093. [CrossRef]

19. Batalha, L.S. *Encapsulation of Bacteriophages in Different Matrices and Evaluation of the Potential for Phagotherapy*; Universidade Federal de Viçosa: Viçosa, MG, Brazil, 2017.

20. Pan, D.; Yu, Z. Intestinal microbiome of poultry and its interaction with host and diet. *Gut Microbes* **2014**, *5*, 108–119. [CrossRef]

21. González-Menéndez, E.; Fernández, L.; Gutiérrez, D.; Pando, D.; Martínez, B.; Rodríguez, A.; García, P. Strategies to Encapsulate the Staphylococcus aureus Bacteriophage phiIPLA-RODI. *Viruses* **2018**, *10*, 495. [CrossRef]

22. Malik, D.J.; Sokolov, I.J.; Vinner, G.K.; Mancuso, F.; Cinquerrui, S.; Vladisavljevic, G.T.; Clokie, M.R.J.; Garton, N.J.; Stapley, A.G.F.; Kirpichnikova, A. Formulation, stabilisation and encapsulation of bacteriophage for phage therapy. *Adv. Colloid Interface Sci.* **2017**, *249*, 100–133. [CrossRef]

23. Balcão, V.M.; Moreli, F.C.; Silva, E.C.; Belline, B.G.; Martins, L.F.; Rossi, F.P.N.; Pereira, C.; Vila, M.M.D.C.; da Silva, A.M. Isolation and Molecular Characterization of a Novel Lytic Bacteriophage That Inactivates MDR Klebsiella pneumoniae Strains. *Pharmaceutics* **2022**, *14*, 1421. [CrossRef] [PubMed]

24. Balcão, V.M.; Belline, B.G.; Silva, E.C.; Almeida, P.F.F.B.; Baldo, D.Â.; Amorim, L.R.P.; Oliveira Júnior, J.M.; Vila, M.M.D.C.; Del Fiol, F.S. Isolation and Molecular Characterization of Two Novel Lytic Bacteriophages for the Biocontrol of Escherichia coli in Uter-ine Infections: In Vitro and Ex Vivo Preliminary Studies in Veterinary Medicine. *Pharmaceutics* **2022**, *14*, 2344. [CrossRef] [PubMed]

25. Harada, L.K.; Silva, E.C.; Rossi, F.P.N.; Cieza, B.; Oliveira, T.J.; Pereira, C.; Tomazetto, G.; Silva, B.B.; Squina, F.M.; Vila, M.M.D.C.; et al. Characterization and in vitro testing of newly isolated lytic bacteriophages for biocontrol of Pseudomonas aeruginosa. *Future Microbiol.* **2022**, *17*, 111–141. [CrossRef] [PubMed]

26. Silva, E.C.; Oliveira, T.J.; Moreli, F.C.; Harada, L.K.; Vila, M.M.D.C.; Balcão, V.M. Newly isolated lytic bacteriophages for Staphylococcus intermedius, structurally and functionally stabilized in a hydroxyethylcellulose gel containing choline geranate: Potential for transdermal permeation in veterinary phage therapy. *Res. Vet. Sci.* **2021**, *135*, 42–58. [CrossRef]

27. Rios, A.C.; Vila, M.M.D.C.; Lima, R.; Del Fiol, F.S.; Tubino, M.; Teixeira, J.A.; Balcão, V.M. Structural and functional stabilization of bacteriophage particles within the aqueous core of a W/O/W multiple emulsion: A potential biotherapeutic system for the inhalational treatment of bacterial pneumonia. *Process Biochem.* **2018**, *64*, 177–192. [CrossRef]

28. Pinheiro, L.A.M.; Pereira, C.; Frazão, C.; Balcão, V.M.; Almeida, A. Efficiency of Phage φ6 for Biocontrol of Pseudomonas syringae pv. syringae: An in Vitro Preliminary Study. *Microorganisms* **2019**, *7*, 286. [CrossRef]

29. Pinheiro, L.A.M.; Pereira, C.; Barreal, M.E.; Gallego, P.P.; Balcão, V.M.; Almeida, A. Use of phage ϕ6 to inactivate Pseudomonas syringae pv. actinidiae in kiwifruit plants: In vitro and ex vivo experiments. *Appl. Microbiol. Biotechnol.* **2019**, *104*, 1319–1330. [CrossRef]

30. Melo, L.D.R.; Sillankorva, S.; Ackermann, H.-W.; Kropinski, A.M.; Azeredo, J.; Cerca, N. Isolation and characterization of a new Staphylococcus epidermidis broad-spectrum bacteriophage. *J. Gen. Virol.* **2014**, *95*, 506–515. [CrossRef]

31. Khan Mirzaei, M.; Nilsson, A.S. Isolation of phages for phage therapy: A comparison of spot tests and efficiency of plating analyses for determination of host range and efficacy. *PLoS ONE* **2015**, *10*, e0118557. [CrossRef]

32. Stuer-Lauridsen, B.; Janzen, T.; Schnabl, J.; Johansen, E. Identification of the host determinant of two prolate-headed phages infecting Lactococcus lactis. *Virology* **2003**, *309*, 10–17. [CrossRef]

33. Storms, Z.J.; Sauvageau, D. Modeling tailed bacteriophage adsorption: Insight into mechanisms. *Virology* **2015**, *485*, 355–362. [CrossRef] [PubMed]

34. Watanabe, K.; Takesue, S.; Ishibashi, K.; Nakahara, S. A computer simulation of the adsorption of Lactobacillus phage PL-1 to host cells: Some factors affecting the process. *Agric. Biol. Chem.* **1982**, *46*, 697–702. [CrossRef]

35. Moldovan, R.; Chapman-McQuiston, E.; Wu, X.L. On kinetics of phage adsorption. *Biophys. J.* **2007**, *93*, 303–315. [CrossRef] [PubMed]

36. Balcão, V.M.; Moreira, A.R.; Moutinho, C.G.; Chaud, M.V.; Tubino, M.; Vila, M.M.D.C. Structural and functional stabilization of phage particles in carbohydrate matrices for bacterial biosensing. *Enzym. Microb. Technol.* **2013**, *53*, 55–69. [CrossRef]

37. Balcão, V.M.; Barreira, S.V.P.; Nunes, T.M.; Chaud, M.V.; Tubino, M.; Vila, M.M.D.C. Carbohydrate hydrogels with stabilized phage particles for bacterial biosensing: Bacterium diffusion studies. *Appl. Biochem. Biotechnol.* **2014**, *172*, 1194–1214. [CrossRef]

38. Harada, L.K.; Bonventi Júnior, W.; Silva, E.C.; Oliveira, T.J.; Moreli, F.C.; Oliveira Júnior, J.M.; Tubino, M.; Vila, M.M.D.C.; Balcão, V.M. Bacteriophage-based biosensing of Pseudomonas aeruginosa: An integrated approach for the putative real-time detection of multi-drug-resistant strains. *Biosensors* **2021**, *11*, 124. [CrossRef]

39. Rocha, L.K.H.; Favaro, L.I.L.; Rios, A.C.; Silva, E.C.; Silva, W.F.; Stigliani, T.P.; Guilger, M.; Lima, R.; Oliveira, J.M.; Aranha, N.; et al. Sericin from Bombyx mori cocoons. Part I: Extraction and physicochemical-biological characterization for biopharmaceutical applications. *Process Biochem.* **2017**, *61*, 163–177. [CrossRef]

40. Oliveira, J.M.; Martins, A.C.G. Construction and test of low cost X-ray tomography scanner for physical-chemical analysis and nondestructive inspections. *AIP Conf. Proc.* **2009**, *1139*, 102–105.

41. Feldkamp, L.A.; Davis, L.C.; Kress, J.W. Practical cone-beam algorithm. *J. Opt. Soc. Am. A* **1984**, *1*, 612–619. [CrossRef]

42. Bates, D.M.; Watts, D.G. *Nonlinear Regression Analysis and Its Applications*, 1st ed.; Wiley-Blackwell: Hoboken, NJ, USA, 1988; ISBN 978-0471816430.

43. Balcão, V.M.; Vieira, M.C.; Malcata, F.X. Adsorption of Protein from Several Commercial Lipase Preparations onto a Hollow-Fiber Membrane Module. *Biotechnol. Prog.* **1996**, *12*, 164–172. [CrossRef]

44. Balcão, V.M.; Oliveira, T.A.; Xavier Malcata, F. Stability of a commercial lipase from Mucor javanicus: Kinetic modelling of pH and temperature dependencies. *Biocatal. Biotransform.* **1998**, *16*, 45–66. [CrossRef]

45. Pires, D.P.; Oliveira, H.; Melo, L.D.R.; Sillankorva, S.; Azeredo, J. Bacteriophage-encoded depolymerases: Their diversity and biotechnological applications. *Appl. Microbiol. Biotechnol.* **2016**, *100*, 2141–2151. [CrossRef] [PubMed]

46. Oliveira, H.; São-José, C.; Azeredo, J. Phage-derived peptidoglycan degrading enzymes: Challenges and future prospects for in vivo therapy. *Viruses* **2018**, *10*, 292. [CrossRef] [PubMed]

47. Kaur, J. Application of UV Light Scattering to Detect Reversible Self-Association and Aggregation of Proteins in Solution. Ph.D. Thesis, University of Connecticut, Mansfield, CT, USA, 2017.

48. Vekshin, N.L. Screening Hypochromism of Chromophores in Macromolecular Biostructures. *Biophysics* **1999**, *44*, 41–51.

49. Manjula, B.; Varaprasad, K.; Sadiku, R.; Raju, K.M. Preparation and characterization of sodium alginate-based hydrogels and their in vitro release studies. *Adv. Polym. Technol.* **2013**, *32*, 21340. [CrossRef]

50. De'Nobili, M.D.; Curto, L.M.; Delfino, J.M.; Pérez, C.D.; Bernhardt, D.; Gerschenson, L.N.; Fissore, E.N.; Rojas, A.M. Alginate utility in edible and nonedible film development and the influence of its macromolecular structure in the antioxidant activity of a pharmaceutical/food interface. In *Alginic Acid: Chemical Structure, Uses and Health Benefits*; Moore, A., Ed.; Nova Science Publishers Inc.: New York, NY, USA, 2014; pp. 119–169. ISBN 978-1-63463-242-3.

51. Rotman, S.G.; Post, V.; Foster, A.L.; Lavigne, R.; Wagemans, J.; Trampuz, A.; Gonzalez Moreno, M.; Metsemakers, W.-J.; Grijpma, D.W.; Richards, R.G.; et al. Alginate chitosan microbeads and thermos-responsive hyaluronic acid hydrogel for phage delivery. *J. Drug Deliv. Sci. Technol.* **2023**, *79*, 103991. [CrossRef]

52. Batalha, L.S.; Gontijo, M.T.P.; Teixeira, A.V.N.C.; Boggione, D.M.G.; Lopez, M.E.S.; Eller, M.R.; Mendonça, R.C.S. Encapsulation in alginate-polymers improves stability and allows controlled release of the UFV-AREG1 bacteriophage. *Food Res. Int.* **2021**, *139*, 109947. [CrossRef]

53. Alves, D.; Cerqueira, M.A.; Pastrana, L.M.; Sillankorva, S. Entrapment of a phage cocktail and cinnamaldehyde on sodium alginate emulsion-based films to fight food contamination by Escherichia coli and Salmonella Enteritidis. *Food Res. Int.* **2020**, *128*, 108791. [CrossRef]

54. Dicastillo, C.L.; Settier-Ramírez, L.; Gavara, R.; Hernández-Muñoz, P.; Carballo, G.L. Development of biodegradable films loaded with phages with antilisterial properties. *Polymers* **2021**, *13*, 327. [CrossRef]

55. Fernandes, R.S.; Moura, M.R.; Glenn, G.M.; Aouada, F.A. Thermal, microstructural, and spectroscopic analysis of Ca^{2+} alginate/clay nanocomposite hydrogel beads. *J. Mol. Liq.* **2018**, *265*, 327–336. [CrossRef]

56. Daemi, H.; Barikani, M. Synthesis and characterization of calcium alginate nanoparticles, sodium homopolymannuronate salt and its calcium nanoparticles. *Sci. Iran.* **2012**, *19*, 2023–2028. [CrossRef]

57. Yallapu, M.M.; Jaggi, M.; Chauhan, S.C. β-Cyclodextrin-curcumin self-assembly enhances curcumin delivery in prostate cancer cells. *Colloids Surf. B Biointerfaces* **2010**, *79*, 113–125. [CrossRef] [PubMed]

58. Chen, X.; Zou, L.-Q.; Niu, J.; Liu, W.; Peng, S.-F.; Liu, C.-M. The Stability, sustained release and cellular antioxidant activity of curcumin nanoliposomes. *Molecules* **2015**, *20*, 14293–14311. [CrossRef]

59. Darandale, S.S.; Vavia, P.R. Cyclodextrin-based nanosponges of curcumin: Formulation and physicochemical characterization. *J. Incl. Phenom. Macrocycl. Chem.* **2012**, *75*, 315–322. [CrossRef]

60. Gangwar, R.K.; Dhumale, V.A.; Kumari, D.; Nakate, U.T.; Gosavi, S.W.; Sharma, R.B.; Kale, S.N.; Datar, S. Conjugation of curcumin with PVP capped gold nanoparticles for improving bioavailability. *Mater. Sci. Eng. C* **2012**, *32*, 2659–2663. [CrossRef]

61. Li, J.; Lee, I.W.; Shin, G.H.; Chen, X.; Park, H.J. Curcumin-Eudragit® E PO solid dispersion: A simple and potent method to solve the problems of curcumin. *Eur. J. Pharm. Biopharm.* **2015**, *94*, 322–332. [CrossRef] [PubMed]

62. Queiroz, M.F.; Melo, K.R.T.; Sabry, D.A.; Sassaki, G.L.; Rocha, H.A.O. Does the Use of Chitosan Contribute to Oxalate Kidney Stone Formation? *Mar. Drugs* **2015**, *13*, 141–158. [CrossRef]

63. Sujima Anbu, A.; Sahi, S.V.; Venkatachalam, P. Synthesis of Bioactive Chemicals Cross-linked Sodium Tripolyphosphate (TPP)—Chitosan Nanoparticles for Enhanced Cytotoxic Activity against Human Ovarian Cancer cell Line (PA-1). *J. Nanomed. Nanotechnol.* **2016**, *7*, 1000418. [CrossRef]

64. Rowbotham, J.S.; Dyer, P.W.; Greenwell, H.C.; Selby, D.; Theodorou, M.K. Copper(II)-mediated thermolysis of alginates: A model kinetic study on the influence of metal ions in the thermochemical processing of macroalgae. *Interface Focus* **2013**, *3*, 20120046. [CrossRef]

65. Lozano-Vazquez, G.; Lobato-Calleros, C.; Escalona-Buendia, H.; Chavez, G.; Alvarez-Ramirez, J.; Vernon-Carter, E.J. Effect of the weight ratio of alginate-modified tapioca starch on the physicochemical properties and release kinetics of chlorogenic acid containing beads. *Food Hydrocoll.* **2015**, *48*, 301–311. [CrossRef]

66. Laia, A.G.S.; Costa, E.S., Jr.; Costa, H.S. A study of sodium alginate and calcium chloride interaction through films for intervertebral disc regeneration uses. In Proceedings of the 21 CBECIMAT: Brazilian Congress of Engineering and Materials Science, Cuiabá, MT, Brazil, 9–13 November 2014.

67. Lopes, S.; Bueno, L.; Aguiar Jr, F.; Finkler, C. Preparation and characterization of alginate and gelatin microcapsules containing Lactobacillus rhamnosus. *An. Acad. Bras. Ciênc.* **2017**, *89*, 1601–1613. [CrossRef] [PubMed]

68. Kim, N.A.; An, I.B.; Lim, D.G.; Lim, J.Y.; Lee, S.Y.; Shim, W.S.; Kang, N.G.; Jeong, S.H. Effects of pH and buffer concentration on the thermal stability of etanercept using DSC and DLS. *Biol. Pharm. Bull.* **2014**, *37*, 808–816. [CrossRef] [PubMed]

69. Vinner, G.K.; Rezaie-Yazdi, Z.; Leppanen, M.; Stapley, A.G.F.; Leaper, M.C.; Malik, D.J. Microencapsulation of Salmonella-Specific Bacteriophage Felix O1 Using Spray-Drying in a pH-Responsive Formulation and Direct Compression Tableting of Powders into a Solid Oral Dosage Form. *Pharmaceuticals* **2019**, *12*, 43. [CrossRef] [PubMed]

70. Petsong, K.; Benjakul, S.; Vongkamjan, K. Optimization of wall material for phage encapsulation via freeze-drying and antimicrobial efficacy of microencapsulated phage against Salmonella. *J. Food Sci. Technol.* **2021**, *58*, 1937–1946. [CrossRef] [PubMed]

71. Chang, R.Y.K.; Kwok, P.C.L.; Khanal, D.; Morales, S.; Kutter, E.; Li, J.; Chan, H.K. Inhalable bacteriophage powders: Glass transition temperature and bioactivity stabilization. *Bioeng. Transl. Med.* **2020**, *5*, e10159. [CrossRef] [PubMed]

72. Fertah, M.; Belfkira, A.; Dahmane, E.M.; Taourirte, M.; Brouillette, F. Extraction and characterization of sodium alginate from Moroccan Laminaria digitata brown seaweed. *Arab. J. Chem.* **2017**, *10*, S3707–S3714. [CrossRef]

73. Theiler, J. Estimating fractal dimension. *J. Opt. Soc. Am. A* **1990**, *7*, 1055–1073. [CrossRef]

74. Hughes, K.A.; Sutherland, I.W.; Clark, J.; Jones, M.V. Bacteriophage and associated polysaccharide depolymerases—Novel tools for study of bacterial biofilms. *J. Appl. Microbiol.* **2002**, *85*, 583–590. [CrossRef]

75. Hyman, P. Phages for Phage Therapy: Isolation, Characterization, and Host Range Breadth. *Pharmaceuticals* **2019**, *12*, 35. [CrossRef]

76. Ross, A.; Ward, S.; Hyman, P. More Is Better: Selecting for Broad Host Range Bacteriophages. *Front. Microbiol.* **2016**, *7*, 1352. [CrossRef]

77. Hendrix, R.W.; Smith, M.C.M.; Burns, R.N.; Ford, M.E.; Hatfull, G.F. Evolutionary relationships among diverse bacteriophages and prophages: All the world's a phage. *Proc. Natl. Acad. Sci. USA* **1999**, *96*, 2192–2197. [CrossRef]

78. Göller, P.C.; Elsener, T.; Lorgé, D.; Radulovic, N.; Bernardi, V.; Naumann, A.; Amri, N.; Khatchatourova, E.; Coutinho, F.H.; Loessner, M.J.; et al. Multi-species host range of staphylococcal phages isolated from wastewater. *Nat. Commun.* **2021**, *12*, 6965. [CrossRef]

79. Letarov, A.V.; Letarova, M.A. The Burden of Survivors: How Can Phage Infection Impact Non-Infected Bacteria? *Int. J. Mol. Sci.* **2023**, *24*, 2733. [CrossRef] [PubMed]

80. Pereira, C.; Moreirinha, C.; Lewicka, M.; Almeida, P.; Clemente, C.; Cunha, Â.; Delgadillo, I.; Romalde, J.L.; Nunes, M.L.; Almeida, A. Bacteriophages with potential to inactivate Salmonella Typhimurium: Use of single phage suspensions and phage cocktails. *Virus Res.* **2016**, *220*, 179–192. [CrossRef] [PubMed]

81. Abedon, S.T.; Culler, R.R. Optimizing bacteriophage plaque fecundity. *J. Theor. Biol.* **2007**, *249*, 582–592. [CrossRef] [PubMed]

82. Mateus, L.; Costa, L.; Silva, Y.J.; Pereira, C.; Cunha, A.; Almeida, A. Efficiency of phage cocktails in the inactivation of Vibrio in aquaculture. *Aquaculture* **2014**, *424–425*, 167–173. [CrossRef]

83. Abedon, S.T. Lysis from without. *Bacteriophage* **2011**, *1*, 46–49. [CrossRef]

84. Letarov, A.V.; Kulikov, E.E. Adsorption of Bacteriophages on Bacterial Cells. *Biochemistry* **2017**, *82*, 1632–1658. [CrossRef]

85. Ahiwale, S.S.; Bankar, A.V.; Tagunde, S.N.; Zinjarde, S.; Ackermann, H.W.; Kapadnis, B.P. Isolation and characterization of a rare waterborne lytic phage of *Salmonella enterica* serovar Paratyphi B. *Can. J. Microbiol.* **2013**, *59*, 318–323. [CrossRef]

86. Hussein, M.N.; Jahell, W.A.; Al-Shekake, T.J.M.; Al-khozai, Z.M.F. Evaluation of Lytic Bacteriophages for Control of Multidrug-Resistant Salmonella Typhi. *Ann. Rom. Soc. Cell Biol.* **2021**, *25*, 11359–11372.

87. Santos, S.B.; Carvalho, C.; Azeredo, J.; Ferreira, E.C. Population dynamics of a Salmonella lytic phage and its host: Implications of the host bacterial growth rate in modelling. *PLoS ONE* **2014**, *9*, e102507. [CrossRef] [PubMed]

88. Tao, C.; Yi, Z.; Zhang, Y.; Wang, Y.; Zhu, H.; Afayibo, D.J.A.; Li, T.; Tian, M.; Qi, J.; Ding, C.; et al. Characterization of a Broad-Host-Range Lytic Phage SHWT1 Against Multidrug-Resistant Salmonella and Evaluation of Its Therapeutic Efficacy in vitro and in vivo. *Front. Vet. Sci.* **2021**, *8*, 683853. [CrossRef] [PubMed]

89. Capparelli, R.; Nocerino, N.; Iannaccone, M.; Ercolini, D.; Parlato, M.; Chiara, M.; Iannelli, D. Bacteriophage Therapy of *Salmonella enterica*: A Fresh Appraisal of Bacteriophage Therapy. *J. Infect. Dis.* **2010**, *201*, 52–61. [CrossRef] [PubMed]

90. Balcão, V.M.; Basu, A.; Cieza, B.; Rossi, F.N.; Pereira, C.; Vila, M.M.D.C.; Setubal, J.C.; Ha, T.; da Silva, A.M. Pseudomonas-tailed lytic phages: Genome mechanical analysis and putative correlation with virion morphogenesis yield. *Future Microbiol.* **2022**, *17*, 1009–1026. [CrossRef]

91. Chi-Hsin, H.; ChongYi, L.; JongKang, L.; Chan-Shing, L. Control of the eel (*Anguilla japonica*) pathogens, Aeromonas hydrophila and Edwardsiella tarda, by bacteriophages. *J. Fish. Soc. Taiwan* **2000**, *27*, 21–31.

92. Pasharawipas, T.; Manopvisetcharean, J.; Flegel, T.W. Phage treatment of Vibrio harveyi: A general concept of protection against bacterial infection. *Res. J. Microbiol.* **2011**, *6*, 560–567. [CrossRef]

93. Prasad, Y.; Arpana; Kumar, D.; Sharma, A.K. Lytic bacteriophages specific to Flavobacterium columnare rescue catfish, Clarias batrachus (Linn.) from columnaris disease. *J. Environ. Biol.* **2011**, *32*, 161–168.

94. Pereira, C.; Moreirinha, C.; Lewicka, M.; Almeida, P.; Clemente, C.; Romalde, J.L.; Nunes, M.L.; Almeida, A. Characterization and in vitro evaluation of new bacteriophages for the biocontrol of Escherichia coli. *Virus Res.* **2017**, *227*, 171–182. [CrossRef]

95. Duarte, J.; Pereira, C.; Moreirinha, C.; Salvio, R.; Lopes, A.; Wang, D.; Almeida, A. New insights on phage efficacy to control Aeromonas salmonicida in aquaculture systems: An in vitro preliminary study. *Aquaculture* **2018**, *495*, 970–982. [CrossRef]

96. Costa, P.; Pereira, C.; Gomes, A.T.P.C.; Almeida, A. Efficiency of single phage suspensions and phage cocktail in the inactivation of Escherichia coli and Salmonella Typhimurium: An in vitro preliminary study. *Microorganisms* **2019**, *7*, 94. [CrossRef]

97. Forti, F.; Roach, D.R.; Cafora, M.; Pasini, M.E.; Horner, D.S.; Fiscarelli, E.V.; Rossitto, M.; Cariani, L.; Briani, F.; Debarbieux, L.; et al. Design of a Broad-Range Bacteriophage Cocktail That Reduces Pseudomonas aeruginosa Biofilms and Treats Acute Infections in Two Animal Models. *Antimicrob. Agents Chemother.* **2018**, *62*, e02573-17. [CrossRef] [PubMed]

98. Campos, W.F.; Silva, E.C.; Oliveira, T.J.; Oliveira, J.M., Jr.; Tubino, M.; Pereira, C.; Vila, M.M.D.C.; Balcão, V.M. Transdermal permeation of bacteriophage particles by choline oleate: Potential for treatment of soft-tissue infections. *Future Microbiol.* **2020**, *15*, 881–896. [CrossRef] [PubMed]

 *antibiotics*

Article

Activity of Phage–Lactoferrin Mixture against Multi Drug Resistant *Staphylococcus aureus* Biofilms

Katarzyna Kosznik-Kwaśnicka [1,2] , Natalia Kaźmierczak [1] and Lidia Piechowicz [1,*]

1 Department of Medical Microbiology, Faculty of Medicine, Medical University of Gdańsk, Dębowa 25, 80-204 Gdansk, Poland
2 Laboratory of Phage Therapy, Institute of Biochemistry and Biophysics, Polish Academy of Sciences, Kładki 24, 80-822 Gdansk, Poland
* Correspondence: lidia.piechowicz@gumed.edu.pl

Abstract: Biofilms are complex bacterial structures composed of bacterial cells embedded in extracellular polymeric substances (EPS) consisting of polysaccharides, proteins and lipids. As a result, biofilms are difficult to eradicate using both mechanical methods, i.e., scraping, and chemical methods such as disinfectants or antibiotics. Bacteriophages are shown to be able to act as anti-biofilm agents, with the ability to penetrate through the matrix and reach the bacterial cells. However, they also seem to have their limitations. After several hours of treatment with phages, the biofilm tends to grow back and phage-resistant bacteria emerge. Therefore, it is now recommended to use a mixture of phages and other antibacterial agents in order to increase treatment efficiency. In our work we have paired staphylococcal phages with lactoferrin, a protein with proven anti-biofilm proprieties. By analyzing the biofilm biomass and metabolic activity, we have observed that the addition of lactoferrin to phage lysate accelerated the anti-biofilm effect of phages and also prevented biofilm re-growth. Therefore, this combination might have a potential use in biofilm eradication procedures in medical settings.

Keywords: phage therapy; phages; lactoferrin; biofilm; *Staphylococcus aureus*; MDRSA; MRSA

Citation: Kosznik-Kwaśnicka, K.; Kaźmierczak, N.; Piechowicz, L. Activity of Phage–Lactoferrin Mixture against Multi Drug Resistant *Staphylococcus aureus* Biofilms. *Antibiotics* **2022**, *11*, 1256. https://doi.org/10.3390/antibiotics11091256

Academic Editors: Aneta Skaradzińska and Anna Nowaczek

Received: 30 August 2022
Accepted: 12 September 2022
Published: 16 September 2022

Publisher's Note: MDPI stays neutral with regard to jurisdictional claims in published maps and institutional affiliations.

1. Introduction

Staphylococcus aureus is a common nosocomial pathogen that can be responsible for wound infections, hospital-acquired pneumonia or sepsis [1,2]. The emergence of antibiotic resistance, especially to methicillin among nosocomial strains, resulted in difficulties in treatment, which is then responsible for prolonged hospital stays, increased mortality and morbidity of infections [3]. The rates of methicillin resistance among clinical isolates vary from country to country, ranging from a small percent in Scandinavian countries to over 50% in the U.S. and Asian countries [3–5]. *S. aureus* can form a biofilm—a complex bacterial structure composed of bacterial cells embedded in extracellular polymeric substance (EPS) that can attach to both organic and inorganic surfaces [6]. The ability to form a biofilm plays an important role in *S. aureus* virulence as cells in biofilms are more resistant to various eradication mechanisms. Furthermore, individual cells can detach from the original biofilm and establish new sites of infection or mediate an acute infection such as sepsis [7]. Bacteriophages, the viruses that infect bacteria, have been shown to be able to successfully eradicate biofilms [8,9]. Phages can prevent biofilm formation and maturation by destroying bacteria in the outer layer of biofilm and planktonic cells. They can also penetrate existing biofilms and eliminate the biofilm structure as phage lytic enzymes, depolymerase and lysins, are being released from the cells upon phage progeny release [10]. However, even phages have their limitations. In some cases, the re-growth of biofilm was observed, and the emergence of resistant bacteria was reported [11,12]. Therefore, it is recommended to pair the phages with other antimicrobials [13,14]. Since the global consensus is to reduce the use of antibiotics, other compounds with antimicrobial activity are also being researched. Lactoferrin is an 80 kDa protein of the transferrin family of non-heme, iron-binding glycoproteins and an important part of the innate immune system. It is present in the blood, at the mucosa, and it is secreted with fluids such as milk, tears, sweat or semen [15]. It has been shown that lactoferrin can act as anti-biofilm agent reducing the biomass, loosening the biofilm structure and enabling its dispersion [15–17]. The detailed mechanisms of lactoferrin's anti-biofilm activity remain to be discovered [15]. However, its potential in treatment should be investigated. Therefore, we have decided to test the combination of phages and lactoferrin against clinical strains of multidrug-resistant *S. aureus* (MDRSA). We have discovered that the phage–lactoferrin mixture significantly reduced biofilm metabolic activity and biomass. Furthermore, the addition of lactoferrin to phage lysate slowed down the process of biofilm re-growth. We believe that combined phage–lactoferrin treatment could be implemented in the eradication of biofilms formed by nosocomial pathogens and should be studied further to fully evaluate its potential.

2. Results

2.1. Lactoferrin Influence on Phage Activity

Bacteriophages vB_SenM-A, vB_SauM-C and vB_SauM-D have previously been characterized and have proven activity against MDRSA biofilms when used alone [8,18]. In order to assess if phage and lactoferrin can be used simultaneously in the form of a cocktail, the phage–lactoferrin mixture was stored at 4 °C for a period of 5 days, with titration performed every 24 h. The phage titer began to drop after 3 days of incubation in case of phages vB_SauM-C and vB_SauM-D (Figure 1). The highest drop in activity (assessed using Plaque Forming Unit—PFU/mL) was observed for phage vB_SauM-C. The titer dropped from an initial 10^9 PFU/mL to 6×10^8 PFU/mL after 4 days of incubation.

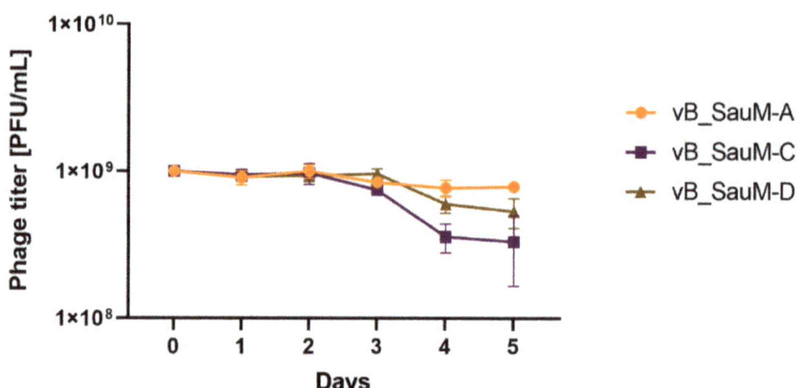

Figure 1. Influence of lactoferrin (10 mg/mL) on phage titer during storage period. Arithmetic mean of triplicates, with error bars representing SD.

2.2. Lactoferrin Influence on Biofilm Formed by MDRSA Strains

To evaluate lactoferrin anti-biofilm activity, we selected appropriate MDRSA strains that were classified as strong biofilm producers during our previous studies [19]. Mature MDRSA biofilms treated with three different concentrations of lactoferrin: 0.1 mg/mL, 1.0 mg/mL and 10 mg/mL, which were chosen based on literature data [16,17]. After 4 h, 12 h of 24 h of incubation with lactoferrin total biofilm biomass (Crystal Violet staining), biofilm metabolic activity (MTT-Formazan assay) and the colony-forming unit (CFU/mL) numbers were assessed. It was found that lactoferrin significantly decreased biofilm biomass and viability in concentrations of 1.0 mg/mL and 10 mg/mL after 12 h and 24 h of incubation, respectively (Figures 2 and 3). In most cases, statistical analysis revealed no difference between 1.0 mg/mL and 10 mg/mL concentrations of lactoferrin on biofilm biomass and viability.

The effect was most pronounced for MDRSA strain no. 121 (more that 50% reduction in both biofilm biomass and viability) (Figures 2 and 3). While strains no. 70, 120 and 203 were the least influenced in case of viability (Figure 3), significant reduction in biofilm biomass was observed (Figure 2). We did not, however, observe a significant reduction in the number of cells creating the biofilm (Figure 4).

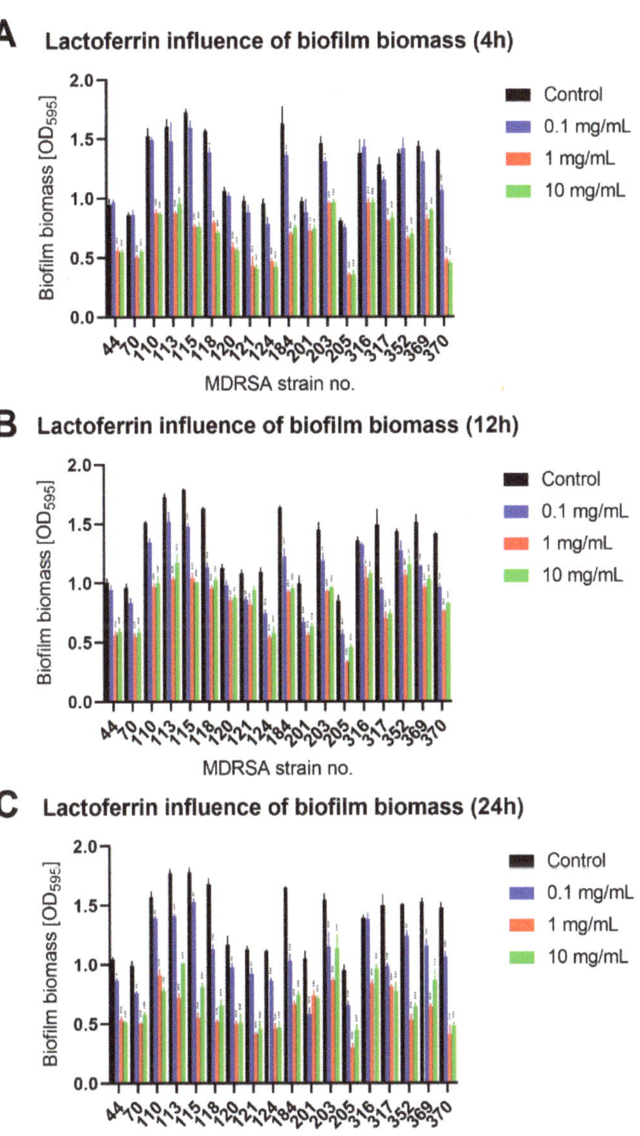

Figure 2. Influence of lactoferrin (0.1 mg/mL, 1.0 mg/mL and 10 mg/mL) on MDRSA biofilm biomass after 4 h (**A**), 12 h (**B**) and 24 h (**C**) of incubation. Mean of triplicates, with error bars representing SD. Statistical analysis was performed using *t*-test, $p < 0.05$ (*), $p < 0.01$ (**), $p < 0.001$ (***).

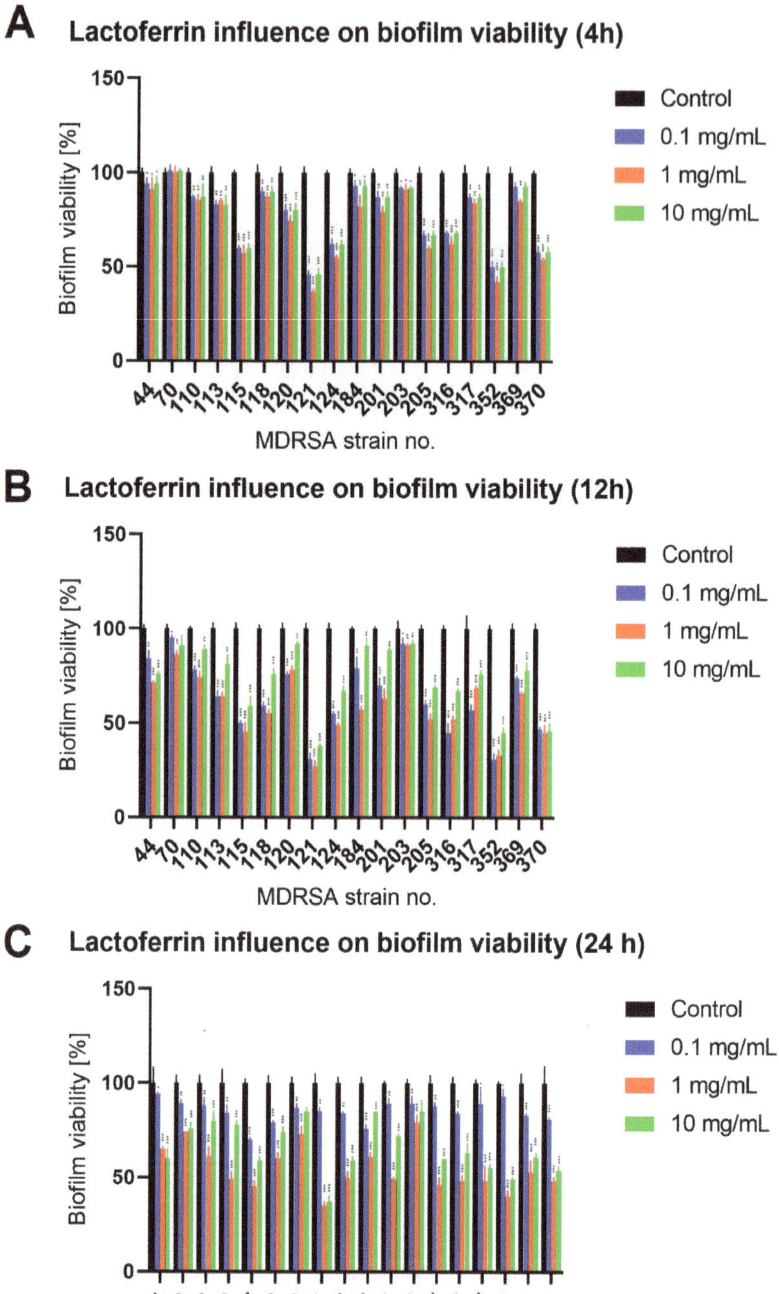

Figure 3. Influence of lactoferrin (0.1 mg/mL, 1.0 mg/mL and 10 mg/mL) on MDRSA biofilm viability after 4 h (**A**), 12 h (**B**) and 24 h (**C**) of incubation. Mean of triplicates, with error bars representing SD. Statistical analysis was performed using *t*-test, $p < 0.05$ (*), $p < 0.01$ (**), $p < 0.001$ (***).

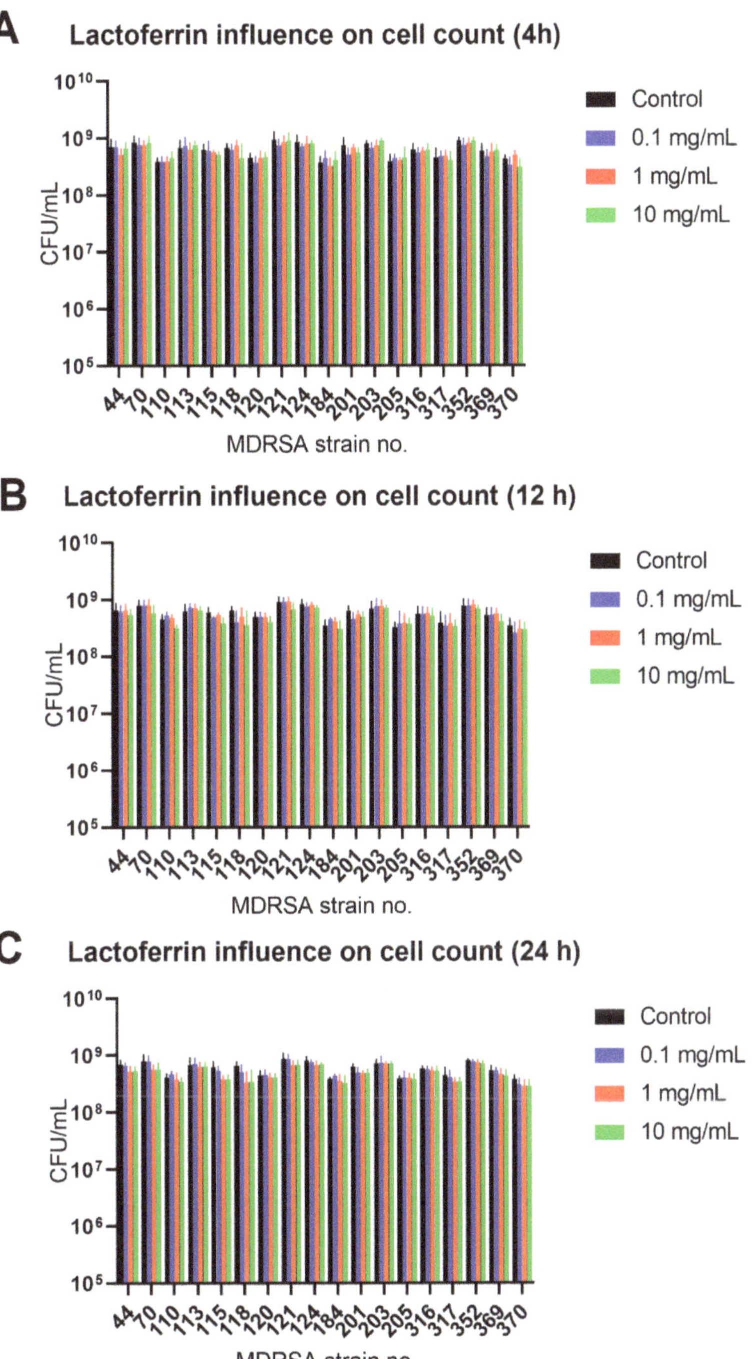

Figure 4. Influence of lactoferrin (0.1 mg/mL, 1.0 mg/mL and 10 mg/mL) on MDRSA biofilm CFU/mL count after 4 h (**A**), 12 h (**B**) and 24 h (**C**) of incubation. Mean of triplicates, with error bars representing SD. Statistical analysis performed using *t*-test showed no significance.

2.3. Biofilm Eradication by Phage–Lactoferrin Mixture

In order to assess if lactoferrin will influence phage anti-biofilm activity, a phage–lactoferrin mixture was prepared. The mixture consisted of one of three previously characterized phages (vB_SauM-A, vB_SauM-C and vB_SauM-D) with proven activity against MDRSA strains [8,18]. Based on our previous studies and presented data on lactoferrin activity, we have chosen the concentration of phage to be 10^9 PFU/mL and 1.0 mg/mL for lactoferrin.

We have observed that the phage–lactoferrin mixture was more efficient against bacterial biofilm in the first hours after administration. The reduction was especially visible in biofilm biomass and metabolic activity for phage vB_SauM-A (Figure 5). In the case of phage vB_SauM-C, we have only observed an increased reduction in biofilm biomass after the first four hours of incubation (Figure 6A). Biofilm biomass and metabolic activity reduction for phage vB_SauM-D (Figure 7) was similar to phage vB_SauM-A. After 12 h, the effectiveness of the phage–lactoferrin cocktail and phage lysate equaled out, and statistical analysis revealed no significant differences. However, after 24 h of incubation in the case of phages vB_SauM-C and vB_SauM-D, we have observed that the CFU/mL would start to increase, signaling the re-growth of biofilm and possible emergence of resistant bacteria (Figures 5I and 7I). When lactoferrin was added to the mixture, this effect was not observed, and in the case of phage vB_SauM-D, there was a statistically significant difference between the phage lysate and phage–lactoferrin treatments (except for strains no. 70 and 110) (Supplementary Materials, Table S1). The difference in CFU/mL between phage vB_SauM-C lysate and the phage–lactoferrin cocktail after 24 h of incubation was statistically significant in the case of strains 70 and 370. In the case of strains 113, 124, 203 and 352, no statistical significance was reported (Supplementary Materials, Table S2).

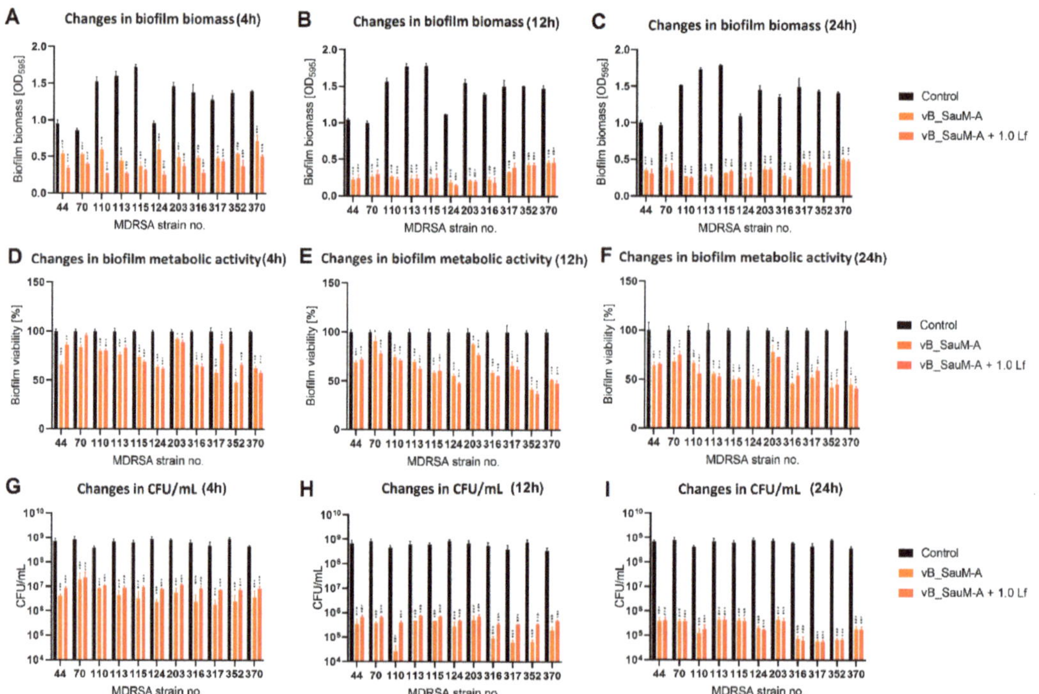

Figure 5. Influence of phage vB_SauM-A and vB_SauM-A+ 1.0 mg/mL lactoferrin (Lf) on MDRSA biofilm: biomass (**A–C**), metabolic activity (**D–F**) and CFU/mL count (**G–I**) after 4 h (**A,D,G**), 12 h (**B,E,H**) and 24 h (**C,F,I**) of incubation. Mean of triplicates, with error bars representing SD. Statistical analysis was performed using *t*-test, $p < 0.05$ (*), $p < 0.01$ (**), $p < 0.001$ (***).

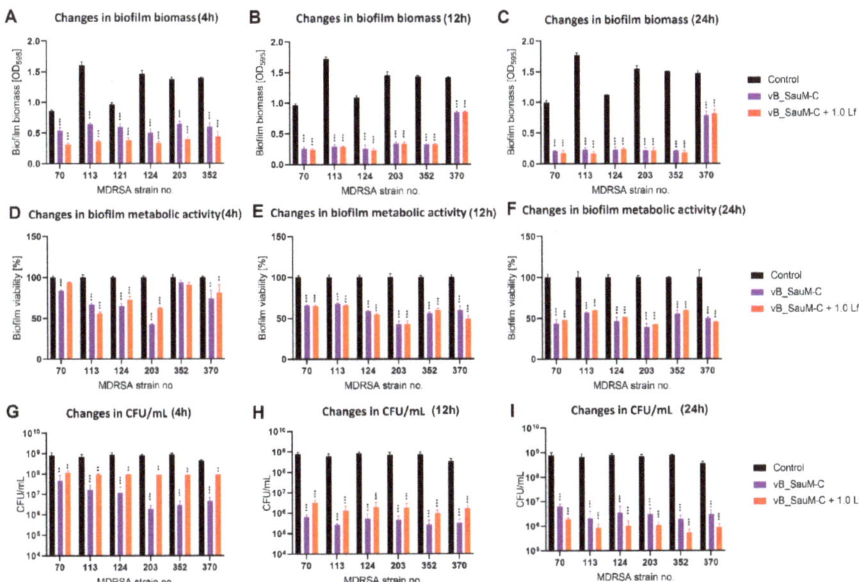

Figure 6. Influence of phage vB_SauM-C and vB_SauM-C+ 1.0 mg/mL lactoferrin (Lf) on MDRSA biofilm: biomass (**A–C**), metabolic activity (**D–F**) and CFU/mL count (**G–I**) after 4 h (**A,D,G**), 12 h (**B,E,H**) and 24 h (**C,F,I**) of incubation. Mean of triplicates, with error bars representing SD. Statistical analysis was performed using *t*-test, $p < 0.01$ (**), $p < 0.001$ (***).

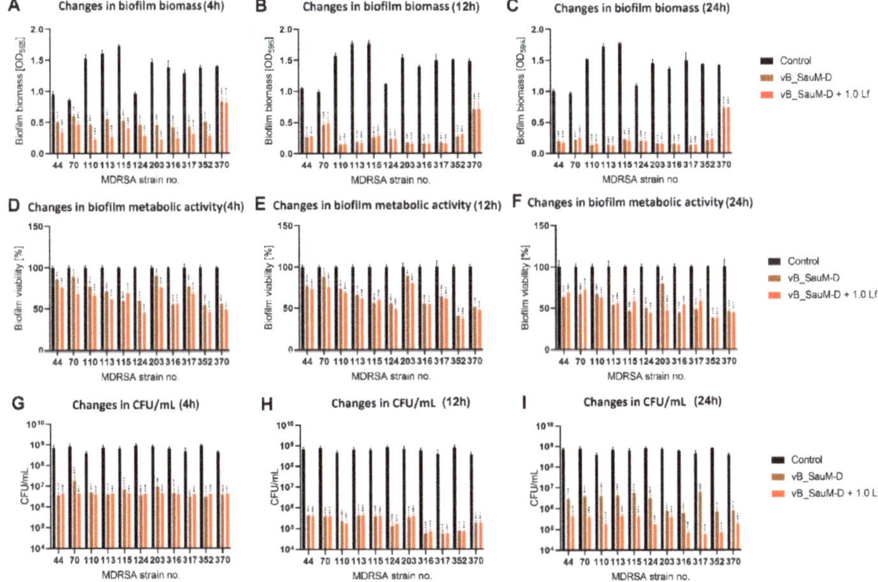

Figure 7. Influence of phage vB_SauM-D and vB_SauM-D+ 1.0 mg/mL lactoferrin (Lf) on MDRSA biofilm: biomass (**A–C**), metabolic activity (**D–F**) and CFU/mL count (**G–I**) after 4 h (**A,D,G**), 12 h (**B,E,H**) and 24 h (**C,F,I**) of incubation. Mean of triplicates, with error bars representing SD. Statistical analysis was performed using *t*-test, $p < 0.05$ (*), $p < 0.01$ (**), $p < 0.001$ (***).

3. Discussion

Bacteriophages are an effective tool against biofilms formed by nosocomial, which are often multi-drug-resistant, strains of bacteria [20,21]. This has been proven by numerous studies by various research groups [8,22,23]. However, even though phages were successful where antibiotics have failed, they also seem to have their limitations [24–26]. Not all bacteriophages can penetrate to the inner layers of the biofilm and are only able to lyse the bacteria from the outermost layers. Additionally, extracellular polymeric substances secreted in biofilm formed by some bacterial genera can immobilize and inactivate the phages [27]. Furthermore, the emergence of phage-resistant bacteria has been reported [26,28]. Therefore, there is a need to find ways to counter these effects and increase the efficacy of phage therapy [26,27]. Currently, pairing bacteriophages and other antimicrobial agents such as antibiotics or essential oils are being investigated with promising results [14,29–31].

Antimicrobial proteins (AMPs), such as lactoferrin, are naturally occurring proteins that act as natural barriers against infection [32]. They are seen as another alternative to antibiotics in combat of antibiotic-resistant strains of bacteria. Lactoferrin has proven antimicrobial activity against various pathogens such as *Herpes simplex* [33], *Papillomavirus* [34], *Pseudomonas aeruginosa* [16], *Salmonella enterica*, *Streptococcus* sp. and *Staphylococcus* sp. [32,35,36].

We have observed that the use of lactoferrin alone had a moderate effect on *S. aureus* biofilm biomass and metabolism. The doses influencing biofilm metabolic state were 1.0 mg/mL and 10 mg/mL, and there was no statistically significant difference between those concentrations. However, the treatment of MDRSA biofilms with lactoferrin did not reduce the number of cells in the biofilm in a statistically significant way. It is therefore possible that lactoferrin acted as a bacteriostatic agent rather than a bactericidal one and prevented further biofilm formation. This was also observed by other research groups. Singh et al. [21] and Ammons et al. [37] have reported that the addition of lactoferrin to the medium prevented *Pseudomonas aeruginosa* from forming biofilms. Additionally, Quinteiri et al. [17] have observed that application of 2.5 mg/mL lactoferrin hydrolysate solution on biofilms attached to glass surfaces caused biofilm dispersion.

Since lactoferrin can influence biofilm dispersion in a significant way, it is therefore proposed to use it as an additive to other antimicrobials to increase their effectiveness. For example, it was reported that lactoferrin increases the inhibitory activity of penicillin up to 4-fold in penicillin-susceptible *S. aureus* strains and up to 16-fold in penicillin-resistant strains by reducing β-lactamase activity [38,39]. Similar results were observed when lactoferrin was paired with other antimicrobials and used against strains of *E. coli* [40], *P. aeruginosa* [37,41,42], *Candida* sp. [43] and *S. epidermilis* [44]. Furthermore, reports by Ammons et al. and Leitch and Willcox suggest that pairing lactoferrin with other compounds (that are not antibiotics) such as xylitol [37,42] or lysozyme [44] have resulted in increased antimicrobial effect. Taking this into account, the pairing of lactoferrin with phages seems to be a logical course of action. However, data on the use of the phage-lactoferrin mixture are very scarce. There are reports that the use of lactoferrin increased phage stability and tolerance to environmental factors [45–47], and there are few in vivo studies of phage–lactoferrin mixture's effectiveness. Experiments performed by Zimecki et al. on mice models reported that a combination of lactoferrin (10 mg i.v.) and T4 phage reduced the bacterial load of *E. coli* in liver more effectively that each of the components separately [48]. Golshahi et al. [45] have observed that the use of lyophilized phages in lactose/lactoferrin in a ration of 60:40 improved phage performance when delivered as inhalable aerosol. However, there are not enough data to conclude whether the phage–lactoferrin mixture can be safely used and whether the use of such a cocktail will increase the effectiveness of the treatment. Therefore, in our work we have decided to analyze if the use of a phage–lactoferrin cocktail will result in increased effectiveness against biofilms formed by clinical strains of MDRSA [19]. Since we have observed that there was no significant difference in activity between 1.0 mg/mL and 10 mg/mL concentrations of lactoferrin, we have chosen to use the lower concentration for our studies. The phage concentration was chosen to be 10^9 PFU/mL based on our previous reports [8]. We have observed that the use of a mixture resulted in significant decrease in all parameters: biomass, metabolic activity and CFU/mL of *S. aureus* biofilms after just 4 h of incubation. The effect of a mixture was more pronounced than the use of phages alone, with statistically significant differences after 12 h and 24 h of incubation. Furthermore, we have observed that the use of lactoferrin prolonged the activity of bacteriophages on the biofilm and prevented its re-growth; this was observed after 24 h of incubation if phages were used alone. This corresponds with the reports of other researchers, which suggests that lactoferrin boosts and prolongs the effects of other antimicrobials [44,45,47]. Therefore, it can be assumed that the use of the phage–lactoferrin cocktail has potential application against biofilms formed by multi-drug-resistant bacteria, though the detailed mechanism remains to be determined [15,16,47]. We believe that more in vitro studies involving other phage types and bacterial genera, followed by in vivo studies, i.e., on *Galleria mellonella* or *Caenorhabditis elegans* models, could deliver more detailed data on phage–lactoferrin synergy and effectivity, helping to answer the question if phage–AMPs mixtures can be used as one of the treatment methods of multi-drug-resistant infections.

4. Materials and Methods

4.1. Bacterial Strains

A total of 18 multi-drug-resistant *Staphylococcus aureus* clinical isolates were chosen from the collection of the Department of Medical Microbiology, the Medical University of Gdańsk. Strains were selected based on their biofilm forming ability and were previously described and characterized [18,19].

4.2. Bacteriophages

Bacteriophages vB_SauM-A, vB_SauM-C and vB_SauM-D were isolated from different wastewater treatment plants and were previously characterized [18]. Their anti-biofilm activity was analyzed and described [8]. Phage propagation, purification and enumeration were performed as described previously [18]. Final phage titer used in this study was 10^9 PFU/mL.

4.3. Lactoferrin

Lactoferrin from bovine milk (Sigma-Aldrich, St. Louis, MO, USA) was dissolved in LB (Luria-Bertani) broth and filtered through 0.22 μm cellulose acetate filter (Merc, Darmstadt, Germany) to form a stock solution of 20 mg/mL and stored at 4 °C. Final concentrations used in the study were 10 mg/mL, 1 mg/mL and 0.1 mg/mL [16,32].

4.4. Lactoferrin Influence on Phage Activity

An amount of 100 μL of phage lysate with titer 10^9 PFU/mL was mixed with 100 μL of lactoferrin at final concentration of 10 mg/mL and then incubated at 4 °C for 5 days. Every 24 h, a 10 μL sample was collected, serial dilutions were made and the mixture was titrated using double agar plate technique. The plates were incubated overnight at 37 °C and then scanned for plaques. The phage titer was calculated based on the number of plaques formed [18,49].

4.5. Assessment of Biofilm Biomass Using Crystal Violet Staining

Biofilms were grown on 96-well plates (Nest Biotechnology, Wuxi, China) in accordance with previously described protocols, with minor modifications [19]. Each well was inoculated with 200 μL of bacterial suspension, and the microtiter plates were incubated for 24 h at 37 °C. After incubation, established biofilms were washed with distilled H_2O, and 200 μL of phage, lactoferrin or phage–lactoferrin mixture in LB was added to a set of wells. After an incubation period (4 h, 12 h or 24 h) at 37 °C, the wells were washed with distilled H_2O, 100 μL of 1% crystal violet (Sigma Aldrich, St. Louis, MO, USA) solution was added to each well and the plate was incubated for 15 min at 37 °C. Excess stain was rinsed off by running tap water until the water was colorless, and the plate was left to air dry.

To solubilize the dye bound to the biofilm, 200 µL of ethanol–acetic acid–water (30:30:40) was added to the wells, and the optical density at 595 nm was measured in the microplate spectrophotometer (BioTek Instruments, Winooski, VT, USA).

4.6. Assessment of Biofilm Metabolic Activity Using MTT

Biofilms were set and treated in accordance with protocol described above. After incubation period, MTT solution in PBS was added to final concentration of 0.5 mg/mL in 100 µL to each well and incubated at 37 °C for 1 h. After staining, the MTT solution was removed, and 200 mL of acidified isopropanol was added to dissolve the MTT formazan product. The absorbance was measured at 540 nm using a microplate spectrophotometer [19].

4.7. Enumeration of Cells in Biofilm Using CFU/mL Count

Biofilms were set and treated in accordance with protocol described at point 4.5. After the incubation period, the number of bacteria adhered to the surface of microplate wells was enumerated in accordance with previously described protocol. Therefore, 200 µL of 0.9% NaCl was added to each well, and biofilm cells were suspended by vigorous pipetting. The 10-fold serial dilutions were immediately performed in 0.9% NaCl and 40 µL of the dilutions were directly plated on LB plates.

4.8. Statistical Analysis

All the experiments were performed in triplicates that were averaged to produce means used for analysis. Mean values were compared using the t-test. Differences were considered statistically significant if $p < 0.05$.

Supplementary Materials: The following supporting information can be downloaded at: https://www.mdpi.com/article/10.3390/antibiotics11091256/s1, Table S1: p values for statistical comparison between influence of phage vB_SauM-D and vB_SauM-D + 0.1 Lf on cell count (CFU/mL) of MDRSA biofilm, Table S2: p values for statistical comparison between influence of phage vB_SauM-C and vB_SauM-C + 0.1 Lf on cell count (CFU/mL) of MDRSA biofilm.

Author Contributions: Conceptualization, K.K.-K.; methodology, K.K.-K.; validation, K.K.-K.; investigation, K.K.-K. and N.K.; writing—original draft preparation, K.K.-K., N.K. and L.P.; writing—review and editing, K.K.-K. and L.P.; visualization, K.K.-K.; supervision, L.P. All authors have read and agreed to the published version of the manuscript.

Funding: This research received no external funding.

Institutional Review Board Statement: Not applicable.

Informed Consent Statement: Not applicable.

Data Availability Statement: Raw data are available from the authors upon request.

Conflicts of Interest: The authors declare no conflict of interest.

References

1. Thompson, R.L.; Cabezudo, I.; Wenzel, R.P. Epidemiology of Nosocomial Infections Caused by Methicillin-Resistant *Staphylococcus aureus*. *Ann. Intern. Med.* **1982**, *97*, 309–317. [CrossRef] [PubMed]
2. Leung, Y.L. Staphylococcus aureus. In *Encyclopedia of Toxicology*, 3rd ed.; Wexler, P., Ed.; Academic Press: Oxford, UK, 2014; pp. 379–380, ISBN 978-0-12-386455-0.
3. Cheung, G.Y.C.; Bae, J.S.; Otto, M. Pathogenicity and Virulence of *Staphylococcus aureus*. *Virulence* **2021**, *12*, 547–569. [CrossRef] [PubMed]
4. Shrestha, B.; Pokhrel, B.; Mohapatra, T. Study of Nosocomial Isolates of *Staphylococcus aureus* with Special Reference to Methicillin Resistant S. Aureus in a Tertiary Care Hospital in Nepal. *Nepal Med. Coll. J. NMCJ* **2009**, *11*, 123–126. [PubMed]
5. Stefani, S.; Chung, D.R.; Lindsay, J.A.; Friedrich, A.W.; Kearns, A.M.; Westh, H.; MacKenzie, F.M. Meticillin-Resistant *Staphylococcus aureus* (MRSA): Global Epidemiology and Harmonisation of Typing Methods. *Int. J. Antimicrob. Agents* **2012**, *39*, 273–282. [CrossRef]
6. Lister, J.L.; Horswill, A.R. *Staphylococcus aureus* Biofilms: Recent Developments in Biofilm Dispersal. *Front. Cell. Infect. Microbiol.* **2014**, *4*, 178. [CrossRef]

7. Costerton, J.W.; Stewart, P.S.; Greenberg, E.P. Bacterial Biofilms: A Common Cause of Persistent Infections. *Science* **1999**, *284*, 1318–1322. [CrossRef]
8. Kaźmierczak, N.; Grygorcewicz, B.; Roszak, M.; Bochentyn, B.; Piechowicz, L. Comparative Assessment of Bacteriophage and Antibiotic Activity against Multidrug-Resistant *Staphylococcus aureus* Biofilms. *Int. J. Mol. Sci.* **2022**, *23*, 1274. [CrossRef]
9. Kifelew, L.G.; Warner, M.S.; Morales, S.; Thomas, N.; Gordon, D.L.; Mitchell, J.G.; Speck, P.G. Efficacy of Lytic Phage Cocktails on *Staphylococcus aureus* and *Pseudomonas aeruginosa* in Mixed-Species Planktonic Cultures and Biofilms. *Viruses* **2020**, *12*, 559. [CrossRef]
10. Chang, C.; Yu, X.; Guo, W.; Guo, C.; Guo, X.; Li, Q.; Zhu, Y. Bacteriophage-Mediated Control of Biofilm: A Promising New Dawn for the Future. *Front. Microbiol.* **2022**, *13*, 825828. [CrossRef]
11. Moons, P.; Faster, D.; Aertsen, A. Lysogenic Conversion and Phage Resistance Development in Phage Exposed *Escherichia coli* Biofilms. *Viruses* **2013**, *5*, 150–161. [CrossRef]
12. Simmons, E.L.; Bond, M.C.; Koskella, B.; Drescher, K.; Bucci, V.; Nadell, C.D. Biofilm Structure Promotes Coexistence of Phage-Resistant and Phage-Susceptible Bacteria. *mSystems* **2020**, *5*, e00877-19. [CrossRef] [PubMed]
13. Ferriol-González, C.; Domingo-Calap, P. Phages for Biofilm Removal. *Antibiotics* **2020**, *9*, 268. [CrossRef] [PubMed]
14. Manohar, P.; Madurantakam Royam, M.; Loh, B.; Bozdogan, B.; Nachimuthu, R.; Leptihn, S. Synergistic Effects of Phage–Antibiotic Combinations against Citrobacter Amalonaticus. *ACS Infect. Dis.* **2022**, *8*, 59–65. [CrossRef]
15. Ammons, M.C.; Copié, V. Mini-Review: Lactoferrin: A Bioinspired, Anti-Biofilm Therapeutic. *Biofouling* **2013**, *29*, 443–455. [CrossRef] [PubMed]
16. Ramamourthy, G.; Vogel, H.J. Antibiofilm Activity of Lactoferrin-Derived Synthetic Peptides against Pseudomonas Aeruginosa PAO11. *Biochem. Cell Biol.* **2021**, *99*, 138–148. [CrossRef]
17. Quintieri, L.; Caputo, L.; Monaci, L.; Cavalluzzi, M.M.; Denora, N. Lactoferrin-Derived Peptides as a Control Strategy against Skinborne Staphylococcal Biofilms. *Biomedicines* **2020**, *8*, 323. [CrossRef]
18. Łubowska, N.; Grygorcewicz, B.; Kosznik-Kwaśnicka, K.; Zauszkiewicz-Pawlak, A.; Węgrzyn, A.; Dołęgowska, B.; Piechowicz, L. Characterization of the Three New Kayviruses and Their Lytic Activity against Multidrug-Resistant *Staphylococcus aureus*. *Microorganisms* **2019**, *7*, 471. [CrossRef]
19. Kaźmierczak, N.; Grygorcewicz, B.; Piechowicz, L. Biofilm Formation and Prevalence of Biofilm-Related Genes among Clinical Strains of Multidrug-Resistant *Staphylococcus aureus*. *Microb. Drug Resist.* **2021**, *27*, 956–964. [CrossRef]
20. Wu, N.; Zhu, T. Potential of Therapeutic Bacteriophages in Nosocomial Infection Management. *Front. Microbiol.* **2021**, *12*, 638094. [CrossRef]
21. Singh, A.; Padmesh, S.; Dwivedi, M.; Kostova, I. How Good Are Bacteriophages as an Alternative Therapy to Mitigate Biofilms of Nosocomial Infections. *Infect. Drug Resist.* **2022**, *15*, 503–532. [CrossRef]
22. Jamal, M.; Andleeb, S.; Jalil, F.; Imran, M.; Nawaz, M.A.; Hussain, T.; Ali, M.; Ur Rahman, S.; Das, C.R. Isolation, Characterization and Efficacy of Phage MJ2 against Biofilm Forming Multi-Drug Resistant Enterobacter Cloacae. *Folia Microbiol.* **2019**, *64*, 101–111. [CrossRef] [PubMed]
23. Tkhilaishvili, T.; Wang, L.; Tavanti, A.; Trampuz, A.; Di Luca, M. Antibacterial Efficacy of Two Commercially Available Bacteriophage Formulations, Staphylococcal Bacteriophage and PYO Bacteriophage, against Methicillin-Resistant *Staphylococcus aureus*: Prevention and Eradication of Biofilm Formation and Control of a Systemic Infection of *Galleria mellonella* Larvae. *Front. Microbiol.* **2020**, *11*, 110. [PubMed]
24. Abedon, S.T.; Kuhl, S.J.; Blasdel, B.G.; Kutter, E.M. Phage Treatment of Human Infections. *Bacteriophage* **2011**, *1*, 66–85. [CrossRef] [PubMed]
25. Górski, A.; Międzybrodzki, R.; Węgrzyn, G.; Jończyk-Matysiak, E.; Borysowski, J.; Weber-Dąbrowska, B. Phage Therapy: Current Status and Perspectives. *Med. Res. Rev.* **2020**, *40*, 459–463. [CrossRef] [PubMed]
26. Dąbrowska, K.; Abedon, S.T. Pharmacologically Aware Phage Therapy: Pharmacodynamic and Pharmacokinetic Obstacles to Phage Antibacterial Action in Animal and Human Bodies. *Microbiol. Mol. Biol. Rev. MMBR* **2019**, *83*, e00012-19. [CrossRef]
27. Principi, N.; Silvestri, E.; Esposito, S. Advantages and Limitations of Bacteriophages for the Treatment of Bacterial Infections. *Front. Pharmacol.* **2019**, *10*, 513. [CrossRef]
28. Mangalea, M.R.; Duerkop, B.A. Fitness Trade-Offs Resulting from Bacteriophage Resistance Potentiate Synergistic Antibacterial Strategies. *Infect. Immun.* **2020**, *88*, e00926-19. [CrossRef]
29. Kamal, F.; Dennis, J.J. Burkholderia Cepacia Complex Phage-Antibiotic Synergy (PAS): Antibiotics Stimulate Lytic Phage Activity. *Appl. Environ. Microbiol.* **2015**, *81*, 1132–1138. [CrossRef]
30. Ghosh, A.; Ricke, S.C.; Almeida, G.; Gibson, K.E. Combined Application of Essential Oil Compounds and Bacteriophage to Inhibit Growth of *Staphylococcus aureus* In Vitro. *Curr. Microbiol.* **2016**, *72*, 426–435. [CrossRef]
31. Jo, A.; Ding, T.; Ahn, J. Synergistic Antimicrobial Activity of Bacteriophages and Antibiotics against *Staphylococcus aureus*. *Food Sci. Biotechnol.* **2016**, *25*, 935–940. [CrossRef]
32. Bruni, N.; Capucchio, M.T.; Biasibetti, E.; Pessione, E.; Cirrincione, S.; Giraudo, L.; Corona, A.; Dosio, F. Antimicrobial Activity of Lactoferrin-Related Peptides and Applications in Human and Veterinary Medicine. *Molecules* **2016**, *21*, 752. [CrossRef] [PubMed]
33. Andersen, J.H.; Jenssen, H.; Gutteberg, T.J. Lactoferrin and Lactoferricin Inhibit Herpes Simplex 1 and 2 Infection and Exhibit Synergy When Combined with Acyclovir. *Antivir. Res.* **2003**, *58*, 209–215. [CrossRef]

34. Mistry, N.; Drobni, P.; Näslund, J.; Sunkari, V.G.; Jenssen, H.; Evander, M. The Anti-Papillomavirus Activity of Human and Bovine Lactoferricin. *Antivir. Res.* **2007**, *75*, 258–265. [CrossRef] [PubMed]

35. Arnold, R.R.; Brewer, M.; Gauthier, J.J. Bactericidal Activity of Human Lactoferrin: Sensitivity of a Variety of Microorganisms. *Infect. Immun.* **1980**, *28*, 893–898. [CrossRef]

36. Jahani, S.; Shakiba, A.; Jahani, L. The Antimicrobial Effect of Lactoferrin on Gram-Negative and Gram-Positive Bacteria. *Int. J. Infect.* **2015**, *2*, iji27594. [CrossRef]

37. Ammons, M.C.B.; Ward, L.S.; Fisher, S.T.; Wolcott, R.D.; James, G.A. In Vitro Susceptibility of Established Biofilms Composed of a Clinical Wound Isolate of Pseudomonas Aeruginosa Treated with Lactoferrin and Xylitol. *Int. J. Antimicrob. Agents* **2009**, *33*, 230–236. [CrossRef]

38. Petitclerc, D.; Lauzon, K.; Cochu, A.; Ster, C.; Diarra, M.S.; Lacasse, P. Efficacy of a Lactoferrin-Penicillin Combination to Treat β-Lactam-Resistant *Staphylococcus aureus* Mastitis. *J. Dairy Sci.* **2007**, *90*, 2778–2787. [CrossRef] [PubMed]

39. Lacasse, P.; Lauzon, K.; Diarra, M.S.; Petitclerc, D. Utilization of Lactoferrin to Fight Antibiotic-Resistant Mammary Gland Pathogens1,2. *J. Anim. Sci.* **2008**, *86*, 66–71. [CrossRef]

40. Chen, P.-W.; Ho, S.-P.; Shyu, C.-L.; Mao, F.C. Effects of Bovine Lactoferrin Hydrolysate on the in Vitro Antimicrobial Susceptibility of *Escherichia coli* Strains Isolated from Baby Pigs. *Am. J. Vet. Res.* **2004**, *65*, 131–137. [CrossRef]

41. Andrés, M.T.; Viejo-Diaz, M.; Pérez, F.; Fierro, J.F. Antibiotic Tolerance Induced by Lactoferrin in Clinical *Pseudomonas aeruginosa* Isolates from Cystic Fibrosis Patients. *Antimicrob. Agents Chemother.* **2005**, *49*, 1613–1616. [CrossRef]

42. Ammons, M.C.B.; Ward, L.S.; Dowd, S.; James, G.A. Combined Treatment of Pseudomonas Aeruginosa Biofilm with Lactoferrin and Xylitol Inhibits the Ability of Bacteria to Respond to Damage Resulting from Lactoferrin Iron Chelation. *Int. J. Antimicrob. Agents* **2011**, *37*, 316–323. [CrossRef] [PubMed]

43. Harris, M.R.; Coote, P.J. Combination of Caspofungin or Anidulafungin with Antimicrobial Peptides Results in Potent Synergistic Killing of Candida Albicans and Candida Glabrata in Vitro. *Int. J. Antimicrob. Agents* **2010**, *35*, 347–356. [CrossRef] [PubMed]

44. Leitch, E.C.; Willcox, M.D. Lactoferrin Increases the Susceptibility of *S. epidermidis* Biofilms to Lysozyme and Vancomycin. *Curr. Eye Res.* **1999**, *19*, 12–19. [CrossRef]

45. Golshahi, L.; Lynch, K.H.; Dennis, J.J.; Finlay, W.H. In Vitro Lung Delivery of Bacteriophages KS4-M and ΦKZ Using Dry Powder Inhalers for Treatment of Burkholderia Cepacia Complex and Pseudomonas Aeruginosa Infections in Cystic Fibrosis. *J. Appl. Microbiol.* **2011**, *110*, 106–117. [CrossRef]

46. Geagea, H.; Gomaa, A.; Remondetto, G.; Moineau, S.; Subirade, M. Molecular Structure of Lactoferrin Influences the Thermal Resistance of Lactococcal Phages. *J. Agric. Food Chem.* **2017**, *65*, 2214–2221. [CrossRef]

47. Wang, X.; Xie, Z.; Zhao, J.; Zhu, Z.; Yang, C.; Liu, Y. Prospects of Inhaled Phage Therapy for Combatting Pulmonary Infections. *Front. Cell. Infect. Microbiol.* **2021**, *11*. [CrossRef]

48. Zimecki, M.; Artym, J.; Kocieba, M.; Weber-Dabrowska, B.; Lusiak-Szelachowska, M.; Górski, A. The Concerted Action of Lactoferrin and Bacteriophages in the Clearance of Bacteria in Sublethally Infected Mice. *Postepy Hig. Med. Doswiadczalnej Online* **2008**, *62*, 42–46.

49. Kosznik-Kwaśnicka, K.; Ciemińska, K.; Grabski, M.; Grabowski, Ł.; Górniak, M.; Jurczak-Kurek, A.; Węgrzyn, G.; Węgrzyn, A. Characteristics of a Series of Three Bacteriophages Infecting Salmonella Enterica Strains. *Int. J. Mol. Sci.* **2020**, *21*, 6152. [CrossRef]

Case Report

Salphage: Salvage Bacteriophage Therapy for Recalcitrant MRSA Prosthetic Joint Infection

James B. Doub [1,*], Vincent Y. Ng [2], Myounghee Lee [3], Andrew Chi [3], Alina Lee [4,5], Silvia Würstle [4,5] and Benjamin Chan [4,5]

1   Division of Clinical Care and Research, Institute of Human Virology, University of Maryland School of Medicine, Baltimore, MD 21201, USA
2   Department of Orthopedic Surgery, University of Maryland School of Medicine, Baltimore, MD 21201, USA; vng@som.umaryland.edu
3   Department of Pharmacy, University of Maryland Medical Center, Baltimore, MD 21201, USA; mlee1@umm.edu (M.L.); andrew.chi@umm.edu (A.C.)
4   Department of Ecology and Evolutionary Biology, Yale University, New Haven, CT 06520, USA; a.lee@yale.edu (A.L.); silvia.wuerstle@yale.edu (S.W.); b.chan@yale.edu (B.C.)
5   Yale Center for Phage Biology & Therapy, Yale University, New Haven, CT 06520, USA
*   Correspondence: jdoub@ihv.umaryland.edu; Tel.: +1-410-706-3454; Fax: +1-410-328-9106

Abstract: Prosthetic joint infections are a devastating complication of joint replacement surgery. Consequently, novel therapeutics are needed to thwart the significant morbidity and enormous financial ramifications that are associated with conventional treatments. One such promising adjuvant therapeutic is bacteriophage therapy given its antibiofilm activity and its ability to self-replicate. Herein we discuss the case of a 70-year-old female who had a recalcitrant MRSA prosthetic knee and femoral lateral plate infection who was successfully treated with adjuvant bacteriophage therapy. Moreover, this case discusses the importance of propagating bacteriophage therapeutics on bacteria that are devoid of toxins and the need to ensure bacteriophage activity to all bacterial morphologies. Overall, this case reinforces the potential benefit of using personalized bacteriophage therapy for recalcitrant prosthetic joint infections, but more translational research is needed to thereby devise effective, reproducible clinical trials.

Keywords: bacteriophage therapy; *Staphylococcus aureus*; prosthetic joint infection; cell surface receptor; transaminitis

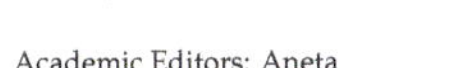

Citation: Doub, J.B.; Ng, V.Y.; Lee, M.; Chi, A.; Lee, A.; Würstle, S.; Chan, B. Salphage: Salvage Bacteriophage Therapy for Recalcitrant MRSA Prosthetic Joint Infection. *Antibiotics* **2022**, *11*, 616. https://doi.org/10.3390/antibiotics11050616

Academic Editors: Aneta Skaradzińska and Anna Nowaczek

Received: 31 March 2022
Accepted: 2 May 2022
Published: 4 May 2022

Publisher's Note: MDPI stays neutral with regard to jurisdictional claims in published maps and institutional affiliations.

1. Introduction

Prosthetic joint infections (PJIs) are the most feared complication of joint replacement surgeries with an estimated 1 to 2% of all knee and hip prosthetics becoming infected during the life of the prosthetic [1]. The gold standard treatment of chronic PJIs is two stage revision surgery whereby the prosthetic is removed and reimplantation of a new arthroplasty is not conducted until after six weeks of antibiotic therapy. This procedure is taxing to patients and has immense financial ramifications [2]. Obstinately, the success rates of this surgical technique have not changed over the past several decades and, consequently, novel therapeutics are drastically needed [3].

One such novel therapeutic for PJI treatments is bacteriophage therapy given these viruses have evolved with bacteria to possess innate anti-biofilm activities, can infect metabolically reduced bacteria and can degrade the biofilm matrices [4]. In addition, bacteriophages can self-replicate on their bacterial hosts, thereby, potentially increasing their numbers to help cure the infection. Therefore, bacteriophages have been proposed to be possible adjuvants with debridement and implant retention surgery to, thereby, cure PJIs without prosthesis removal [5]. However, bacteriophages also hold promise as an adjuvant in revision surgeries when retention of prosthetics is not feasible. Herein we discuss a case of a patient who had a recalcitrant methicillin resistant *Staphylococcus aureus* (MRSA) PJI of her knee and femoral plate that conventional treatments were not able to cure. Rather, only after adjuvant bacteriophage therapy was a sustained clinical and microbiological cure achieved

2. Case

A 70-year-old female with a past medical history of hypertension, diabetes and chronic lymphedema of the right leg presented to the University of Maryland for a second opinion of her recalcitrant MRSA PJI. The patient initially underwent bilateral knee arthroplasties in 2008 for progressive osteoarthritis. Unfortunately, her right knee arthroplasty was complicated by an MRSA PJI requiring revision surgery and intravenous vancomycin and then indefinite oral suppression doxycycline therapy given her chronic lymphedema. She then underwent right hip arthroplasty for progressive osteoarthritis, but this was complicated by a periprosthetic femur fracture requiring the insertion of an extensive lateral femoral plate, cerclage wires and single staged revision of right knee arthroplasty for MRSA PJI followed by 6 weeks of vancomycin therapy and then indefinite oral doxycycline therapy (Figure 1). Three months later, she had worsening pain in her knee and a draining sinus tract. Arthrocentesis culture again grew MRSA. A repeat revision surgery was deemed unlikely to be successful given her chronic lymphedema and extensive chronic MRSA infection. Subsequently, she was recommended for fusion or amputation.

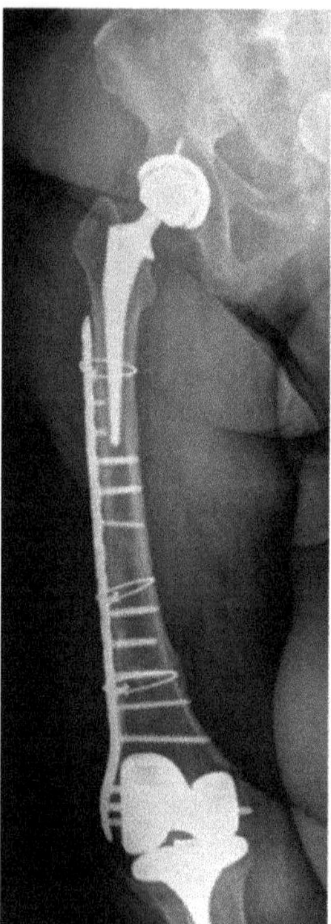

Figure 1. Knee and hip prosthetics with lateral plate with three cerclage wires.

Given the sinus tract and the extent of hardware infection (knee prosthetic, lateral plate and hip prosthetic) in correlation with erosion of the medial condyle, salvage of her prosthesis was not feasible as the prosthetic was loose. Therefore, explantation of all hardware followed by reconstruction with a right revision hip replacement and knee replacement with total femur implant was deemed the best chance to eradicate the infection, salvage her lower extremity and allow for ambulation. However, given the extent and recalcitrant nature of her infection, adjuvant bacteriophage therapy was discussed with this patient. She agreed to this experimental therapy and a repeat arthrocentesis of her knee was obtained. Again, only MRSA was cultured, and the clinical isolate was sent to Dr. Benjamin Chan to create a personalized bacteriophage therapy.

Her clinical isolate was matched to the bacteriophage Mallokai, which had adequate growth inhibition and plaque formation. This bacteriophage was then amplified on her clinical isolate to titers of 1×10^{10} PFU/mL. Evaluation of the bacteriophage therapeutic did not reveal any endotoxins and USP-71 sterility testing was negative. Expanded access was granted by the FDA (IND 27250) and approval by the University of Maryland of Baltimore Institutional Review Board (HP-00094883EA) was obtained.

She then underwent removal of her knee prosthetic, lateral plate and hip prosthetic. At the end of the surgery, an intraoperative dose of bacteriophage (1×10^{10} PFU/mL) was diluted in 10 mL of normal saline, with resulting titers administered being 1×10^9 PFU/mL. Intravenous daptomycin, 500 mg every day, was started after the operation. The next day intravenous bacteriophage therapy (1×10^{10} PFU/mL) was diluted in 50 mL of normal saline and infused over 30 min, with resulting titers administered being 2×10^8 PFU/mL. She received three doses of intravenous bacteriophage therapy and was planned to receive two more days of intravenous bacteriophage therapy. However, before her fourth dose she developed an increase in aspartate and alanine transferase up to 2.5 times the upper limit of normal and further doses were held. Four days after stopping bacteriophage her transaminitis returned to normal. Operative cultures grew two different distinct MRSA colony morphologies that were spatially separated, with one distinct colony morphology obtained from knee tissues and another colony morphology from the proximal end of the femoral lateral plate adjacent to the distal hip prosthetic. We retrospectively tested both morphologies to ensure bacteriophage activity (Figure 2).

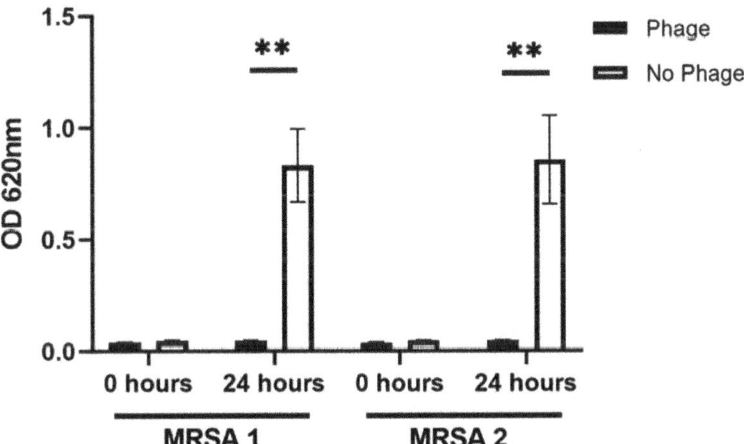

Figure 2. Bacteriophage growth inhibtion assays for the two different MRSA clinical isolates: Significant bacteriophage (Mallokai) induced growth inhibition was observed for both MRSA clinical isolates at 24 h (Wilcoxon test, ** $p < 0.005$). Experiment was conducted with six replicates and was reproduced in triplicate. Error bars are SD.

She then completed six weeks of intravenous ceftaroline therapy, 600 mg every 12 h, as daptomycin was cost prohibitive. After six weeks, all antibiotics were stopped, and after two months off antibiotic therapy, a repeat arthrocentesis was obtained that showed no evidence of infection in the hip or knee joints. One month later, she underwent reconstruction of her hip and knee joints with an intramedullary total femur implant (Figure 3). No MRSA could be cultured from operating room cultures and no infection was seen in the operating room, but a rare growth of Streptococcus pyogenes was seen in one culture. While this was thought to be a contaminate, especially given the lack of any signs of infection in the operating room, we elected to treat for four weeks with intravenous ceftriaxone therapy, 2 g every day, to be conservative. In addition, as a result of her chronic lower extremity lymphedema and the extensive reconstruction, we elected to use oral Cephalexin, 500 mg twice daily, to prevent further PJIs after the ceftriaxone therapy. Given the prolonged period in which she had not ambulated, she is currently receiving rehab to increase her mobility but has had no signs of recurrence of her infection 12 months later.

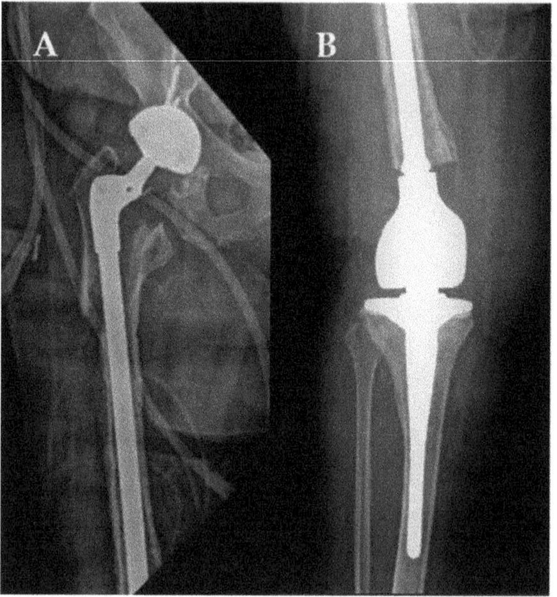

Figure 3. After cure of recalcitrant MRSA PJI and implantation of right total hip arthroplasty (**A**) extending in continuity with a total knee arthroplasty (**B**).

3. Discussion

Two stage revision surgery is the gold standard for the treatment of chronic PJIs, but this intervention is associated with significant morbidity and financial ramifications [1,2]. When PJIs fail, two stage revision surgery limited standardized options are available. This is in part because the biofilm laden prosthetic was removed and, consequently, all theoretical niduses of infection were eliminated. Therefore, when immunocompetent patients have recurrence with the same pathogen, this suggests that a deep-seated infection may be present. Bacterial factors that cause these deep-seated infections include: persister cells, small colony variants, plasma protein aggregates and microscopic abscesses in cortical canaliculi [6–8]. Furthermore, when additional infected hardware is present beyond the infected prosthetic, this complicates treatment and makes eradication even more difficult as seen here.

In this case the patient had failed standard of care revision surgeries and a draining sinus tract was present over her knee, indicating a chronic MRSA infection. A previous periprosthetic femur fracture required a large lateral plate that was also infected with MRSA and, obstinately, this was directly adjacent to the distal end of her hip prosthetic. Unfortunately, her hardware could not be salvaged given the instability of the knee prosthetic, but it was paramount to cure her infection to thereby implant a megaprosthesis to give her the best chance to ambulate again on her lower extremity (Figure 3). Therefore, the use of intraoperative and intravenous bacteriophage therapy was used as an adjuvant to help sterilize the joint. The benefits of using bacteriophage therapy with surgical interventions have been discussed and have been used by other researchers [5,9–11]. Moreover, surgical interventions are a central dogma of PJI treatments to achieve infection source control. No standard of care PJI treatment is recommended without some form of surgery [12,13]. However, the use of surgical intervention does hinder the ability to definitively prove effectiveness of the bacteriophage therapeutic, but in this case the intraoperative cultures off antibiotics for many months supports sterilization of her infection. However, only randomized clinical trials will be able to definitively prove if using bacteriophages as therapy are effective adjuvants in PJI treatments.

While the bacteriophage therapy helped cure her recalcitrant MRSA infection, we did observe a mild transaminitis with our protocol of intraoperative phage and then intravenous phage therapy. As seen in a previous reported case, it was not until after the third intravenous dose that we observed a mild transaminitis, which further supports the need for close monitoring of liver enzymes when using bacteriophage therapy for PJIs [14]. Additionally, since her bacteriophage therapeutic was grown on her MRSA clinical isolate, we retrospectively evaluated her therapeutic for possible enterotoxins in which we observed low levels of Staphylococcal enterotoxin A (3 ng/mL). It is unknown if this low level of enterotoxin caused her to have mild liver inflammation or if the inflammation was cytokine driven for reasons we state elsewhere [14,15]. Nonetheless, this case reinforces that bacteriophage therapies can be contaminated with other elements beyond endotoxins and supports the need to grow bacteriophage therapeutics on bacteria that are devoid of these contaminants to mitigate the potential risks [15].

What was also interesting was that the patient had two distinct different MRSA clinical isolates, which had different colony morphologies. Nonetheless, given the narrow spectrum of bacteriophage activity, the bacteriophage used here was tested to ensure growth inhibition to both morphologies. Figure 2 shows the results in which no differences in growth inhibition were seen for the two morphologies. Even though the bacteriophage had activity to both isolates, this case reinforces the need to ensure bacteriophage activity to all isolates and morphologies given that assuming activity is what has led to failed clinical trials [16]. Furthermore, most Staphylococcal bacteriophages bind to teichoic acid, but this receptor can have different glycosylation patterns in different environments, which can have ramifications for bacteriophage activity [17–19]. Therefore, at this nascent stage when using bacteriophages for PJI, there is ample evidence supporting bacteriophage therapy use as an adjuvant to surgical interventions instead of in lieu of surgery. This is to thereby allow for effective and reproducible treatments as well as to isolate all bacteria and all potential phenotypes of the causative bacteria to ensure activity of the bacteriophage therapeutic.

4. Materials and Methods

4.1. Bacteriophage Screening, Amplification and Purification

The screening, amplification and purification followed the same procedure as previously described [20]. To note, the concentrated bacteriophage was buffered in plasmalyte. The final therapeutic was quality control tested for titers, sterility (USP <71>) and endotoxin levels. The results from quality control testing can be found in Table 1. Retrospectively, we also tested for levels of Staphylococcal enterotoxin A-C with use of commercially available ELISA testing (Thermo Fisher Scientific, Waltham, MA, USA), which can be seen in Table 1.

Table 1. Titer, endotoxin, sterility and exotoxin levels of the bacteriophage used in this case.

Phage ID	Titer (PFU/mL)	Endotoxin (EU/Dose)	USP <71> Sterility	Staphylococcal Enterotoxin A (ng/mL)
Mallokai	1×10^{10}	<1	No Growth	3

4.2. Testing for Bacteriophage Activity to the Two MRSA Morphologies

Overnight cultures of the two MRSA morphologies were grown in tryptic soy broth (TSB) to optical density 0.30–0.60, representing exponential growth (OD 620 nm). Bacterial cultures were then diluted to OD of 0.024 with TSB and placed into wells of microtiter plates. Negative control included TSB without bacteria or bacteriophages (not shown in Figure 2), but no changes in OD were seen. Positive control included bacteria in TSB without bacteriophages. Wells of bacteria in TSB were infected with 0.05 mL of bacteriophage with the same titers as used for the patient (1×10^{10} PFU/mL). OD was read at time zero and again after microwell plates were incubated at 37 degrees Celsius for 24 h. Results were reproduced in triplicate.

5. Conclusions

In conclusion, this case adds to the growing data supporting the potential use of bacteriophage therapy as an adjuvant to surgical interventions in PJI treatment. Bacteriophage therapy may be a promising agent in treating PJI to either circumvent the need for revision surgery or to enhance the efficacy of revision surgery, especially in complex cases that have high risk of recurrence. However, more translational research is needed to clarify many aspects of this therapeutic to devise effective, reproducible protocols before efficacy clinical trials are conducted.

Author Contributions: J.B.D. and V.Y.N. carried out the clinical experimental treatment. J.B.D. wrote the manuscript with contributions from B.C. and S.W. Isolation, amplification, and purification of the bacteriophage was conducted by B.C., A.L. and S.W. Doses for clinical use were prepared by M.L. and A.C. All authors edited the manuscript and all authors have read and agreed to the published version of the manuscript.

Funding: This research received no external funding.

Institutional Review Board Statement: Expanded access was granted by the FDA (IND-27250) and University of Maryland Institutional Review Board (HP-00094883EA).

Informed Consent Statement: Written informed consent was obtained from the subject involved in the study to publish this manuscript.

Data Availability Statement: Not applicable.

Acknowledgments: We appreciate the help of other members of the University of Maryland IDS pharmacy team.

Conflicts of Interest: J.B.D. has a patent pending with respect to use of bacteriophage therapy with surgery for prosthetic joint infections. All other authors declare no conflict of interest.

References

1. Natsuhara, K.M.; Shelton, T.J.; Meehan, J.P.; Lum, Z.C. Mortality During Total Hip Periprosthetic Joint Infection. *J. Arthroplast.* **2019**, *34*, S337–S342. [CrossRef] [PubMed]
2. Premkumar, A.; Kolin, D.A.; Farley, K.X.; Wilson, J.M.; McLawhorn, A.S.; Cross, M.B.; Sculco, P.K. Projected Economic Burden of Periprosthetic Joint Infection of the Hip and Knee in the United States. *J. Arthroplast.* **2021**, *36*, 1484–1489.e3. [CrossRef] [PubMed]
3. Xu, C.; Goswami, K.; Li, W.T.; Tan, T.L.; Yayac, M.; Wang, S.H.; Parvizi, J. Is Treatment of Periprosthetic Joint Infection Improving Over Time? *J. Arthroplast.* **2020**, *35*, 1696–1702.e1. [CrossRef] [PubMed]
4. Chan, B.K.; Abedon, S.T. Bacteriophages and their enzymes in biofilm control. *Curr. Pharm. Des.* **2015**, *21*, 85–99. [CrossRef] [PubMed]
5. Doub, J.B.; Ng, V.Y.; Johnson, A.; Amoroso, A.; Kottilil, S.; Wilson, E. Potential Use of Adjuvant Bacteriophage Therapy With Debridement, Antibiotics, and Implant Retention Surgery to Treat Chronic Prosthetic Joint Infections. *Open Forum Infect. Dis.* **2021**, *8*, ofab277. [CrossRef] [PubMed]
6. Del Pozo, J.L. Biofilm-related disease. *Expert Rev. Anti-Infect. Ther.* **2018**, *16*, 51–65. [CrossRef] [PubMed]
7. Pestrak, M.J.; Gupta, T.T.; Dusane, D.H.; Guzior, D.V.; Staats, A.; Harro, J.; Horswill, A.R.; Stoodley, P. Investigation of synovial fluid induced Staphylococcus aureus aggregate development and its impact on surface attachment and biofilm formation. *PLoS ONE* **2020**, *15*, e0231791.

8. de Mesy Bentley, K.L.; Trombetta, R.; Nishitani, K.; Bello-Irizarry, S.N.; Ninomiya, M.; Zhang, L.; Chung, H.L.; McGrath, J.L.; Daiss, J.L.; Awad, H.A.; et al. Evidence of Staphylococcus Aureus Deformation, Proliferation, and Migration in Canaliculi of Live Cortical Bone in Murine Models of Osteomyelitis. *J. Bone Miner. Res.* **2017**, *32*, 985–990. [CrossRef] [PubMed]

9. Ferry, T.; Kolenda, C.; Batailler, C.; Gustave, C.A.; Lustig, S.; Malatray, M.; Fevre, C.; Josse, J.; Petitjean, C.; Chidiac, C.; et al. Phage Therapy as Adjuvant to Conservative Surgery and Antibiotics to Salvage Patients With Relapsing *S. aureus* Prosthetic Knee Infection. *Front. Med.* **2020**, *7*, 570572. [CrossRef] [PubMed]

10. Onsea, J.; Soentjens, P.; Djebara, S.; Merabishvili, M.; Depypere, M.; Spriet, I.; De Munter, P.; Debaveye, Y.; Nijs, S.; Vanderschot, P.; et al. Bacteriophage Application for Difficult-to-treat Musculoskeletal Infections: Development of a Standardized Multidisciplinary Treatment Protocol. *Viruses* **2019**, *11*, 891. [CrossRef] [PubMed]

11. Ferry, T.; Kolenda, C.; Batailler, C.; Gaillard, R.; Gustave, C.A.; Lustig, S.; Fevre, C.; Petitjean, C.; Leboucher, G.; Laurent, F.; et al. Case Report: Arthroscopic "Debridement Antibiotics and Implant Retention" With Local Injection of Personalized Phage Therapy to Salvage a Relapsing *Pseudomonas Aeruginosa* Prosthetic Knee Infection. *Front. Med.* **2021**, *8*, 569159. [CrossRef] [PubMed]

12. Osmon, D.R.; Berbari, E.F.; Berendt, A.R.; Lew, D.; Zimmerli, W.; Steckelberg, J.M.; Rao, N.; Hanssen, A.; Wilson, W.R. Diagnosis and management of prosthetic joint infection: Clinical practice guidelines by the Infectious Diseases Society of America. *Clin. Infect. Dis.* **2013**, *56*, e1–e25. [CrossRef] [PubMed]

13. Izakovicova, P.; Borens, O.; Trampuz, A. Periprosthetic joint infection: Current concepts and outlook. *EFORT Open Rev.* **2019**, *4*, 482–494. [CrossRef] [PubMed]

14. Doub, J.B.; Wilson, E. Observed transaminitis with a unique bacteriophage therapy protocol to treat recalcitrant Staphylococcal biofilm infections. *Infection* **2022**, *50*, 281–283. [CrossRef] [PubMed]

15. Doub, J.B. Risk of Bacteriophage Therapeutics to Transfer Genetic Material and Contain Contaminants Beyond Endotoxins with Clinically Relevant Mitigation Strategies. *Infect. Drug Resist.* **2021**, *14*, 5629–5637. [CrossRef] [PubMed]

16. Jault, P.; Leclerc, T.; Jennes, S.; Pirnay, J.P.; Que, Y.A.; Resch, G.; Rousseau, A.F.; Ravat, F.; Carsin, H.; Le Floch, R.; et al. Efficacy and tolerability of a cocktail of bacteriophages to treat burn wounds infected by Pseudomonas aeruginosa (PhagoBurn): A randomised, controlled, double-blind phase 1/2 trial. *Lancet Infect. Dis.* **2019**, *19*, 35–45. [CrossRef]

17. Doub, J.B.; Urish, K.; Lee, M.; Fackler, J. Impact of Bacterial Phenotypic Variation with Bacteriophage therapy: A Pilot Study with Prosthetic Joint Infection Isolates. *Int. J. Infect. Dis.* **2022**, *119*, 44–46. [CrossRef] [PubMed]

18. Mistretta, N.; Brossaud, M.; Telles, F.; Sanchez, V.; Talaga, P.; Rokbi, B. Glycosylation of Staphylococcus aureus cell wall teichoic acid is influenced by environmental conditions. *Sci. Rep.* **2019**, *9*, 3212. [CrossRef] [PubMed]

19. Azam, A.H.; Tanji, Y. Peculiarities of Staphylococcus aureus phages and their possible application in phage therapy. *Appl. Microbiol. Biotechnol.* **2019**, *103*, 4279–4289. [CrossRef] [PubMed]

20. Doub, J.B.; Ng, V.Y.; Wilson, E.; Corsini, L.; Chan, B.K. Successful Treatment of a Recalcitrant *Staphylococcus epidermidis* Prosthetic Knee Infection with Intraoperative Bacteriophage Therapy. *Pharmaceuticals* **2021**, *14*, 231. [CrossRef] [PubMed]

Review

Enterococcal Phages: Food and Health Applications

Carlos Rodríguez-Lucas [1,2,]* and Victor Ladero [3,4,]*

1 Microbiology Laboratory, Hospital Universitario Central de Asturias, 33011 Oviedo, Spain
2 Translational Microbiology Group, Instituto de Investigación Sanitaria del Principado de Asturias (ISPA), 33011 Oviedo, Spain
3 Department of Technology and Biotechnology of Dairy Products, Dairy Research Institute, IPLA CSIC, 33300 Villaviciosa, Spain
4 Molecular Microbiology Group, Instituto de Investigación Sanitaria del Principado de Asturias (ISPA), 33011 Oviedo, Spain
* Correspondence: carlos.rodriguezlu@sespa.es (C.R.-L.); ladero@ipla.csic.es (V.L.)

Abstract: *Enterococcus* is a diverse genus of Gram-positive bacteria belonging to the lactic acid bacteria (LAB) group. It is found in many environments, including the human gut and fermented foods. This microbial genus is at a crossroad between its beneficial effects and the concerns regarding its safety. It plays an important role in the production of fermented foods, and some strains have even been proposed as probiotics. However, they have been identified as responsible for the accumulation of toxic compounds—biogenic amines—in foods, and over the last 20 years, they have emerged as important hospital-acquired pathogens through the acquisition of antimicrobial resistance (AMR). In food, there is a need for targeted measures to prevent their growth without disturbing other LAB members that participate in the fermentation process. Furthermore, the increase in AMR has resulted in the need for the development of new therapeutic options to treat AMR enterococcal infections. Bacteriophages have re-emerged in recent years as a precision tool for the control of bacterial populations, including the treatment of AMR microorganism infections, being a promising weapon as new antimicrobials. In this review, we focus on the problems caused by *Enterococcus faecium* and *Enterococcus faecalis* in food and health and on the recent advances in the discovery and applications of enterococcus-infecting bacteriophages against these bacteria, with special attention paid to applications against AMR enterococci.

Keywords: *Enterococcus faecalis*; *Enterococcus faecium*; antimicrobial resistance; bacteriophage; food; health

Citation: Rodríguez-Lucas, C.; Ladero, V. Enterococcal Phages: Food and Health Applications. *Antibiotics* **2023**, *12*, 842. https://doi.org/10.3390/antibiotics12050842

Academic Editors: Aneta Skaradzińska and Anna Nowaczek

Received: 28 March 2023
Revised: 24 April 2023
Accepted: 30 April 2023
Published: 2 May 2023

1. Introduction

The discovery of antibiotics in the mid-20th century is one of the scientific advances that has had the most significant impacts on increasing life expectancy. The prescription of antibiotics and other antimicrobials to treat bacterial infections, which can often cause permanent damage and even end a patient's life, has become a routine treatment. However, in recent decades, the misuse of antimicrobials has led to a rapid increase in the isolation of antimicrobial-resistant (AMR) bacteria. Today, AMR bacteria pose a great health problem; in Europe alone, up to 133,000 deaths in 2019 were attributable to infections caused by AMR bacteria [1], with an estimated cost for health services of over EUR 1000 million every year [2]. Seven pathogenic species were responsible for most of the deaths registered, namely, *Enterococcus faecium*, *Staphylococcus aureus*, *Klebsiella pneumoniae*, *Acinetobacter baumannii*, *Pseudomonas aeruginosa* and *Enterobacter* spp., listed together as the ESKAPE group by the World Health Organization (WHO, Geneva, Switzerland) [1,3]. Because of the great impact of these pathogens in terms of nosocomial infections, deaths and the economic losses of health services, the WHO has encouraged the scientific community to search for new ways to combat them [3]. In the context of this problem, bacteriophages have re-emerged as a potential weapon to fight AMR bacteria.

In this review, we focus on the advances in the characterization and application of enterococcal phages as rediscovered weapons against AMR *E. faecium* and *Enterococcus faecalis*.

2. *E. faecium* and *E. faecalis*

Enterococci are a diverse group of Gram-positive bacteria belonging to the lactic acid bacteria (LAB) group. The members of this genus are Gram-positive coccus-shaped bacteria that possess a versatile metabolism allowing them to adapt to very diverse environments and to resist rough conditions [4,5]. *E. faecium* and *E. faecalis* are the most studied species from the genus *Enterococcus* due to their role in human health [6]. These species are considered commensal bacteria of the gastrointestinal tract in mammals, including that of human beings [6–8]. They have also been isolated from a wide variety of environments that are mostly, but not exclusively, related to animal and human facilities, for example, cattle facilities [7], farms [9], hospitals [10] and wastewater facilities [11]. In fact, due to their persistent presence in the intestinal habitat, their robustness and endurance and the ease of their cultivation in the laboratory, enterococci are used as indicators of fecal contamination [12]. In addition, due to their presence in the gut, feces and milk of animals [13], they are also commonly found in foods of animal origin, such as meat and dairy products [11,14].

Although *E. faecium* and *E. faecalis* are considered harmless commensal bacteria, some strains are used as safe and effective probiotics, and they are present in certain cheeses in which they participate in the elimination of foodborne pathogens via the production of bacteriocins [15,16]. They can behave as opportunistic pathogens, and in recent years, they have been established as one of the major nosocomial pathogens. *E. faecalis* is considered the most pathogenic, but *E. faecium* has gained more concern due to the increasing acquisition of AMR [6,16].

2.1. *E. faecium* and *E. faecalis* in Food

As previously mentioned, their presence in the gut, feces and milk of mammals results in their presence in raw materials of animal origin, such as meat and dairy products. Their resistance to adverse environmental conditions allows them to grow in a wide range of pH values, temperatures and salt concentrations and to colonize foods, including fermented foods [17]. Enterococci have an ambiguous status regarding food safety. In fact, although these microorganisms belong to the LAB group (safe bacteria involved in the production of fermented foods), they have been granted with neither the Generally Regarded as Safe (GRAS) status nor the Qualified Presumption of Safety (QPS) status. Enterococci can be considered a valuable asset in cheese making, as some strains can be used as adjunct starter cultures [18,19]. The role of enterococci during cheese making is based on the large variety of technologically interesting enzymatic activities, such as protease, peptidase and lipolytic activities, which contribute to the organoleptic properties during the maturation process [18]. In addition, some strains are able to produce enterocins—so-called bacteriocins—and these produced by different strains of enterococcus that can inhibit the growth of several foodborne pathogens, such as *Staphylococcus aureus*, *Listeria monocytogenes* and *Salmonella enterica* [20]. However, their presence in foods has also been associated with the production of biogenic amines (BAs), toxic compounds that can cause food poisoning [21,22]. In fact, *E. faecalis* has been identified as the main species responsible for the accumulation of elevated concentrations of tyramine and putrescine [23–25], two of the most frequent BAs in dairy products [26,27]. Moreover, due to the increase in recent years of multidrug-resistant (MDR) bacteria, including enterococci, and the coining of the One Health concept to prevent their proliferation [28], concern regarding MDR enterococci in food has emerged [29]. These AMR enterococci can reach the food chain, where they can be transmitted directly or indirectly to humans and act as reservoirs of AMR genes that can be transferred to human-adapted strains or to other pathogenic bacteria [29]. Of special concern is the increase in vancomycin-resistant enterococci (VRE) from different food sources [30–33]. Thus, although the presence of enterococci in food could be considered beneficial in some scenarios, in general, they are considered a potential health threat.

2.2. E. faecium and E. faecalis in Human Health

Enterococci are considered a commensal organism of the human gastrointestinal (GI) tract, and they can be found in the GI microbiota of more than 90% of healthy people [34]. The first description of an enterococcal human infection was in 1899, when MacCallum and Hastings reported infective endocarditis (IE) caused by a bacterium that they called *Micrococcus zymogenes*, later identified as a member of the *Enterococcus* genus [35,36]. Enterococci were subsequently shown to be the cause of several kinds of infections, both community- (e.g., including pelvic infections, urinary tract infections and IE) and healthcare-associated infections (HAIs) (e.g., including surgical site infections, and urinary and bloodstream catheter-related infections) [37]. Therefore, enterococci usually display low levels of virulence, but they can also act as an opportunistic pathogen, causing severe infections, mainly in vulnerable patients, such as those who are immunocompromised, have undergone invasive procedures (e.g., the insertion of urinary or blood catheters) and have previously received antimicrobial treatments [34].

2.2.1. Epidemiology of Enterococcal Infections

E. faecium and *E. faecalis* account for more than 90% of the enterococci recovered from clinical samples in humans. Among them, *E. faecalis* is the most frequent species (80–90%) causing human infection, followed by *E. faecium* (5–10%) and other species (less than 10%) [38]. In recent decades, enterococci have become a first-rate clinical problem, being one of the most common microorganisms of HAIs around the world [39,40]. Several factors related to the host and to the microorganism have contributed to this conversion of a commensal pathogen into one of the major causes of HAIs. The most relevant factors associated with enterococci are their intrinsic resistance to some antimicrobials (e.g., aminoglycosides, cephalosporins and clindamycin); their ability to acquire and disseminate AMR determinants (e.g., linezolid and vancomycin resistance); and the plasticity of their genome, which may contribute to improve their adaptation to harsh environments. Moreover, the increasing number of patients undergoing immunomodulatory therapies, undergoing invasive procedures or receiving multiple antimicrobial treatments, all of which are factors associated with the host, favors the role of enterococci to cause disease [34,41].

2.2.2. Antimicrobial Resistance in Enterococci
Resistance to ß-Lactams

Enterococci are intrinsically resistant to cephalosporins, and they present a natural reduced susceptibility to penicillin due to the expression of low-affinity penicillin binding proteins (PBPs), designated PBP4 in *E. faecalis* and PBP5 in *E. faecium* [34,42]. Moreover, many enterococci strains show tolerance to the bactericidal activity of ß-lactams, with the minimal bactericidal concentrations being higher than the minimum inhibitory concentrations (MICs) [5]. This situation can be solved with the addition of an aminoglycoside (typically streptomycin or gentamicin) to an active ß-lactam, which results in bactericidal synergism [43,44].

A higher-level resistance to penicillin or ampicillin resistance in enterococci can be due to the overexpression of chromosomal PBP4 and PBP5 in *E. faecalis* and *E. faecium*, respectively, or through acquired mechanisms [45–47]. The former is anecdotic, as acquired mechanisms are the most frequent cause of ampicillin resistance. Acquired mechanisms include ß-lactamase production and mutation acquisition in low-affinity PBP4 and PBP5 [48,49]. Currently, ampicillin resistance in enterococci is mainly mediated by the acquisition of mutations in PBP, and it is far more prevalent in *E. faecium* than in *E. faecalis* [49]. Ampicillin-resistant *E. faecium* due to acquired mutations in the PBP5-encoding gene has been linked to a hospital-associated (HA) clade, and it emerged in the late 1970s in the United States (US) [34]. Today, there is a high rate of ampicillin resistance in *E. faecium* strains, and it exceeds 70% in many countries [5,50].

Resistance to Aminoglycosides

Enterococci are intrinsically resistant to clinically achievable concentrations of aminoglycosides due to the poor penetration of these agents through the bacterial cell wall in *E. faecalis* and due to two chromosomally encoded genes, namely 6'-N-aminoglycoside acetyltransferase (*aac(6')-Ii*) and rRNA methyltransferase (*efmM*) in *E. faecium* [5,34]. As previously mentioned, this type of resistance can be overcome with the addition of an agent that disrupts cell wall synthesis, such as ß-lactams. Some strains can also exhibit a high level of aminoglycoside resistance (MIC > 500 mg/L for gentamycin and MIC > 2000 mg/L for streptomycin) through the acquisition of aminoglycoside-modifying enzymes (phosphotransferases, acetyltransferases and nucleotidyltransferases), which inhibit the aforementioned synergic effect [5,51].

Resistance to Glycopeptides

Vancomycin, the main member of the glycopeptide family, was the first-line treatment of ampicillin-resistant *E. faecium* for decades, without reports of VRE strains until the 1980s [52–54]. Glycopeptide resistance in enterococci is mediated by the acquisition of eight different genes of the *van* operon (*vanA*, *vanB*, *vanD*, *vanE*, *vanG*, *vanL*, *vanM* and *vanN*). Moreover, *E. casseliflavus* and *E. gallinarum* exhibit intrinsic low-level resistance to glycopeptides through the presence of a *vanC* gene in their chromosome [55,56]. These genes code for the terminal amino acids of peptidoglycan precursors different from the original form (D-Ala-D-Ala). Thus, the modified amino acids D-Ala-D-Lactate and D-Ala-D-Serine present a lower affinity to glycopeptides, leading to high-level and low-level resistance to glycopeptides, respectively [57]. The *vanA* and *vanB* genes are the main mechanism of resistance to glycopeptides in enterococci, mainly being present in *E. faecium* [55,58]. The prevalence of glycopeptide-resistant *E. faecium* varies widely between continents and countries. Accordingly, the percentage of resistance to glycopeptides in *E. faecium* invasive isolates is more than 60% in the US, 37% in Australia and 16.8% in European countries (with national percentages ranging from 0.0 to 56.6%) [49,50,59–61]. HA ampicillin-resistant *E. faecium* clones often acquire resistance to glycopeptides, highlighting the importance of *E. faecium* as a nosocomial pathogen [62,63]. The scarce active antimicrobials available to treat infections caused by this MDR microorganism are a global cause for concern.

Resistance to Linezolid

Although linezolid resistance in enterococci remains uncommon, the number of linezolid-resistant enterococci (LRE) has increased in recent years. The main mechanism of linezolid resistance in Gram-positive bacteria is point mutations in the central loop of domain V of the *23S rRNA* gene, among which the G2576T (*Escherichia coli* numbering) nucleotide mutation is the most described [64,65]. Other point mutations in the genes *rplC*, *rplD* and *rplV*, which code for the L3, L4 and L22 ribosomal proteins, respectively, are also associated with a decreased susceptibility to linezolid; however, they play a minor role [66,67]. Moreover, the acquisition and dissemination of transferable linezolid resistance genes, namely, *cfr*-like, *optrA* and *poxtA* genes, have been increasingly reported in linezolid-resistant Gram-positive bacteria in recent years [10,67–70]. The *cfr*-like genes encode a 23S rRNA methyltransferase, which confers resistance to phenicols, lincosamides, oxazolidinones, pleuromutilins and streptogramin A (PhLOPS$_A$ phenotype) [71,72]. However, the *optrA* and *poxtA* genes code for the ribosomal protection proteins of the ABC-F family, and they confer resistance to oxazolidinones and phenicols, as well as tetracyclines in the case of *poxtA* [73,74]. Nowadays, the *cfr*, *cfr(B)*, *cfr(D)*, *optrA* and *poxtA* genes have been described among enterococci from different sources (animal, human and environmental samples) and countries [75]. The main linezolid resistance mechanisms are mutations in the 23S rRNA in *E. faecium* and the *optrA* gene in *E. faecalis* [68]. The spread of these transferable linezolid resistance genes to difficult-to-treat bacteria, such as VRE, is a cause for concern. Unfortunately, outbreaks caused by *E. faecium* strains that are resistant to vancomycin and linezolid (*optrA*-positive) have already been reported [76,77].

Resistance to Daptomycin

Daptomycin-resistant enterococci (previously called daptomycin-nonsusceptible enterococci) are uncommon, and they have often been associated with prior exposure to the drug [67]. Daptomycin resistance in enterococci is mainly mediated by structural alterations of the cell envelope through a variety of mutations, mainly in the three-component regulatory system LiaFSR [5,78]. This alteration of the cell envelope produces a repulsion of daptomycin from the membrane. Moreover, daptomycin resistance in *E. faecium* is also associated with mutations in the *cls* gene [49]. Daptomycin resistance is more common in *E. faecium* than in *E. faecalis*, which is probably related to the use of this drug to treat vancomycin-resistant *E. faecium* infections [78].

3. Bacteriophages of *E. faecium* and *E. faecalis*

Bacteriophages, or phages, have emerged in recent years as a potential bioweapon to combat MDR bacteria [79,80]. Phages are viruses that infect and kill bacteria. They are the most abundant entities on Earth and the most genetically diverse biological entities due to their mosaic genome structure and ability to mutate and recombine [81,82]. In addition, they are ubiquitous in all types of environments, from the sea to the human gut [83]. As they are natural predators of bacteria, they have been suggested to be one of the most promising alternative therapeutic agents against MDR bacterial infections [79]. The use of phages as therapeutic agents (phage therapy) was suggested immediately after their discovery in the early XX century by Frederick Twort and Félix d'Herelle. However, the discovery of antibiotics, with a broader spectrum of action, meant that their use declined rapidly [79], except for in some countries in Eastern Europe and in the former Soviet Union where phage therapy was active, as in the Eliava Institute in Georgia [84], a reference center for phage therapy worldwide. The global problem of MDR bacteria and their consequences in terms of lives and health system costs [2] have led to a renewed interest in the study of phages and their application in phage therapy. Moreover, under the umbrella of phage therapy, the use of phages has been proposed in other fields, including food safety [85–87].

In this context, in recent years several enterococcal-infecting phages have been isolated and characterized—genetically and functionally (Table S1). It is remarkable that the number of *E. faecalis*-infecting phages that have been characterized is higher than that of *E. faecium*-infecting phages [88], as, in the last year, the number of *E. faecium* phages has increased. In Table S1, we can see that 101 genomes of *E. faecalis*-infecting phages are available, whereas only 16 of *E. faecium* can be found. Whether or not this bias is related to abundance, the ease of isolation under laboratory conditions or different searching pressures is unclear. It is astonishing that although there is a large number of molecular techniques available for the precise identification of bacterial species, there is still a large number of phage genomes (24) identified as infecting *Enterococcus* spp. (Table S1). However, it is remarkable that some phages are able to infect strains of both species, that is, *E. faecalis* and *E. faecium* [89–91]. This could be considered an advantage if a general phage cocktail designed to treat enterococcal infections is intended.

The *Enterococcus*-infecting phages that have been isolated to date are taxonomically widely diverse, as there are phages belonging to eleven different genera from four families (Figure 1; Table S1). Regarding the genus, the most abundant genera, accounting for almost half of the phages, are *Efquatrovirus* (represented by 39 phages) and *Saphexavirus* (represented by 25 phages), both belonging to the *Siphoviridae* family. Siphoviridae is the most abundant morphology, with three times more isolates than the Myoviridae and Podoviridae morphologies. The genome size distribution has a wide range, from approximately 18 kbp to 150 Kbp, but this heterogeneity is mostly related to taxonomic differences rather than genome diversity. Small genomes are typical of the *Rountreeviridae* family (previously known as *Podoviridae*), whereas large genomes are characteristic of the *Herelleviridae* family (previously known as *Myoviridae*) [92,93]. Nevertheless, within the same genera, heterogeneity in the genome size is observed, thus indicating differences among their members (Table S1).

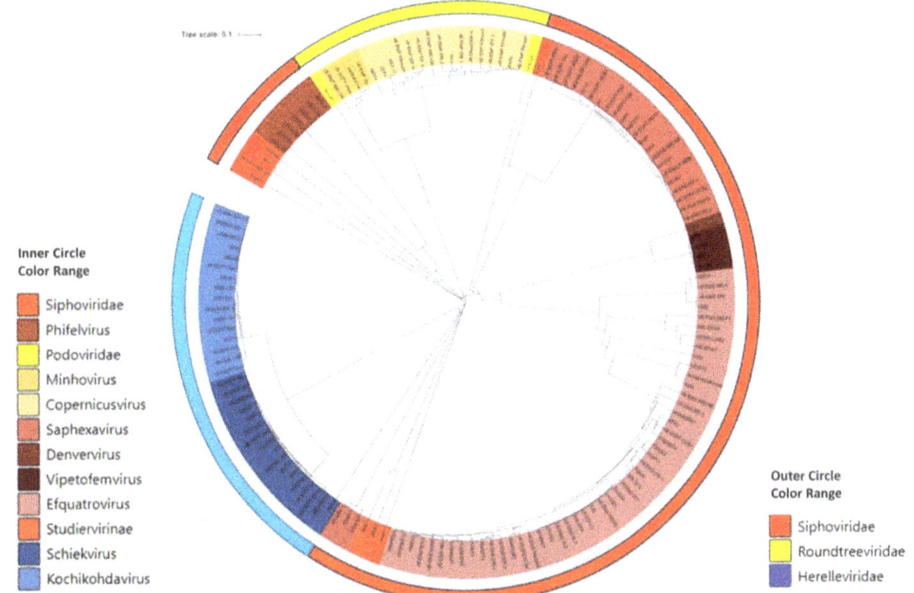

Figure 1. Phylogenetic tree of *Enterococcus*-infecting bacteriophages based on the major capsid proTable S1. The tree was generated by using the unweighted pair group method with arithmetic means (UPGMA) and by employing MAFFT v.7 software (https://mafft.cbrc.jp/alignment/server/ (accessed on 12 January 2023)). The generated phylogenetic tree was visualized using the iTOL web server (https://itol.embl.de/ (accessed on 7 March 2023)).

Most of the reported enterococcal-infecting phages are virulent, at least those included in databases as single entries. In general, temperate phages are described as prophages that are identified and characterized as part of an analysis of one strain or isolate genome [94]. This is linked to the fact that temperate phages are not a good option for phage therapy due to their known involvement in the phenomenon of horizontal gene transfer, one of the main mechanisms involved in the spread of virulence and AMR genes [95]. In addition, after entering a new infecting cell, if a temperate phage undergoes the lysogenic cycle, the infected bacteria do not die; they can continue to spread and become resistant to infection by that same phage. Nevertheless, in some cases, if no alternative exists, temperate phages can be converted into virulent ones by selecting or constructing mutant phage variants [96,97], as has been achieved in the case of the *E. faecalis* ΦEf11 prophage, resulting in an additional increase in host range and progeny [98]. As previously mentioned, the lifestyle (virulent or temperate) of the described phages could be biased by the fact that this is an exclusion criterion for phage therapy due to their putative role in the transference of antimicrobial resistance genes [95]. In this sense, the absence of these genes is also a requirement. Most probably, this is related to the lack of such genes in the genomes of the described phages.

Although enterococci have been documented in many different ecosystems, most of the reported isolation sources are sewage and wastewater (Table S1). This could be related to the role of *Enterococcus* as an indicator of fecal contamination in water [12]. Nevertheless, other sources, in the search for increased phage diversity or specific target applications, have been assayed, such as human stools [89] and cheese [87,88] (Table S1).

4. Food Applications of Enterococcal Bacteriophages

As mentioned in a previous section, *Enterococcus* plays a yin–yang role in foods, where it can be considered a beneficial player responsible for the accumulation of toxic compounds or reservoirs of AMR genes.

E. faecalis has been identified as the main cause of the accumulation of tyramine [23,99] and putrescine [24]—together with *Lactococcus lactis* [100]—in dairy products, one of the food matrixes in which BAs can reach the highest concentration [27,101]. Strategies for the reduction of BAs in food have been proposed, for example, eliminating BAs after they have formed and accumulated via the addition of BA-degrading microorganisms [102–104] or reducing the number of BA-producing microorganisms via different treatments, such as the use of pasteurization or high-pressure technologies [105,106], with the latter being the technical process most employed during food production. However, these methods work by generally reducing the microorganisms present in the food matrix, thus affecting other bacteria that participate in the fermentation process. As enterococci belong to the LAB group, the methodologies applied to reduce their presence also act on other LAB, affecting the development of the organoleptic characteristic of the final product. Thus, tailored measures targeting only the *Enterococcus* population are needed. In this sense, phages infecting *E. faecalis* have been proposed as highly specific tools to reduce the content of BAs in dairy products [88]. The *E. faecalis* Q69 phage has been applied to reduce the presence of tyramine, one of the most toxic BAs found in cheese [107,108], in an experimental cheese model [87]. The phage was added directly to milk [multiplicity of infection (MOI 0.1)] used for cheese making, and after 60 days of ripening, reductions in tyramine concentrations of about 85% were achieved [87], and most importantly, the concentration of tyramine was reduced below the safety threshold level proposed [109]. In another assay, the *E. faecalis*-infecting phage 156 was applied (MOI 0.1) to reduce tyramine and putrescine, another toxic BA frequently found in cheese [21,26,110], taking advantage of the fact that *E. faecalis* is responsible for the accumulation of elevated concentrations of both BAs. In this case, significative reductions in tyramine and putrescine of 95% and 77%, respectively, were achieved [111] after 60 days of ripening. Interestingly, both phages showed the ability to control the population of *E. faecalis* from the early stages of cheese making, and both were partially resistant to the pasteurization process, allowing for both technologies to be applied if desired. The fact that some of the phages proposed as a tool to reduce the content of BAs in dairy products can infect MDR enterococci, including VRE [88,111], suggests that bacteriophages could also be applied to reduce the presence of MDR enterococci in food [112], thus contributing to the One Health strategy's aim of reducing the amount of MDR bacteria in the environment.

Biofilms in the food industry also present a food safety threat, as they can act as reservoirs of foodborne spoilage or pathogenic bacteria [113,114], including BA-producing ones [115]. Different phages have been described as potential tools to eliminate biofilms formed by *E. faecalis* or *E. faecium* [91,116–118]. Although the great potential of enterococcal phages as biofilm elimination agents in food facilities surfaces, to the best of our knowledge, there are no reports regarding this interesting application, thus opening the opportunity for further research.

5. Human Health Applications of Enterococcal Bacteriophages

The continued increase in AMR among enterococci and their abilities to form biofilms and survive in harsh environments have been related to poor clinical outcomes in some cases [119]. Different studies have reported the use of enterococcal phages in the treatment of *Enterococcus* infections through in vitro and in vivo studies, including the use of biofilm, root canal and animal models. However, the use of phage therapy in human patients is limited to some case reports, and, at present, no clinical trials are being carried out.

5.1. In Vitro Models

5.1.1. Biofilm Models

The use of a single phage, a phage cocktail or their combination with antimicrobials in the treatment of bacterial biofilms has been explored in several studies [120,121]. Biofilm-associated infections are related to poor microbiological and clinical outcomes when antimicrobials are used. Phage therapy, in some cases, has been proven to be more effective against MDR biofilm infections than antimicrobials [91]. Several studies have reported the ability of different phages to infect and disrupt biofilms [120,121]. Anti-biofilm activity is mainly tested using microtiter plates with the crystal violet method and confocal laser scanning microscopy [121]. Using these methodologies, the anti-biofilm activity of several phages belonging to different families (*Herelleviridae*: vB_EfaH_EF1TV; *Siphoviridae*: Efa02, EfaS-SRH2, SHEF2 and vB_EfsS_V583; and *Podoviridae*: vB_ZEFP) has been demonstrated against *E. faecalis* [119,122–125]. Moreover, phage therapy has been shown to be a potential weapon against the biofilms formed by MDR *Enterococcus*. In a previous study, it was found that the EFDG1 phage was able to infect vancomycin-resistant *E. faecium* and *E. faecalis* strains. Moreover, EFDG1 significantly reduced a 2-week-old biofilm formed by a vancomycin-resistant *E. faecalis* strain V583 [91]. Furthermore, the vB_EfsS_V583 phage inhibited the biofilm formation by a vancomycin-resistant *E. faecalis* strain for 7 days. However, a poor ability to eradicate mature biofilms was revealed, with significant disruption only being observed in 1- to 2-day-old biofilms [124].

Some authors have developed models closely resembling biofilm formation during in vivo infections to avoid the possible limitations of microtiter plate studies. Thus, El-Atrees et al. studied the effects of the EPA, EPC and EPE phages on *E. faecalis* (EF104, EF134 and EF151) adherence to urinary catheter surfaces [126]. Their findings proved the ability to prevent biofilm formation by reducing the number of cells adhering to the catheter surface to a range of 30.8–43.8%. Moreover, they were also able to eradicate the number of cells pre-adhering to the catheter surface to a range of 48.2–71.1%. Nevertheless, when the anti-biofilm activities of the same phages were evaluated on microtiter plates, they showed more efficacy in both the prevention of biofilm formation and the eradication of the preformed biofilm, achieving ranges of 38–39.9% and 71–78.4%, respectively [126]. In a similar study, silicone Foley catheters were covered with an *E. faecalis* biofilm, and then they were exposed to the vB_EfaS-271 phage for 3, 6 or 24 h. A significant decrease in the number of viable *E. faecalis* cells was observed after three hours when a higher MOI was used. However, lower MOI ratios needed a longer time (6 h) for considerable effects to be observed. Unexpected results were observed at 24 h, with a large number of *E. faecalis* cells surviving in samples treated with 10 MOI compared to those treated with 0.0001 or 0.01 MOI. The authors speculated that a greater selection of phage-resistant mutants could occur under high-MOI conditions. However, such mutants seemed to be less competitive than wild-type cells [127]. In another approximation of biofilm formation during in vivo infections, Melo et al. developed an in vitro collagen wound model (CWM) of biofilm formation with two phages: vB_EfaS-Zip (*Rountreeviridae*) and vB_EfaP-Max (*Siphoviridae*). Both phages showed lytic activity against *E. faecium* and *E. faecalis*. In the CWM, vB_EfaP-Max and vB_EfaS-Zip were able to reduce the number of viable cells of *E. faecalis* and *E. faecium*, respectively, during the first eight hours. However, in both cases, the number of cells in the control and phage-treated biofilms in the CWM was similar at 24 h. In a new CWM, a cocktail comprising the two phages was used to infect a dual-species (*E. faecium* and *E. faecalis*) biofilm. In this last assay, a statistically significant reduction in the concentrations of the cells in the treated biofilms was observed at 3, 6 and 8 h compared to those in the control, and a residual reduction was also detected at 24 h. The emergence of phage resistance might be related to the loss of or a reduction in anti-biofilm activity when a phage alone or a cocktail phage is applied, respectively [90]. Although the selection of phage-resistant mutants is an issue in phage therapy, in some cases, this selection may be an opportunity. Liu et al. described the strong lytic activity of the EFap02 phage against the *E. faecalis* strain Efa02 and identified the glycosyltransferase gene Group 2 (*gtr2*) as its receptor. Unfortunately, the rapid emergence of phage-resistant mutants was observed by the authors. The phage-resistant strain EFa02R had loss-of-function mutations in the *gtr2* gene, responsible for the biosynthesis of capsular polysaccharides. Not only does the loss of receptors in EFa02R prevent phage adsorption, but it also impairs the biofilm formation ability of these mutants. Therefore, capsular polysaccharide loss could revert the inactivation of some antimicrobials caused by the biofilm [125].

A phage alone or a phage cocktail combined with antimicrobials could prevent the emergence of phage resistance and enhance their activities separately [128]. Phage–antimicrobial synergy has been reported, even against bacteria resistant to the antimicrobial used in the combination [129]. Similar to this, daptomycin plus a phage cocktail (113 and 9184) showed synergic bactericidal activity against daptomycin-nonsusceptible *E. faecium* [130]. Likewise, phage–antimicrobial synergy against vancomycin-resistant *E. faecalis* V583 was also observed when a combined treatment of vancomycin and the EFLK1 phage was applied. This combination was able to reduce viable bacterial counts by nearly 8 logs in a well-established biofilm, whereas treatment with the phage alone only achieved a reduction of 4 logs, and vancomycin alone failed entirely [131].

5.1.2. Human Root Canal Model (Ex Vivo)

E. faecalis is frequently detected in asymptomatic and persistent endodontic infections, with prevalence ranging from 24 to 77% [132]. Its ability to invade the dentinal tubes of the root canal walls of human teeth, their aforementioned ability to form biofilms and their ability to survive in harsh environments allow this microorganism to cause persistent infections, and it is difficult to treat them [121,132,133]. The treatment of these infections includes mechanical debridement and chemical agents, such as chlorhexidine and sodium hypochlorite (NaOCl), which are generally effective [132,133]. However, several treatment failures have been reported, and therefore, the development of new therapeutic alternatives is necessary [132]. In this context, enterococcal phage therapy has been studied in ex vivo models of root canal infection. In a previous study, it was found that the EFDG1 phage was able to prevent *E. faecalis* root canal infection in an ex vivo model performed with human-extracted teeth [91]. Likewise, the vB_ZEFP phage showed a greater ability to reduce bacterial leakage from the root apex than other treatments (NaOCl and NaOCl plus EDTA) [123]. Similar studies have demonstrated the efficacy of different phages to destroy *E. faecalis* biofilms in root canal systems: vB_Efa29212_2e (*Siphoviridae*), vB_Efa29212_3e (*Herelleviriae*) and vB_EfaS_HEf13 (*Siphoviridae*) [134–137]. With regard to MDR *Enterococcus*, Tinoco et al. evaluated the activity of ΦEf11/ΦFL1C(Δ36)P^{nisA}, an engineered phage, against vancomycin-resistant *E. faecalis* V583 in an ex vivo model of root canal [138]. The treatment with the phage generated a reduction of 99% for the V583-infected models. In contrast, a scarce reduction of 18% was observed in biofilms formed by the *E. faecalis* JH2-2 strain (a fusidic acid- and rifampicin-resistant, vancomycin-susceptible strain) [138].

5.2. In Vivo Models

In vivo studies using animal models are essential to evaluate the safety and efficiency of new therapies, including phage therapy. Among them, the most frequently used include models performed in *Galleria mellonella*, zebrafish embryos and mice [139]. The *G. mellonella* animal model has previously been used to assess the virulence of VRE [89]. Although this model is cost-effective in the evaluation of the potential of phage therapy, to date, only two studies have assessed the efficacy of phage therapy against larvae infected by *Enterococcus* [89,140]. In the first study, the administration of a phage cocktail comprising the MDA1 (*Rountreeviridae*) and MDA2 (*Herelleviridae*) phages was effective against larvae infected with a vancomycin-resistant *E. faecium* strain (VRE004). The larvae were injected with 10^7 colony-forming units (CFU)/10 μL of the VRE004 strain, and two groups were employed. These groups were injected with the phage cocktail at a concentration of 2×10^6 plaque-forming units (PFU)/10^6 μL, one of them 1 h prior to (the prophylactic group) and the other 1 h after (the treatment group) the VRE injection. After 48 h of follow-up, both groups demonstrated efficacy, being 3.7 (the treatment group) and 6.5 (the prophylactic group) times more likely to survive than the larvae injected with VRE only [140]. In the second study, the activity of the phage vB_EfaH_163 (*Herelleviridae*) against larvae infected with a vancomycin-resistant *E. faecium* (VRE-13) strain was studied. The larvae were injected with the VRE-13 strain at a concentration of 10^5 CFU/larva. After 1 h, the larvae were injected with PBS (the control group) or a phage suspension at an MOI of 0.1, and the number of deaths was monitored for five days. Treatment with the vB_EfaH_163 phage increased larval survival by 20% compared with the control group, although no statistically significant differences were observed [89]. In another assay, the therapeutic potential of phage SHEF2 (*Siphoviridae*) in treating systemic *E. faecalis* infections in an in vivo zebrafish embryo infection model was studied. In this model, the zebrafish embryos were infected with *E. faecalis* (OS16 strain), and two hours later, they were injected with SHEF2 or a heat-killed sample of SHEF2 at an MOI of 20 (with respect to the *E. faecalis* inoculum). The zebrafish infected with OS16 alone or with heat-killed SHEF2 showed a mortality rate of 73%, whereas those injected with SHEF2 showed a mortality rate of 16% ($p < 0.0001$) [119]. Nowadays, a number of studies are evaluating enterococcal phage therapy against different kinds of *E. faecalis* infection using mice models. Enterococcal phage therapy has been shown to be a promising candidate for the treatment of *E. faecalis* endophthalmitis (a rare cause of postoperative infection) in mice models. Thus, Kishimoto et al. demonstrated a decrease in the number of viable bacteria and the infiltration of neutrophils in mice eyes infected with vancomycin-susceptible and -resistant *E. faecalis* when they were treated with phages [ΦEF24C-P2, ΦEF7H, ΦEF14H1 and ΦEF19G (*Herelleviridae*)] [141,142]. Several studies have evaluated the efficacy of phage therapy in an *E. faecalis* sepsis mice model using intraperitoneal injections. The intraperitoneal administration of phage IME-EF1 or its endolysin, at a 10 MOI, 30 min after *E. faecalis* 002 inoculation resulted in survival rates of 60% and 80%, respectively [143]. A single injection of 3×10^8 PFU of the phage ENB6, administered 45 min after a vancomycin-resistant *E. faecium* (CRMEN 44) challenge, was able to rescue 100% of the mice [144]. A single intraperitoneal administration of other phages (at different doses and times following bacterium inoculation) was also enough to protect all the mice infected with enterococcus (including VRE): ΦEF24C (MOI 0.01/20 min), EF-P29 (4×10^5 PFU/1 h), SSsP-1 and GVEsP-1 (3×10^9 phage stock/3 h) and a phage cocktail (comprising the phages EFDG1 and EFLK1) (2×10^8 PFU/0 h and 1 h) [145–148].

5.3. Phage Therapy in Humans

To date, only a few case reports have described the use of phage therapy against enterococcus infections in humans (Table 1) [149–153]. Three patients suffering from chronic bacterial prostatitis caused by *E. faecalis*, previously unsuccessfully treated with long-term targeted antimicrobials, autovaccines and laser bio-stimulation, were selected for phage therapy. Phage treatment was rectally applied, twice daily, with 10 mL of bacterial phage lysate (with a phage titter between 10^7 and 10^9 PFU/mL) for 30 days. Encouraging results were obtained in the three patients regarding bacterial eradication, the abatement of clinical symptoms and the lack of early disease recurrence [152]. In another case of chronic bacterial prostatitis, this time polymicrobial (with different staphylococcal species, *E. faecalis* and *Streptococcus mitis*), three phage preparations from Eliava Institute (Pyo, Intesti and Staphylococcal Bacteriophage preparations), with an approximate phage titter between 10^5 and 10^7 PFU/mL, were used. These preparations were applied via three routes: the oral route (20 mL of Pyo and Intesti Bacteriophage per day for 14 days), the rectal route (Staphylococcal Bacteriophage suppositories twice a day for 10 days) and the urethral route (Intesti Bacteriophage instillations once a day for 10 days). A significant improvement in symptoms was observed after phage therapy, and the patient was considered in full remission [149]. A commercial preparation of Pyo Bacteriophage (with an unknown phage titter) (Eliava Institute, Tbilisi, Georgia) was also applied in a case of a recurrent femur osteomyelitis infection after multiple failed medical and surgical therapy regimens. The infected bone was rinsed with 40 mL of the phage solution after the debridement surgery (intraoperative), followed three times per day with 10–20 mL for seven days using a draining system (the draining system was closed to allow a contact time of 10 min). The patient was concomitantly treated with amoxicillin for three months. A follow-up of the patient after eight months showed no signs of clinical or radiological recurrence, and the patient was considered infection-free [151]. Pyo and Intesti Bacteriophages (with an unknown phage titter) were also used in a case of an *E. faecalis* hip prosthetic joint infection. In this case, the phages were administered orally (10 mL), with the first one administered in the morning and the second one administered in the evening, for two periods of 19 days with a pause of 2 weeks between both periods. The patient was treated concomitantly with amoxicillin in the first phage treatment period and doxycycline in the second one. Antimicrobial therapy was suspended after a final course of doxycycline for four months. Over the next two years, the patient recovered and had no hip complaints [153]. Lastly, Paul et al. reported the first case of a vancomycin-resistant *E. faecium* abdominal infection treated with intravenous injections of cocktail phages (comprising the EFgrKN and EFgrNG phages, with a joint titter between 10^7 and 10^8 PFU/mL). Phage therapy was intravenously administered (2 mL/Kg/12 h over 2 h) for 20 days in a one-year-old girl, critically ill, needing three successive liver transplants with a persistent vancomycin-resistant *E. faecium* infection. Although the disease course was complex, the authors linked the clinical improvement to the phage application [150].

Table 1. Characteristics of available studies of phage therapy against *Enterococcus* in humans.

Type of Infection and No of Subjects (n)	Target Strain	Phage	Application Route	Concomitant Antimicrobial Use	Outcomes	Reference
Chronic bacterial prostatitis (n = 3)	*E. faecalis*	No data	Rectal	No	Bacterial eradication Abatement of symptoms Lack of early disease recurrence	[146]
Chronic bacterial prostatitis (n = 1)	*E. faecalis* [a]	Pyo [b], Intesti [b] and Staphylococcal bacteriophage [b]	Oral, rectal and urethral	No	Bacterial eradication Significant improvement in symptoms	[147]
Femur osteomyelitis (n = 1)	*E. faecalis*	Pyo bacteriophage [b]	Direct rise of the infection site	Yes (amoxicillin)	No signs of clinical or radiological recurrence	[148]
Hip prosthetic joint infection (n = 1)	*E. faecalis*	Pyo [b] and Intesti bacteriophage [b]	Oral	Yes (amoxicillin and doxycycline)	Not hip complaints	[149]
Intrabdominal infection (n=1)	VR *E. faecium*	EFgrKN and EFgrNG	Intravenous	Yes (linezolid)	Clinical improvement	[150]

VR: vancomycin-resistant. [a] Polymicrobial infection which include members of the Staphylococcal species (including *S. aureus*), *Streptococcus mitis* and *E. faecalis*. [b] Standard phage preparations made by Eliava Institute (Tbilisi, Georgia).

As can be seen in these cases, the use of phage therapy carried out on a compassionate-use basis has provided encouraging results in the healing of patients who suffered from difficult-to-treat infections and did not have other alternative treatment options. To date, clinical trials have assessed the effect of phage therapy on a few pathogens or different types of infections (e.g., urinary tract infections) with discordant results, but no clinical trials focusing on Enterococcus have been performed [120,154]. The safety and feasibility of phage therapy, including when it is administrated intravenously, have been demonstrated in several reports [155,156]. Therefore, the development of double-blind randomized clinical trials is necessary to assess the true efficacy of phage therapy in human health. Moreover, these studies must answer some questions and assess the contribution of various factors to the outcome of phage therapy, such as the quality of phage preparation; their titter, dosage and route of administration; and the concomitant use of antimicrobials.

6. Conclusions

AMR in pathogenic microorganisms is a major global threat and a leading and increasing cause of mortality worldwide. The interest in the possibilities of phage therapy as a new weapon to fight AMR microorganisms has boosted research on bacteriophages, particularly that on *Enterococcus*-infecting phages. As shown in this review, there are a large number of characterized phages that fulfill the requirements for application in phage therapy and feed. However, the lack of legislation regulating their use in food limits their possible application in most countries. Similarly, the lack of procedures, as well as the lack of a definition of good production practices for phage suspensions, limits their application to compassionate use in patients for whom there is no other treatment alternative. Thus, a step forward is still needed to standardize procedures that allow for their systematic use in practical clinical applications beyond compassionate use.

Supplementary Materials: The following is available online at https://www.mdpi.com/article/10.3 390/antibiotics12050842/s1, Table S1: Complete bacteriophage genomes used for the construction of the phylogenetic tree.

Author Contributions: Conceptualization, writing—original draft and writing—review and editing, C.R.-L. and V.L. All authors have read and agreed to the published version of the manuscript.

Funding: This research was funded by the Plan for Science, Technology and Innovation of the Principality of Asturias 2018–2022, co-financed by FEDER (AYUD/2021/50916).

Institutional Review Board Statement: Not applicable.

Informed Consent Statement: Not applicable.

Data Availability Statement: Data are contained within the article.

Conflicts of Interest: The authors declare no conflict of interest.

References

1. Mestrovic, T.; Aguilar, G.R.; Swetschinski, L.R.; Ikuta, K.S.; Gray, A.P.; Weaver, N.D.; Han, C.; Wool, E.E.; Hayoon, A.G.; Hay, S.I.; et al. The burden of bacterial antimicrobial resistance in the WHO European region in 2019: A cross-country systematic analysis. *Lancet Public Health* **2022**, *7*, e897–e913. [CrossRef]
2. O'Neill, J. *Tackling Drug-Resistant Infections Globally: Final Report and Recomendations*; HM Government: London, UK, 2016.
3. Mulani, M.S.; Kamble, E.E.; Kumkar, S.N.; Tawre, M.S.; Pardesi, K.R. Emerging Strategies to Combat ESKAPE Pathogens in the Era of Antimicrobial Resistance: A Review. *Front. Microbiol.* **2019**, *10*, 539. [CrossRef] [PubMed]
4. Gaca, A.O.; Lemos, J.A. Adaptation to Adversity: The Intermingling of Stress Tolerance and Pathogenesis in Enterococci. *Microbiol. Mol. Biol. Rev.* **2019**, *83*, e00008-19. [CrossRef] [PubMed]
5. García-Solache, M.; Rice, L.B. The Enterococcus: A Model of Adaptability to Its Environment. *Clin. Microbiol. Rev.* **2019**, *32*, e00058-18. [CrossRef]
6. Ramos, S.; Silva, V.; Dapkevicius, M.L.E.; Igrejas, G.; Poeta, P. Enterococci, from Harmless Bacteria to a Pathogen. *Microorganisms* **2020**, *8*, 1118. [CrossRef] [PubMed]
7. Zaheer, R.; Cook, S.R.; Barbieri, R.; Goji, N.; Cameron, A.; Petkau, A.; Polo, R.O.; Tymensen, L.; Stamm, C.; Song, J.; et al. Surveillance of *Enterococcus* spp. reveals distinct species and antimicrobial resistance diversity across a One-Health continuum. *Sci. Rep.* **2020**, *10*, 3937. [CrossRef] [PubMed]
8. Ladero, V.; Fernandez, M.; Alvarez, M.A. Isolation and identification of tyramine-producing enterococci from human fecal samples. *Can. J. Microbiol.* **2009**, *55*, 215–218. [CrossRef]
9. Kim, S.-H.; Kim, D.-H.; Lim, H.-W.; Seo, K.-H. High prevalence of non-faecalis and non-faecium Enterococcus spp. in farmstead cheesehouse and their applicability as hygiene indicators. *LWT* **2020**, *126*, 109271. [CrossRef]
10. Rodríguez-Lucas, C.; Fernández, J.; Vázquez, X.; de Toro, M.; Ladero, V.; Fuster, C.; Rodicio, R.; Rodicio, M.R. Detection of the *optrA* Gene Among Polyclonal Linezolid-Susceptible Isolates of *Enterococcus faecalis* Recovered from Community Patients. *Microb. Drug Resist.* **2022**, *28*, 773–779. [CrossRef]
11. Vec, P.; Franz, C.M.A.P. The Family Enterococcaceae. In *Lactic Acid Bacteria: Biodiversity and Taxonomy*; Holzapfel, W.H., Wood, B.B.J., Eds.; John Wiley & Sons, Incorporated: Berlin/Heidelberg, Germany, 2014; pp. 171–244, ISBN 9781118655252.

12. Gowda, H.N.; Kido, H.; Wu, X.; Shoval, O.; Lee, A.; Lorenzana, A.; Madou, M.; Hoffmann, M.; Jiang, S.C. Development of a proof-of-concept microfluidic portable pathogen analysis system for water quality monitoring. *Sci. Total Environ.* **2022**, *813*, 152556. [CrossRef]

13. Jiménez, E.; Ladero, V.; Chico, I.; Maldonado-Barragán, A.; López, M.; Martín, V.; Fernández, L.; Fernández, M.; Álvarez, A.M.; Torres, C.; et al. Antibiotic resistance, virulence determinants and production of biogenic amines among enterococci from ovine, feline, canine, porcine and human milk. *BMC Microbiol.* **2013**, *13*, 288. [CrossRef] [PubMed]

14. Martín-Platero, A.M.; Valdivia, E.; Maqueda, M.; Martínez-Bueno, M. Characterization and safety evaluation of enterococci isolated from Spanish goats' milk cheeses. *Int. J. Food Microbiol.* **2009**, *132*, 24–32. [CrossRef] [PubMed]

15. Khan, H.; Flint, S.; Yu, P.-L. Enterocins in food preservation. *Int. J. Food Microbiol.* **2010**, *141*, 1–10. [CrossRef]

16. Hanchi, H.; Mottawea, W.; Sebei, K.; Hammami, R. The Genus Enterococcus: Between Probiotic Potential and Safety Concerns—An Update. *Front. Microbiol.* **2018**, *9*, 1791. [CrossRef] [PubMed]

17. Krawczyk, B.; Wityk, P.; Gałęcka, M.; Michalik, M. The Many Faces of *Enterococcus* spp.—Commensal, Probiotic and Opportunistic Pathogen. *Microorganisms* **2021**, *9*, 1900. [CrossRef]

18. Moreno, M.F.; Sarantinopoulos, P.; Tsakalidou, E.; De Vuyst, L. The role and application of enterococci in food and health. *Int. J. Food Microbiol.* **2006**, *106*, 1–24. [CrossRef] [PubMed]

19. Giraffa, G. Functionality of enterococci in dairy products. *Int. J. Food Microbiol.* **2003**, *88*, 215–222. [CrossRef]

20. Giraffa, G. Enterococci from foods. *FEMS Microbiol. Rev.* **2002**, *26*, 163–171. [CrossRef]

21. Wu, Y.; Pang, X.; Wu, Y.; Liu, X.; Zhang, X. Enterocins: Classification, Synthesis, Antibacterial Mechanisms and Food Applications. *Molecules* **2022**, *27*, 2258. [CrossRef] [PubMed]

22. Ladero, V.; Calles-Enriquez, M.; Fernandez, M.; Alvarez, M.A. Toxicological Effects of Dietary Biogenic Amines. *Curr. Nutr. Food Sci.* **2010**, *6*, 145–156. [CrossRef]

23. Ladero, V.; Fernández, M.; Calles-Enríquez, M.; Sánchez-Llana, E.; Cañedo, E.; Martín, M.C.; Alvarez, M.A. Is the production of the biogenic amines tyramine and putrescine a species-level trait in enterococci? *Food Microbiol.* **2012**, *30*, 132–138. [CrossRef]

24. Ladero, V.; Martínez, N.; Martín, M.C.; Fernández, M.; Alvarez, M.A. qPCR for quantitative detection of tyramine-producing bacteria in dairy products. *Food Res. Int.* **2010**, *43*, 289–295. [CrossRef]

25. Ladero, V.; Cañedo, E.; Pérez, M.; Martin, M.C.; Fernandez, M.; Alvarez, M.A. Multiplex qPCR for the detection and quantification of putrescine-producing lactic acid bacteria in dairy products. *Food Control* **2012**, *27*, 307–313. [CrossRef]

26. O'sullivan, D.J.; Fallico, V.; O'sullivan, O.; McSweeney, P.L.H.; Sheehan, J.J.; Cotter, P.D.; Giblin, L. High-throughput DNA sequencing to survey bacterial histidine and tyrosine decarboxylases in raw milk cheeses. *BMC Microbiol.* **2015**, *15*, 266. [CrossRef]

27. Fernández, M.; Linares, D.M.; del Río, B.; Ladero, V.; Alvarez, M.A. HPLC quantification of biogenic amines in cheeses: Correlation with PCR-detection of tyramine-producing microorganisms. *J. Dairy Res.* **2007**, *74*, 276–282. [CrossRef]

28. EFSA Panel on Biological Hazards (BIOHAZ). EFSA Scientific Opinion on risk based control of biogenic amine formation in fermented foods. *EFSA J.* **2011**, *9*, 2393–2486. [CrossRef]

29. Hernando-Amado, S.; Coque, T.M.; Baquero, F.; Martínez, J.L. Defining and combating antibiotic resistance from One Health and Global Health perspectives. *Nat. Microbiol.* **2019**, *4*, 1432–1442. [CrossRef]

30. Economou, V.; Gousia, P. Agriculture and food animals as a source of antimicrobial-resistant bacteria. *Infect. Drug Resist.* **2015**, *8*, 49–61. [CrossRef]

31. Gaglio, R.; Couto, N.; Marques, C.; Lopes, M.D.F.S.; Moschetti, G.; Pomba, C.; Settanni, L. Evaluation of antimicrobial resistance and virulence of enterococci from equipment surfaces, raw materials, and traditional cheeses. *Int. J. Food Microbiol.* **2016**, *236*, 107–114. [CrossRef] [PubMed]

32. Silvetti, T.; Morandi, S.; Brasca, M. Does *Enterococcus faecalis* from Traditional Raw Milk Cheeses Serve as a Reservoir of Antibiotic Resistance and Pathogenic Traits? *Foodborne Pathog. Dis.* **2019**, *16*, 359–367. [CrossRef]

33. Klein, G.; Pack, A.; Reuter, G. Antibiotic Resistance Patterns of Enterococci and Occurrence of Vancomycin-Resistant Enterococci in Raw Minced Beef and Pork in Germany. *Appl. Environ. Microbiol.* **1998**, *64*, 1825–1830. [CrossRef] [PubMed]

34. Lawpidet, P.; Tengjaroenkul, B.; Saksangawong, C.; Sukon, P. Global Prevalence of Vancomycin-Resistant Enterococci in Food of Animal Origin: A Meta-Analysis. *Foodborne Pathog. Dis.* **2021**, *18*, 405–412. [CrossRef]

35. Arias, C.A.; Murray, B.E. The rise of the Enterococcus: Beyond vancomycin resistance. *Nat. Rev. Microbiol.* **2012**, *10*, 266–278. [CrossRef] [PubMed]

36. Maccallum, W.G.; Hastings, T.W. A case of acute endocarditis caused by micrococcus zymogenes (Nov. spec.), with a description of the microorganism. *J. Exp. Med.* **1899**, *4*, 521–534. [CrossRef]

37. Frobisher, M.; Denny, E.R. A study of micrococcus zymogenes. *J. Bacteriol.* **1928**, *16*, 301–314. [CrossRef]

38. Murray, B.E. The life and times of the Enterococcus. *Clin. Microbiol. Rev.* **1990**, *3*, 46–65. [CrossRef]

39. Růžičková, M.; Vítězová, M.; Kushkevych, I. The characterization of *Enterococcus* genus: Resistance mechanisms and inflammatory bowel disease. *Open Med.* **2020**, *15*, 211–224. [CrossRef] [PubMed]

40. Ruiz-Garbajosa, P.; Regt, M.; Bonten, M.; Baquero, F.; Coque, T.M.; Cantón, R.; Harmsen, H.J.M.; Willems, R.J.L. High-density fecal *Enterococcus faecium* colonization in hospitalized patients is associated with the presence of the polyclonal subcluster CC17. *Eur. J. Clin. Microbiol. Infect. Dis.* **2012**, *31*, 519–522. [CrossRef]

41. Mendes, R.E.; Castanheira, M.; Farrell, D.J.; Flamm, R.K.; Sader, H.S.; Jones, R.N. Longitudinal (2001–14) analysis of enterococci and VRE causing invasive infections in European and US hospitals, including a contemporary (2010–13) analysis of oritavancin in vitro potency. *J. Antimicrob. Chemother.* **2016**, *71*, 3453–3458. [CrossRef]
42. Malani, P.N.; Kauffman, C.A.; Zervos, M.J. Enterococcal disease, epidemiology, and treatment. In *The Enterococci*; ASM Press: Washington, DC, USA, 2014; pp. 385–408.
43. Fontana, R.; Ligozzi, M.; Pittaluga, F.; Satta, G. Intrinsic Penicillin Resistance in Enterococci. *Microb. Drug Resist.* **1996**, *2*, 209–213. [CrossRef]
44. Bisno, A.L. Antimicrobial Treatment of Infective Endocarditis due to Viridans Streptococci, Enterococci, and Staphylococci. *J. Am. Med. Assoc.* **1989**, *261*, 1471. [CrossRef]
45. Moellering, R.C.; Wennersten, C.; Weinberg, A.N. Synergy of Penicillin and Gentamicin against Enterococci. *J. Infect. Dis.* **1971**, *124*, S207–S213. [CrossRef]
46. Zhang, X.; Paganelli, F.L.; Bierschenk, D.; Kuipers, A.; Bonten, M.J.M.; Willems, R.J.L.; van Schaik, W. Genome-Wide Identification of Ampicillin Resistance Determinants in *Enterococcus faecium*. *PLoS Genet.* **2012**, *8*, e1002804. [CrossRef]
47. Duez, C.; Zorzi, W.; Sapunaric, F.; Amoroso, A.; Thamm, I.; Coyette, J. The penicillin resistance of Enterococcus faecalis JH2-2r results from an overproduction of the low-affinity penicillin-binding protein PBP4 and does not involve a psr-like gene. *Microbiology* **2001**, *147*, 2561–2569. [CrossRef]
48. Infante, V.H.P.; Conceição, N.; de Oliveira, A.G.; Darini, A.L.D.C. Evaluation of polymorphisms in pbp4 gene and genetic diversity in penicillin-resistant, ampicillin-susceptible *Enterococcus faecalis* from hospitals in different states in Brazil. *FEMS Microbiol. Lett.* **2016**, *363*, fnw044. [CrossRef] [PubMed]
49. Murray, B.E. Beta-lactamase-producing enterococci. *Antimicrob. Agents Chemother.* **1992**, *36*, 2355–2359. [CrossRef] [PubMed]
50. Prieto, A.M.G.; van Schaik, W.; Rogers, M.R.C.; Coque, T.M.; Baquero, F.; Corander, J.; Willems, R.J.L. Global Emergence and Dissemination of Enterococci as Nosocomial Pathogens: Attack of the Clones? *Front. Microbiol.* **2016**, *7*, 788. [CrossRef]
51. ECDC. Antimicrobial resistance in the EU in 2012. *Vet. Rec.* **2014**, *174*, 341. [CrossRef]
52. Hegstad, K.; Mikalsen, T.; Coque, T.M.; Werner, G.; Sundsfjord, A. Mobile genetic elements and their contribution to the emergence of antimicrobial resistant Enterococcus faecalis and *Enterococcus faecium*. *Clin. Microbiol. Infect.* **2010**, *16*, 541–554. [CrossRef]
53. Leclercq, R.; Derlot, E.; Duval, J.; Courvalin, P. Plasmid-Mediated Resistance to Vancomycin and Teicoplanin in *Enterococcus faecium*. *N. Engl. J. Med.* **1988**, *319*, 157–161. [CrossRef]
54. Uttley, A.H.C.; George, R.C.; Naidoo, J.; Woodford, N.; Johnson, A.P.; Collins, C.H.; Morrison, D.; Gilfillan, A.J.; Fitch, L.E.; Heptonstall, J. High-level vancomycin-resistant enterococci causing hospital infections. *Epidemiol. Infect.* **1989**, *103*, 173–181. [CrossRef] [PubMed]
55. Kirst, H.A.; Thompson, D.G.; Nicas, T.I. Historical Yearly Usage of Vancomycin. *Antimicrob. Agents Chemother.* **1998**, *42*, 1303–1304. [CrossRef] [PubMed]
56. Cattoir, V.; Leclercq, R. Twenty-five years of shared life with vancomycin-resistant enterococci: Is it time to divorce? *J. Antimicrob. Chemother.* **2013**, *68*, 731–742. [CrossRef]
57. Murray, B.E. Vancomycin-Resistant Enterococcal Infections. *N. Engl. J. Med.* **2000**, *342*, 710–721. [CrossRef]
58. Cetinkaya, Y.; Falk, P.; Mayhall, C.G. Vancomycin-Resistant Enterococci. *Clin. Microbiol. Rev.* **2000**, *13*, 686–707. [CrossRef] [PubMed]
59. Torres, C.; Alonso, C.A.; Ruiz-Ripa, L.; Leon-Sampedro, R.; Del Campo, R.; Coque, T.M.; León-Sampedro, R.; Del Campo, R.; Coque, T.M. Antimicrobial Resistance in Enterococcus spp. of animal origin. *Microbiol. Spectr.* **2018**, *6*, 185–227. [CrossRef]
60. Greenstein, R.J.; Collins, M.T. Emerging pathogens: Is Mycobacterium avium subspecies paratuberculosis zoonotic? *Lancet* **2004**, *364*, 396–397. [CrossRef]
61. Weiner-Lastinger, L.M.; Abner, S.; Benin, A.L.; Edwards, J.R.; Kallen, A.J.; Karlsson, M.; Magill, S.S.; Pollock, D.; See, I.; Soe, M.M.; et al. Antimicrobial-resistant pathogens associated with pediatric healthcare-associated infections: Summary of data reported to the National Healthcare Safety Network, 2015–2017. *Infect. Control Hosp. Epidemiol.* **2020**, *41*, 19–30. [CrossRef]
62. Rodríguez-Lucas, C.; Fernández, J.; Raya, C.; Bahamonde, A.; Quiroga, A.; Muñoz, R.; Rodicio, M.R. Establishment and Persistence of Glycopeptide-Resistant *Enterococcus faecium* ST80 and ST117 Clones Within a Health Care Facility Located in a Low-Prevalence Geographical Region. *Microb. Drug Resist.* **2022**, *28*, 217–221. [CrossRef]
63. Leavis, H.L.; Bonten, M.J.; Willems, R.J. Identification of high-risk enterococcal clonal complexes: Global dispersion and antibiotic resistance. *Curr. Opin. Microbiol.* **2006**, *9*, 454–460. [CrossRef]
64. Tedim, A.P.; Ruíz-Garbajosa, P.; Rodríguez, M.C.; Rodríguez-Baños, M.; Lanza, V.F.; Derdoy, L.; Zurita, G.C.; Loza, E.; Cantón, R.; Baquero, F.; et al. Long-term clonal dynamics of *Enterococcus faecium* strains causing bloodstream infections (1995–2015) in Spain. *J. Antimicrob. Chemother.* **2017**, *72*, 48–55. [CrossRef]
65. Mendes, R.E.; Deshpande, L.; Streit, J.M.; Sader, H.S.; Castanheira, M.; Hogan, P.A.; Flamm, R.K. ZAAPS programme results for 2016: An activity and spectrum analysis of linezolid using clinical isolates from medical centres in 42 countries. *J. Antimicrob. Chemother.* **2018**, *73*, 1880–1887. [CrossRef] [PubMed]
66. Pfaller, M.A.; Mendes, R.E.; Streit, J.M.; Hogan, P.A.; Flamm, R.K. Five-Year Summary of In Vitro Activity and Resistance Mechanisms of Linezolid against Clinically Important Gram-Positive Cocci in the United States from the LEADER Surveillance Program (2011 to 2015). *Antimicrob. Agents Chemother.* **2017**, *61*, e00609-17. [CrossRef]

67. Long, K.S.; Vester, B. Resistance to Linezolid Caused by Modifications at Its Binding Site on the Ribosome. *Antimicrob. Agents Chemother.* **2012**, *56*, 603–612. [CrossRef] [PubMed]

68. Bender, J.K.; Cattoir, V.; Hegstad, K.; Sadowy, E.; Coque, T.M.; Westh, H.; Hammerum, A.M.; Schaffer, K.; Burns, K.; Murchan, S.; et al. Update on prevalence and mechanisms of resistance to linezolid, tigecycline and daptomycin in enterococci in Europe: Towards a common nomenclature. *Drug Resist. Updat.* **2018**, *40*, 25–39. [CrossRef]

69. Deshpande, L.M.; Castanheira, M.; Flamm, R.K.; Mendes, R.E. Evolving oxazolidinone resistance mechanisms in a worldwide collection of enterococcal clinical isolates: Results from the SENTRY Antimicrobial Surveillance Program. *J. Antimicrob. Chemother.* **2018**, *73*, 2314–2322. [CrossRef] [PubMed]

70. Elghaieb, H.; Freitas, A.R.; Abbassi, M.S.; Novais, C.; Zouari, M.; Hassen, A.; Peixe, L. Dispersal of linezolid-resistant enterococci carrying poxtA or optrA in retail meat and food-producing animals from Tunisia. *J. Antimicrob. Chemother.* **2019**, *74*, 2865–2869. [CrossRef]

71. Moure, Z.; Lara, N.; Marín, M.; Sola-Campoy, P.J.; Bautista, V.; Gómez-Bertomeu, F.; Gómez-Dominguez, C.; Pérez-Vázquez, M.; Aracil, B.; Campos, J.; et al. Interregional spread in Spain of linezolid-resistant Enterococcus spp. isolates carrying the optrA and poxtA genes. *Int. J. Antimicrob. Agents* **2020**, *55*, 105977. [CrossRef]

72. Schwarz, S.; Werckenthin, C.; Kehrenberg, C. Identification of a Plasmid-Borne Chloramphenicol-Florfenicol Resistance Gene in *Staphylococcus sciuri*. *Antimicrob. Agents Chemother.* **2000**, *44*, 2530–2533. [CrossRef]

73. Shen, J.; Wang, Y.; Schwarz, S. Presence and dissemination of the multiresistance gene cfr in Gram-positive and Gram-negative bacteria. *J. Antimicrob. Chemother.* **2013**, *68*, 1697–1706. [CrossRef]

74. Wang, Y.; Lv, Y.; Cai, J.; Schwarz, S.; Cui, L.; Hu, Z.; Zhang, R.; Li, J.; Zhao, Q.; He, T.; et al. A novel gene, *optrA*, that confers transferable resistance to oxazolidinones and phenicols and its presence in *Enterococcus faecalis* and *Enterococcus faecium* of human and animal origin. *J. Antimicrob. Chemother.* **2015**, *70*, 2182–2190. [CrossRef] [PubMed]

75. Antonelli, A.; D'Andrea, M.M.; Brenciani, A.; Galeotti, C.L.; Morroni, G.; Pollini, S.; Varaldo, P.E.; Rossolini, G.M. Characterization of poxtA, a novel phenicol–oxazolidinone–tetracycline resistance gene from an MRSA of clinical origin. *J. Antimicrob. Chemother.* **2018**, *73*, 1763–1769. [CrossRef] [PubMed]

76. Ruiz-Ripa, L.; Feßler, A.T.; Hanke, D.; Eichhorn, I.; Azcona-Gutiérrez, J.M.; Pérez-Moreno, M.O.; Seral, C.; Aspiroz, C.; Alonso, C.A.; Torres, L.; et al. Mechanisms of Linezolid Resistance Among Enterococci of Clinical Origin in Spain—Detection of *optrA*- and *cfr*(D)-Carrying E. *faecalis*. *Microorganisms* **2020**, *8*, 1155. [CrossRef] [PubMed]

77. Egan, S.; Corcoran, S.; McDermott, H.; Fitzpatrick, M.; Hoyne, A.; McCormack, O.; Cullen, A.; Brennan, G.; O'Connell, B.; Coleman, D. Hospital outbreak of linezolid-resistant and vancomycin-resistant ST80 *Enterococcus faecium* harbouring an optrA-encoding conjugative plasmid investigated by whole-genome sequencing. *J. Hosp. Infect.* **2020**, *105*, 726–735. [CrossRef] [PubMed]

78. Lazaris, A.; Coleman, D.; Kearns, A.M.; Pichon, B.; Kinnevey, P.; Earls, M.R.; Boyle, B.; O'connell, B.; Brennan, G.I.; Shore, A.C. Novel multiresistance cfr plasmids in linezolid-resistant methicillin-resistant Staphylococcus epidermidis and vancomycin-resistant *Enterococcus faecium* (VRE) from a hospital outbreak: Co-location of cfr and optrA in VRE. *J. Antimicrob. Chemother.* **2017**, *72*, 3252–3257. [CrossRef]

79. Palmer, K.L.; Daniel, A.; Hardy, C.; Silverman, J.; Gilmore, M.S. Genetic Basis for Daptomycin Resistance in Enterococci. *Antimicrob. Agents Chemother.* **2011**, *55*, 3345–3356. [CrossRef]

80. Hesse, S.; Adhya, S. Phage Therapy in the Twenty-First Century: Facing the Decline of the Antibiotic Era; Is It Finally Time for the Age of the Phage? *Annu. Rev. Microbiol.* **2019**, *73*, 155–174. [CrossRef]

81. Maimaiti, Z.; Li, Z.; Xu, C.; Chen, J.; Chai, W. Global trends and hotspots of phage therapy for bacterial infection: A bibliometric visualized analysis from 2001 to 2021. *Front. Microbiol.* **2023**, *13*, 1067803. [CrossRef]

82. Abedon, S.T. Chapter 1 Phage Evolution and Ecology. *Adv. Appl. Microbiol.* **2009**, *67*, 1–45. [CrossRef]

83. Pedulla, M.L.; Ford, M.E.; Houtz, J.M.; Karthikeyan, T.; Wadsworth, C.; Lewis, J.A.; Jacobs-Sera, D.; Falbo, J.; Gross, J.; Pannunzio, N.R.; et al. Origins of Highly Mosaic Mycobacteriophage Genomes. *Cell* **2003**, *113*, 171–182. [CrossRef]

84. Dion, M.B.; Oechslin, F.; Moineau, S. Phage diversity, genomics and phylogeny. *Nat. Rev. Microbiol.* **2020**, *18*, 125–138. [CrossRef] [PubMed]

85. Kutateladze, M. Experience of the Eliava Institute in bacteriophage therapy. *Virol. Sin.* **2015**, *30*, 80–81. [CrossRef]

86. Moye, Z.D.; Woolston, J.M.; Sulakvelidze, A. Bacteriophage Applications for Food Production and Processing. *Viruses* **2018**, *10*, 205. [CrossRef] [PubMed]

87. García, P.; Martínez, B.; Obeso, J.; Rodríguez, A. Bacteriophages and their application in food safety. *Lett. Appl. Microbiol.* **2008**, *47*, 479–485. [CrossRef]

88. Eladero, V.; Egomez-Sordo, C.; Esanchez-Llana, E.; del Rio, B.; Redruello, B.; Efernandez, M.; Martin, M.C.; Alvarez, M.A. Q69 (an E. faecalis-Infecting Bacteriophage) As a Biocontrol Agent for Reducing Tyramine in Dairy Products. *Front. Microbiol.* **2016**, *7*, 445. [CrossRef]

89. Del Rio, B.; Sánchez-Llana, E.; Martínez, N.; Fernández, M.; Ladero, V.; Alvarez, M.A. Isolation and Characterization of Enterococcus faecalis-Infecting Bacteriophages From Different Cheese Types. *Front. Microbiol.* **2021**, *11*, 592172. [CrossRef] [PubMed]

90. Pradal, I.; Casado, A.; del Rio, B.; Rodriguez-Lucas, C.; Fernandez, M.; Alvarez, M.A.; Ladero, V. *Enterococcus faecium* Bacteriophage vB_EfaH_163, a New Member of the *Herelleviridae* Family, Reduces the Mortality Associated with an *E. faecium* vanR Clinical Isolate in a *Galleria mellonella* Animal Model. *Viruses* **2023**, *15*, 179. [CrossRef] [PubMed]

91. Melo, L.D.R.; Ferreira, R.; Costa, A.R.; Oliveira, H.; Azeredo, J. Efficacy and safety assessment of two enterococci phages in an in vitro biofilm wound model. *Sci. Rep.* **2019**, *9*, 6643. [CrossRef]

92. Khalifa, L.; Brosh, Y.; Gelman, D.; Coppenhagen-Glazer, S.; Beyth, S.; Poradosu-Cohen, R.; Que, Y.-A.; Beyth, N.; Hazan, R. Targeting Enterococcus faecalis Biofilms with Phage Therapy. *Appl. Environ. Microbiol.* **2015**, *81*, 2696–2705. [CrossRef] [PubMed]

93. Barylski, J.; Kropinski, A.M.; Alikhan, N.-F.; Adriaenssens, E.M. ICTV Report Consortium ICTV Virus Taxonomy Profile: Herelleviridae. *J. Gen. Virol.* **2020**, *101*, 362–363. [CrossRef] [PubMed]

94. Adriaenssens, E.M.; Sullivan, M.B.; Knezevic, P.; van Zyl, L.J.; Sarkar, B.L.; Dutilh, B.E.; Alfenas-Zerbini, P.; Łobocka, M.; Tong, Y.; Brister, J.R.; et al. Taxonomy of prokaryotic viruses: 2018–2019 update from the ICTV Bacterial and Archaeal Viruses Subcommittee. *Arch. Virol.* **2020**, *165*, 1253–1260. [CrossRef] [PubMed]

95. Matos, R.C.; Lapaque, N.; Rigottier-Gois, L.; Debarbieux, L.; Meylheuc, T.; Gonzalez-Zorn, B.; Repoila, F.; Lopes, M.D.F.; Serror, P. Enterococcus faecalis Prophage Dynamics and Contributions to Pathogenic Traits. *PLoS Genet.* **2013**, *9*, e1003539. [CrossRef]

96. Fernández, L.; Gutiérrez, D.; García, P.; Rodríguez, A. The Perfect Bacteriophage for Therapeutic Applications—A Quick Guide. *Antibiotics* **2019**, *8*, 126. [CrossRef] [PubMed]

97. Dedrick, R.M.; Guerrero-Bustamante, C.A.; Garlena, R.A.; Russell, D.A.; Ford, K.; Harris, K.; Gilmour, K.C.; Soothill, J.; Jacobs-Sera, D.; Schooley, R.T.; et al. Engineered bacteriophages for treatment of a patient with a disseminated drug-resistant Mycobacterium abscessus. *Nat. Med.* **2019**, *25*, 730–733. [CrossRef] [PubMed]

98. Ladero, V.; García, P.; Bascarán, V.; Herrero, M.; Alvarez, M.A.; Suárez, J.E. Identification of the Repressor-Encoding Gene of the *Lactobacillus* Bacteriophage A2. *J. Bacteriol.* **1998**, *180*, 3474–3476. [CrossRef]

99. Zhang, H.; Fouts, D.E.; DePew, J.; Stevens, R.H. Genetic modifications to temperate Enterococcus faecalis phage φEf11 that abolish the establishment of lysogeny and sensitivity to repressor, and increase host range and productivity of lytic infection. *Microbiology* **2013**, *159*, 1023–1035. [CrossRef]

100. Ladero, V.; Fernández, M.; Cuesta, I.; Alvarez, M.A. Quantitative detection and identification of tyramine-producing enterococci and lactobacilli in cheese by multiplex qPCR. *Food Microbiol.* **2010**, *27*, 933–939. [CrossRef]

101. Del Rio, B.; Linares, D.M.; Ladero, V.; Redruello, B.; Fernández, M.; Martin, M.C.; Alvarez, M.A. Putrescine production via the agmatine deiminase pathway increases the growth of Lactococcus lactis and causes the alkalinization of the culture medium. *Appl. Microbiol. Biotechnol.* **2015**, *99*, 897–905. [CrossRef]

102. Linares, D.M.; Martín, M.; Ladero, V.; Alvarez, M.A.; Fernández, M. Biogenic Amines in Dairy Products. *Crit. Rev. Food Sci. Nutr.* **2011**, *51*, 691–703. [CrossRef]

103. Herrero-Fresno, A.; Martínez, N.; Sánchez-Llana, E.; Díaz, M.; Fernández, M.; Martin, M.C.; Ladero, V.; Alvarez, M.A. Lactobacillus casei strains isolated from cheese reduce biogenic amine accumulation in an experimental model. *Int. J. Food Microbiol.* **2012**, *157*, 297–304. [CrossRef]

104. Capozzi, V.; Russo, P.; Ladero, V.; Fernández, M.; Fiocco, D.; Alvarez, M.A.; Grieco, F.; Spano, G. Biogenic Amines Degradation by Lactobacillus plantarum: Toward a Potential Application in Wine. *Front. Microbiol.* **2012**, *3*, 122. [CrossRef]

105. Alvarez, M.A.; Moreno-Arribas, M.V. The problem of biogenic amines in fermented foods and the use of potential biogenic amine-degrading microorganisms as a solution. *Trends Food Sci. Technol.* **2014**, *39*, 146–155. [CrossRef]

106. Calzada, J.; Del Olmo, A.; Picon, A.; Gaya, P.; Nuñez, M. Proteolysis and biogenic amine buildup in high-pressure treated ovine milk blue-veined cheese. *J. Dairy Sci.* **2013**, *96*, 4816–4829. [CrossRef]

107. Linares, D.M.; Del Río, B.; Ladero, V.; Martínez, N.; Fernández, M.; Martín, M.C.; Álvarez, M.A. Factors Influencing Biogenic Amines Accumulation in Dairy Products. *Front. Microbiol.* **2012**, *3*, 180. [CrossRef]

108. Del Rio, B.; Redruello, B.; Ladero, V.; Cal, S.; Obaya, A.J.; Alvarez, M.A. An altered gene expression profile in tyramine-exposed intestinal cell cultures supports the genotoxicity of this biogenic amine at dietary concentrations. *Sci. Rep.* **2018**, *8*, 17038. [CrossRef]

109. Linares, D.M.; del Rio, B.; Redruello, B.; Ladero, V.; Martin, M.C.; Fernández, M.; Ruas-Madiedo, P.; Alvarez, M.A. Comparative analysis of the in vitro cytotoxicity of the dietary biogenic amines tyramine and histamine. *Food Chem.* **2016**, *197*, 658–663. [CrossRef]

110. Brink, B.T.; Damink, C.; Joosten, H.; Veld, J.H.I. Occurrence and formation of biologically active amines in foods. *Int. J. Food Microbiol.* **1990**, *11*, 73–84. [CrossRef]

111. Del Rio, B.; Redruello, B.; Linares, D.M.; Ladero, V.; Ruas-Madiedo, P.; Fernandez, M.; Martin, M.C.; Alvarez, M.A. The biogenic amines putrescine and cadaverine show in vitro cytotoxicity at concentrations that can be found in foods. *Sci. Rep.* **2019**, *9*, 120. [CrossRef]

112. Del Rio, B.; Sánchez-Llana, E.; Redruello, B.; Magadan, A.H.; Fernández, M.; Martin, M.C.; Ladero, V.; Alvarez, M.A. Enterococcus faecalis Bacteriophage 156 Is an Effective Biotechnological Tool for Reducing the Presence of Tyramine and Putrescine in an Experimental Cheese Model. *Front. Microbiol.* **2019**, *10*, 566. [CrossRef]

113. Gołaś-Prądzyńska, M.; Łuszczyńska, M.; Rola, J.G. Dairy Products: A Potential Source of Multidrug-Resistant *Enterococcus faecalis* and *Enterococcus faecium* Strains. *Foods* **2022**, *11*, 4116. [CrossRef]

114. Alvarez-Ordóñez, A.; Coughlan, L.M.; Briandet, R.; Cotter, P.D. Biofilms in Food Processing Environments: Challenges and Opportunities. *Annu. Rev. Food Sci. Technol.* **2019**, *10*, 173–195. [CrossRef]
115. Galiè, S.; García-Gutiérrez, C.; Miguélez, E.M.; Villar, C.J.; Lombó, F. Biofilms in the Food Industry: Health Aspects and Control Methods. *Front. Microbiol.* **2018**, *9*, 898. [CrossRef]
116. Diaz, M.; Ladero, V.; Del Rio, B.; Redruello, B.; Fernández, M.; Martin, M.C.; Alvarez, M.A. Biofilm-Forming Capacity in Biogenic Amine-Producing Bacteria Isolated from Dairy Products. *Front. Microbiol.* **2016**, *7*, 591. [CrossRef]
117. Topka-Bielecka, G.; Bloch, S.; Nejman-Faleńczyk, B.; Grabski, M.; Jurczak-Kurek, A.; Górniak, M.; Dydecka, A.; Necel, A.; Węgrzyn, G.; Węgrzyn, A. Characterization of the Bacteriophage vB_EfaS-271 Infecting *Enterococcus faecalis. Int. J. Mol. Sci.* **2020**, *21*, 6345. [CrossRef]
118. Yang, D.; Chen, Y.; Sun, E.; Hua, L.; Peng, Z.; Wu, B. Characterization of a Lytic Bacteriophage vB_EfaS_PHB08 Harboring Endolysin Lys08 Against *Enterococcus faecalis* Biofilms. *Microorganisms* **2020**, *8*, 1332. [CrossRef]
119. Coyne, A.J.K.; Stamper, K.; Kebriaei, R.; Holger, D.J.; El Ghali, A.; Morrisette, T.; Biswas, B.; Wilson, M.; Deschenes, M.V.; Canfield, G.S.; et al. Phage Cocktails with Daptomycin and Ampicillin Eradicates Biofilm-Embedded Multidrug-Resistant *Enterococcus faecium* with Preserved Phage Susceptibility. *Antibiotics* **2022**, *11*, 1175. [CrossRef]
120. Al-Zubidi, M.; Widziolek, M.; Court, E.K.; Gains, A.F.; Smith, R.E.; Ansbro, K.; Alrafaie, A.; Evans, C.; Murdoch, C.; Mesnage, S.; et al. Identification of Novel Bacteriophages with Therapeutic Potential That Target *Enterococcus faecalis. Infect. Immun.* **2019**, *87*, e00512-19. [CrossRef]
121. Canfield, G.S.; Duerkop, B.A. Molecular mechanisms of enterococcal-bacteriophage interactions and implications for human health. *Curr. Opin. Microbiol.* **2020**, *56*, 38–44. [CrossRef]
122. Bolocan, A.S.; Upadrasta, A.; Bettio, P.H.A.; Clooney, A.G.; Draper, L.A.; Ross, R.P.; Hill, C. Evaluation of Phage Therapy in the Context of *Enterococcus faecalis* and Its Associated Diseases. *Viruses* **2019**, *11*, 366. [CrossRef]
123. D'andrea, M.M.; Frezza, D.; Romano, E.; Marmo, P.; De Angelis, L.H.; Perini, N.; Thaller, M.C.; Di Lallo, G. The lytic bacteriophage vB_EfaH_EF1TV, a new member of the Herelleviridae family, disrupts biofilm produced by Enterococcus faecalis clinical strains. *J. Glob. Antimicrob. Resist.* **2020**, *21*, 68–75. [CrossRef]
124. El-Telbany, M.; El-Didamony, G.; Askora, A.; Ariny, E.; Abdallah, D.; Connerton, I.F.; El-Shibiny, A. Bacteriophages to Control Multi-Drug Resistant *Enterococcus faecalis* Infection of Dental Root Canals. *Microorganisms* **2021**, *9*, 517. [CrossRef]
125. Goodarzi, F.; Hallajzadeh, M.; Sholeh, M.; Talebi, M.; Mahabadi, V.P.; Amirmozafari, N. Anti-biofilm Activity of a Lytic Phage Against Vancomycin-Resistant Enterococcus faecalis. *Iran. J. Pathol.* **2022**, *17*, 285–293. [CrossRef]
126. Liu, J.; Zhu, Y.; Li, Y.; Lu, Y.; Xiong, K.; Zhong, Q.; Wang, J. Bacteriophage-Resistant Mutant of Enterococcus faecalis Is Impaired in Biofilm Formation. *Front. Microbiol.* **2022**, *13*, 913023. [CrossRef]
127. El-Atrees, D.M.; El-Kased, R.F.; Abbas, A.M.; Yassien, M.A. Characterization and anti-biofilm activity of bacteriophages against urinary tract Enterococcus faecalis isolates. *Sci. Rep.* **2022**, *12*, 13048. [CrossRef]
128. Topka-Bielecka, G.; Nejman-Faleńczyk, B.; Bloch, S.; Dydecka, A.; Necel, A.; Węgrzyn, A.; Węgrzyn, G. Phage–Bacteria Interactions in Potential Applications of Bacteriophage vB_EfaS-271 against *Enterococcus faecalis. Viruses* **2021**, *13*, 318. [CrossRef]
129. Song, M.; Wu, D.; Hu, Y.; Luo, H.; Li, G. Characterization of an Enterococcus faecalis Bacteriophage vB_EfaM_LG1 and Its Synergistic Effect With Antibiotic. *Front. Cell. Infect. Microbiol.* **2021**, *11*, 698807. [CrossRef]
130. Lev, K.; Coyne, A.J.K.; Kebriaei, R.; Morrisette, T.; Stamper, K.; Holger, D.J.; Canfield, G.S.; Duerkop, B.A.; Arias, C.A.; Rybak, M.J. Evaluation of Bacteriophage-Antibiotic Combination Therapy for Biofilm-Embedded MDR *Enterococcus faecium. Antibiotics* **2022**, *11*, 392. [CrossRef]
131. Morrisette, T.; Lev, K.L.; Canfield, G.S.; Duerkop, B.A.; Kebriaei, R.; Stamper, K.C.; Holger, D.; Lehman, S.M.; Willcox, S.; Arias, C.A.; et al. Evaluation of Bacteriophage Cocktails Alone and in Combination with Daptomycin against Daptomycin-Nonsusceptible *Enterococcus faecium. Antimicrob. Agents Chemother.* **2022**, *66*, e01623-21. [CrossRef]
132. Shlezinger, M.; Coppenhagen-Glazer, S.; Gelman, D.; Beyth, N.; Hazan, R. Eradication of Vancomycin-Resistant Enterococci by Combining Phage and Vancomycin. *Viruses* **2019**, *11*, 954. [CrossRef]
133. Stuart, C.H.; Schwartz, S.A.; Beeson, T.J.; Owatz, C.B. Enterococcus faecalis: Its Role in Root Canal Treatment Failure and Current Concepts in Retreatment. *J. Endod.* **2006**, *32*, 93–98. [CrossRef]
134. Voit, M.; Trampuz, A.; Moreno, M.G. In Vitro Evaluation of Five Newly Isolated Bacteriophages against *E. faecalis* Biofilm for Their Potential Use against Post-Treatment Apical Periodontitis. *Pharmaceutics* **2022**, *14*, 1779. [CrossRef]
135. Moryl, M.; Palatyńska-Ulatowska, A.; Maszewska, A.; Grzejdziak, I.; de Oliveira, S.D.; Pradebon, M.C.; Steier, L.; Różalski, A.; de Figueiredo, J.A.P. Benefits and Challenges of the Use of Two Novel vB_Efa29212_2e and vB_Efa29212_3e Bacteriophages in Biocontrol of the Root Canal *Enterococcus faecalis* Infections. *J. Clin. Med.* **2022**, *11*, 6494. [CrossRef] [PubMed]
136. Lee, D.; Im, J.; Na, H.; Ryu, S.; Yun, C.-H.; Han, S.H. The Novel Enterococcus Phage vB_EfaS_HEf13 Has Broad Lytic Activity Against Clinical Isolates of Enterococcus faecalis. *Front. Microbiol.* **2019**, *10*, 2877. [CrossRef]
137. Paisano, A.F.; Spira, B.; Cai, S.; Bombana, A.C. In vitro antimicrobial effect of bacteriophages on human dentin infected with Enterococcus faecalis ATCC 29212. *Oral Microbiol. Immunol.* **2004**, *19*, 327–330. [CrossRef] [PubMed]
138. Tinoco, J.M.; Buttaro, B.; Zhang, H.; Liss, N.; Sassone, L.M.; Stevens, R. Effect of a genetically engineered bacteriophage on Enterococcus faecalis biofilms. *Arch. Oral Biol.* **2016**, *71*, 80–86. [CrossRef]

139. Tinoco, J.M.; Liss, N.; Zhang, H.; Nissan, R.; Gordon, W.; Tinoco, E.; Sassone, L.; Stevens, R. Antibacterial effect of genetically-engineered bacteriophage φEf11/φFL1C(Δ36)P nisA on dentin infected with antibiotic-resistant Enterococcus faecalis. *Arch. Oral Biol.* **2017**, *82*, 166–170. [CrossRef]

140. Cieślik, M.; Bagińska, N.; Górski, A.; Jończyk-Matysiak, E. Animal Models in the Evaluation of the Effectiveness of Phage Therapy for Infections Caused by Gram-Negative Bacteria from the ESKAPE Group and the Reliability of Its Use in Humans. *Microorganisms* **2021**, *9*, 206. [CrossRef] [PubMed]

141. El Haddad, L.; Angelidakis, G.; Clark, J.R.; Mendoza, J.F.; Terwilliger, A.L.; Chaftari, C.P.; Duna, M.; Yusuf, S.T.; Harb, C.P.; Stibich, M.; et al. Genomic and Functional Characterization of Vancomycin-Resistant Enterococci-Specific Bacteriophages in the *Galleria mellonella* Wax Moth Larvae Model. *Pharmaceutics* **2022**, *14*, 1591. [CrossRef]

142. Kishimoto, T.; Ishida, W.; Fukuda, K.; Nakajima, I.; Suzuki, T.; Uchiyama, J.; Matsuzaki, S.; Todokoro, D.; Daibata, M.; Fukushima, A. Therapeutic Effects of Intravitreously Administered Bacteriophage in a Mouse Model of Endophthalmitis Caused by Vancomycin-Sensitive or -Resistant Enterococcus faecalis. *Antimicrob. Agents Chemother.* **2019**, *63*, e01088-19. [CrossRef]

143. Kishimoto, T.; Ishida, W.; Nasukawa, T.; Ujihara, T.; Nakajima, I.; Suzuki, T.; Uchiyama, J.; Todokoro, D.; Daibata, M.; Fukushima, A.; et al. In Vitro and In Vivo Evaluation of Three Newly Isolated Bacteriophage Candidates, phiEF7H, phiEF14H1, phiEF19G, for Treatment of *Enterococcus faecalis* Endophthalmitis. *Microorganisms* **2021**, *9*, 212. [CrossRef]

144. Zhang, W.; Mi, Z.; Yin, X.; Fan, H.; An, X.; Zhang, Z.; Chen, J.; Tong, Y. Characterization of Enterococcus faecalis Phage IME-EF1 and Its Endolysin. *PLoS ONE* **2013**, *8*, e80435. [CrossRef]

145. Biswas, B.; Adhya, S.; Washart, P.; Paul, B.; Trostel, A.N.; Powell, B.; Carlton, R.; Merril, C.R. Bacteriophage Therapy Rescues Mice Bacteremic from a Clinical Isolate of Vancomycin-Resistant *Enterococcus faecium*. *Infect. Immun.* **2002**, *70*, 1664. [CrossRef]

146. Cheng, M.; Liang, J.; Zhang, Y.; Hu, L.; Gong, P.; Cai, R.; Zhang, L.; Zhang, H.; Ge, J.; Ji, Y.; et al. The Bacteriophage EF-P29 Efficiently Protects against Lethal Vancomycin-Resistant Enterococcus faecalis and Alleviates Gut Microbiota Imbalance in a Murine Bacteremia Model. *Front. Microbiol.* **2017**, *8*, 837. [CrossRef]

147. Uchiyama, J.; Rashel, M.; Takemura, I.; Wakiguchi, H.; Matsuzaki, S. In Silico and In Vivo Evaluation of Bacteriophage φEF24C, a Candidate for Treatment of *Enterococcus faecalis* Infections. *Appl. Environ. Microbiol.* **2008**, *74*, 4149–4163. [CrossRef]

148. Tkachev, P.V.; Pchelin, I.M.; Azarov, D.V.; Gorshkov, A.N.; Shamova, O.V.; Dmitriev, A.V.; Goncharov, A.E. Two Novel Lytic Bacteriophages Infecting *Enterococcus* spp. Are Promising Candidates for Targeted Antibacterial Therapy. *Viruses* **2022**, *14*, 831. [CrossRef]

149. Gelman, D.; Beyth, S.; Lerer, V.; Adler, K.; Poradosu-Cohen, R.; Coppenhagen-Glazer, S.; Hazan, R. Combined bacteriophages and antibiotics as an efficient therapy against VRE Enterococcus faecalis in a mouse model. *Res. Microbiol.* **2018**, *169*, 531–539. [CrossRef]

150. Johri, A.V.; Johri, P.; Hoyle, N.; Pipia, L.; Nadareishvili, L.; Nizharadze, D. Case Report: Chronic Bacterial Prostatitis Treated With Phage Therapy After Multiple Failed Antibiotic Treatments. *Front. Pharmacol.* **2021**, *12*, 6926148. [CrossRef]

151. Paul, K.; Merabishvili, M.; Hazan, R.; Christner, M.; Herden, U.; Gelman, D.; Khalifa, L.; Yerushalmy, O.; Coppenhagen-Glazer, S.; Harbauer, T.; et al. Bacteriophage Rescue Therapy of a Vancomycin-Resistant *Enterococcus faecium* Infection in a One-Year-Old Child following a Third Liver Transplantation. *Viruses* **2021**, *13*, 1785. [CrossRef]

152. Onsea, J.; Soentjens, P.; Djebara, S.; Merabishvili, M.; Depypere, M.; Spriet, I.; De Munter, P.; Debaveye, Y.; Nijs, S.; Vanderschot, P.; et al. Bacteriophage Application for Difficult-To-Treat Musculoskeletal Infections: Development of a Standardized Multidisciplinary Treatment Protocol. *Viruses* **2019**, *11*, 891. [CrossRef]

153. Letkiewicz, S.; Międzybrodzki, R.; Fortuna, W.; Weber-Dabrowska, B.; Gorski, A. Eradication of Enterococcus faecalis by phage therapy in chronic bacterial prostatitis—Case report. *Folia Microbiol.* **2009**, *54*, 457–461. [CrossRef]

154. Neuts, A.-S.; Berkhout, H.J.; Hartog, A.; Goosen, J.H.M. Bacteriophage therapy cures a recurrent *Enterococcus faecalis* infected total hip arthroplasty? A case report. *Acta Orthop.* **2021**, *92*, 678–680. [CrossRef]

155. Górski, A.; Borysowski, J.; Międzybrodzki, R. Phage Therapy: Towards a Successful Clinical Trial. *Antibiotics* **2020**, *9*, 827. [CrossRef] [PubMed]

156. Aslam, S.; Lampley, E.; Wooten, D.; Karris, M.; Benson, C.; Strathdee, S.; Schooley, R.T. Lessons Learned From the First 10 Consecutive Cases of Intravenous Bacteriophage Therapy to Treat Multidrug-Resistant Bacterial Infections at a Single Center in the United States. *Open Forum Infect. Dis.* **2020**, *7*, ofaa389. [CrossRef] [PubMed]

Review

Recent Advances in the Application of Bacteriophages against Common Foodborne Pathogens

Kinga Hyla, Izabela Dusza and Aneta Skaradzińska *

Department of Biotechnology and Food Microbiology, Faculty of Biotechnology and Food Science, Wrocław University of Environmental and Life Sciences, Chełmońskiego 37, 51-630 Wrocław, Poland
* Correspondence: aneta.skaradzinska@upwr.edu.pl

Abstract: Bacteriophage potential in combating bacterial pathogens has been recognized nearly since the moment of discovery of these viruses at the beginning of the 20th century. Interest in phage application, which initially focused on medical treatments, rapidly spread throughout different biotechnological and industrial fields. This includes the food safety sector in which the presence of pathogens poses an explicit threat to consumers. This is also the field in which commercialization of phage-based products shows the greatest progress. Application of bacteriophages has gained special attention particularly in recent years, presumably due to the potential of conventional antibacterial strategies being exhausted. In this review, we present recent findings regarding phage application in fighting major foodborne pathogens, including *Salmonella* spp., *Escherichia coli*, *Yersinia* spp., *Campylobacter jejuni* and *Listeria monocytogenes*. We also discuss advantages of bacteriophage use and challenges facing phage-based antibacterial strategies, particularly in the context of their widespread application in food safety.

Keywords: bacteriophages; food safety; phage biocontrol; foodborne pathogens; bacteriophage application; phage commercial products; foodborne illness

Citation: Hyla, K.; Dusza, I.; Skaradzińska, A. Recent Advances in the Application of Bacteriophages against Common Foodborne Pathogens. *Antibiotics* **2022**, *11*, 1536. https://doi.org/10.3390/antibiotics11111536

Academic Editor: Grzegorz Wegrzyn

Received: 13 October 2022
Accepted: 28 October 2022
Published: 2 November 2022

Publisher's Note: MDPI stays neutral with regard to jurisdictional claims in published maps and institutional affiliations.

1. Introduction

Foodborne illness has been affecting people's lives unceasingly, impacting human welfare and contributing to significant economic losses in many countries and populations.

Food products may become contaminated at different stages along the food chain including slaughtering, milking, fermentation, processing, storage, packaging or finally consumption of the product [1]. The most popular strategies for elimination of the pathogens are implementation of high standard hygiene procedures, rational running of the process line, and use of biocides and disinfectants [2]. However, currently applied methods for elimination of foodborne pathogens are unreliable. For instance, use of steam, dry heat or UV light leads to changes in organoleptic properties of the product. Moreover, there are some limitations in the use of certain antimicrobial approaches, in particular products such as fresh fruits, vegetables and ready-to-eat (RTE) products. A major problem is that extensive use of sanitizers leads to the development of microbial resistance [1].

Bacteriophages, or phages for short, are viruses that selectively infect and replicate in bacterial cells. Due to their unique properties, phages have been perceived as promising tools in combating bacterial pathogens not only in human treatments (phage therapy) but also in various industrial fields. This includes the food production sector in which, for the sake of the consumer's safety, all implemented antimicrobial procedures must be selected with special care. In contrast to routinely used antibacterial approaches, phage application does not change the properties of the product, viruses may be applied on a variety of matrices and the problem of resistance may be overcome much more easily compared to chemical antibacterials [3–5]. Phages are isolated from a variety of foods, which indicates their natural contact with humans with this route [6]. Furthermore, phage-based strategies are cost-effective and consumer-friendly, making them an important alternative to standard antibacterial procedures.

The ability of bacteriophages to effectively eliminate foodborne pathogens has been reported in numerous scientific articles [7–10]. Since the latest years have brought a significant increase in bacteriophage studies, in this review, we summarize the recent findings regarding efficiency of phages against the main food-related pathogens and potential application of these viruses in approaches of microbiological control in food production. We believe that ensuring safety of food products is a key global concern, and searching for novel solutions that will allow high standards of food production to be maintained is urgently needed. In this aspect, the use of phages as tools providing safety for consumers should be seen as an important alternative to currently applied methods.

1.1. Bacteriophages

Bacteriophages are the most ubiquitous biological entities on Earth, with an estimated total global population of 10^{31} particles [11]. They are also known as phages, from the Greek *phagein* meaning "to eat" [12]. They can be found in a variety of ecosystems, including extreme environments, such as the Sahara desert, hot springs and cold polar waters [13]. They are characterized by a great diversity in structure, size and organization of the genome [14].

Bacteriophages were first mentioned in 1896, when the British bacteriologist Ernest Hankin reported unusual antibacterial activity in the waters of the Yamuna and Ganga rivers. Furthermore, he suggested that this unknown factor inhibited the development of the epidemic of cholera caused by *Vibrio cholerae*. Notably, his hypothesis has never been confirmed by the scientific community [15]. The researcher who first observed translucencies on a bacterial lawn caused by bacteriophages was Frederick Twort. However, at that time he considered it to be the action of a "transmissible vitreous transformation", the amount of which increased after the death of the cell. In 1917, the French-Canadian microbiologist Félix d'Herelle published results of his research, describing the phenomenon of bright "zonas", which he eventually called plaques. He was also the first to propose the hypothesis that the translucencies could be caused by a virus that parasitizes bacteria. He called it a "bacteriophage" [13].

One of the most characteristic features of bacteriophages is their high specificity, regarding one species, or a specific strain of bacteria they infect [14]. This high specificity is based on selective binding of the virus receptor with the ligand at the bacterial surface. Proteins, polysaccharides, lipopolysaccharides (LPS) and carbohydrate moieties as well as outer membrane proteins, pili and flagella may be used by phages as keys to the entrance to bacterial cells [16]. Bacteriophages multiply with two basic replication cycles. In the lytic cycle, once the genetic material of the phage is inside the relevant host, it is replicated with the molecular apparatus of the bacterium. In a relatively short time after phage penetration, new virions are assembled and released to the environment. This sequence of molecular events naturally leads to lysis of the bacterial cell [17]. Bacteriophages amplifying with the lytic cycle are called lytic or virulent phages [18]. In the lysogenic cycle, performed by temperate phages, nucleic acid of the virus integrates with the bacterial genome and multiplies with the host as a prophage. However, as a result of unfavorable changes in environmental conditions, the phage can revert to the lytic cycle [13]. Clearly, due to the lack of lytic activity, temperate phages are not relevant candidates for practical use.

Bacteriophage efficacy in disrupting bacterial cells is an excellent feature for their use as microbiological tools in antibacterial strategies. However, within years of phage application, other characteristics of bacterial viruses have been perceived as beneficial, particularly compared to standard chemical antibacterial agents. Bacteriophages recognize and infect only a particular bacterial host and therefore they do not affect the natural microflora of the organism. As a result, they are better tolerated by the human body than antibiotics and thus are considered a safer treatment option [19,20]. Moreover, frequently, a single dose of bacteriophage preparation is sufficient to achieve the therapeutic effect [21,22]. This results from the unique phage ability of "auto-dosing", which means that phages are capable of increasing in number specifically where their hosts are located [23]. Bacteriophages are inhabitants of humans as they are found in the respiratory or digestive tract. It is thought that the first viruses enter the intestines within 4 days of birth [20]. This natural contact with phages also indirectly supports the safety of the intended phage application. It is not without significance that, compared to antibiotics, phage acquisition is easier, faster and cheaper [24]. For applications in food safety of particular importance is that phages do not impact organoleptic, rheological and nutritional properties of the product [25].

There are surely also limitations of phage use, such as the narrow spectrum of activity, which may be a serious obstacle in production of universal preparations intended for use on a large scale, or development of phage-resistant strains. However, both limitations may be overcome by the use of phage cocktails, which may be composed of phages with different specificity, broadening the lytic spectrum of the preparation, or with phages with similar activity so that when bacteria acquire resistance to one phage from the preparation they likely remain susceptible to another [24,26]. It is noteworthy that the effectivity of phage cocktails is multifaceted and other factors, including the mobilization of virulence or antibiotic resistance genes or phage coinfections have to be considered [27]. Limitations of phage application are discussed more specifically later in this review.

1.2. Foodborne Pathogens

The first food infections were reported in the 5th century BC, when it was observed that illnesses occurring at that time may be related to the consumed food [28]. Since then, scientists have shown that pathogens that contribute to food contamination include viruses, parasites and bacteria of which the latter are considered the most common cause of foodborne infections [29]. Among the bacterial species of utmost importance are frequently listed *Salmonella* spp., *Listeria monocytogenes*, *Escherichia coli* and *Campylobacter jejuni* [30,31].

Bacterial food-related pathogens are mainly mesophilic micro-organisms tolerating temperatures in the range of 20–45 °C, so they can easily survive in the human body. Of note, some pathogens, e.g., *Y. enterocolitica*, persist at temperatures lower than 10 °C, which entails the need to adjust preventive methods to remove potentially occurring bacterial cells. Production of spores by many bacterial species also hinders decontamination.

However, one of the most significant challenges is the ability of pathogens to form a biofilm structure [28]. In a biofilm, cells are embedded in the extracellular polymeric substance (EPS), also known as the extracellular matrix, providing micro-organisms resistance to environmental factors [32]. Food is a particularly favorable matrix for biofilm development. In the food production sector, biofilm structures may form directly on food products as well as on equipment that may come into contact with food [33]. Foodborne pathogens able to form biofilms are of particular concern for food producers as the use of disinfectants and other antimicrobial agents is inefficient due to the limited penetration of the product within the structure [34]. Even if the biofilm is cleared once, the contamination likely returns [33]. Many foodborne pathogens, e.g., *L. monocytogenes*, *S. enterica*, *C. jejuni* and *E. coli*, are able to form biofilms, so implementing antimicrobial strategies developed to efficiently remove these highly resistant structures is urgently needed.

There are two routes for the development of a foodborne infection. The first route, which is referred to as intoxication, is that a pathogen on the food surface or inside the food product produces a toxin, which then enters the organism with the meal and affects its metabolism. In the second case, a pathogen that has directly entered the digestive system with food is able to adapt and multiply within cells [28].

The most comprehensive data related to food infections date back to 2016 when the EFSA (European Food Safety Authority) published a summary report on trends and sources of zoonoses, zoonotic agents and food-borne outbreaks in the preceding year [35]. According to presented data, in 2015, there were 45,874 cases of food-related illness within the countries of the European Union. The number of outbreaks in which two or more people were infected after consuming the same food was 4362 of which 33.7% were caused by bacteria, mainly *Campylobacter* spp. and *Salmonella* spp. Reported foodborne infections were mainly related to contamination of products of animal origin, such as pork and eggs, but also other products as shellfish, milk, and fish [28]. In the same year, the World Health Organization (WHO) estimated the number of foodborne outbreaks across the globe. A total of 31 infectious agents were identified to contribute to the illnesses of 600 million people and the deaths of 420,000 people worldwide. Both reports conclude that special attention should be paid to food safety at every stage of production in the so-called "farm to fork" approach [36].

The possibility of food contamination depends on numerous factors, such as the type of the product and the method of its production, the water content in food, or the amount of accessible oxygen during the process which inhibits development of anaerobic bacteria but at the same time creates conditions for development of aerobic strains. There are various methods which are used as a standard to eliminate undesirable bacteria. These include thermal treatment at low (chilling, freezing) and high temperatures (pasteurization, sterilization), drying, reduction of water activity, fermentation, use of microbial growth inhibitors or ozonation. However, use of the available sanitization methods frequently influences sensory and organoleptic characteristics of the product (Figure 1) [37].

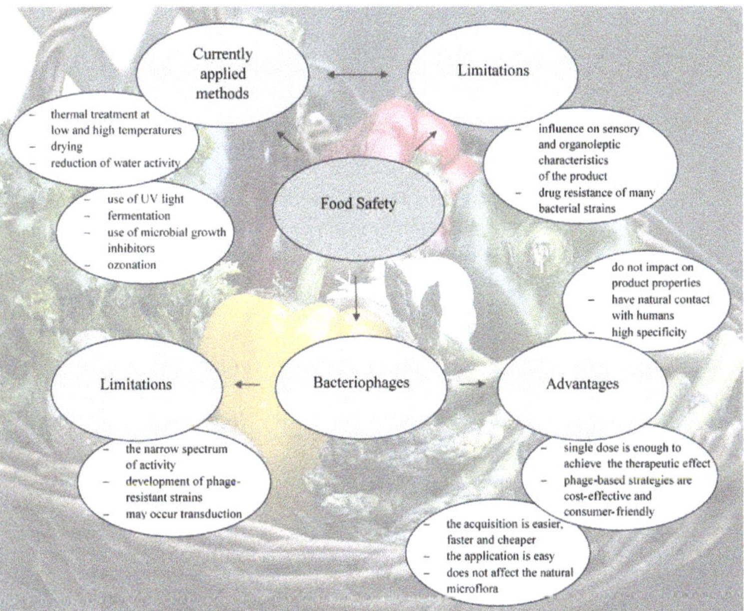

Figure 1. Currently used antimicrobial methods in food safety compared to bacteriophage application.

It is worth emphasizing that a number of system procedures have been developed and implemented to maintain standards of safety within food processing. These include good manufacturing practices (GMP), sanitation standard operating procedures (SSOP) and hazard analysis and critical control points (HACCP). The latter is the most widely used strategy in maintaining production safety [38]. Nevertheless, despite implementing different safety protocols, the final products must still be tested for microbial contamination to ensure consumer safety and reduce the risk of disease [39].

Especially noteworthy is the problem of the industrial breeding of livestock, particularly on large farms in which there is an increased risk of microbial contamination and eventually a disease outbreak due to hindered waste management control. It is commonly known that abuse of antibiotics in husbandry has become a relevant global problem. The amount of these drugs applied in farming exceeds 50 million kilograms each year worldwide [40]. Replacement of antibiotics is a critical challenge as abundant and virtually uncontrolled release of these medicaments into the environment contributes to the rapid development of resistant bacterial strains. Considering the demanding environment of livestock farms and the specificity of animal breeding, development of new preventive methods against microbial contaminations is critically needed.

1.3. Salmonella

Salmonella is a rod-shaped, Gram-negative bacterium belonging to the Enterobacteriaceae family. It is the cause of one of the most common foodborne diseases—salmonellosis. The illness usually manifests with abdominal pain, vomiting, diarrhea, fever and headache. Based on the EFSA report from 2020, the pathogen is transmitted through the soil, water, feed, and feces, among others. Bacteria are typically found in meat and animal products, which become the source of infection for humans. Chickens are the most frequently infected animals, followed by cattle, turkeys, pigs, ducks, and geese. The most common cause of infection are the two serotypes *Salmonella* Enteritidis and *Salmonella* Typhimurium [41].

The potential of bacteriophages in combating *Salmonella* infections has been confirmed in numerous studies [25,42–44]. A recent example is the research of Yan et al. (2020), who investigated the efficacy of the LYPSET phage in elimination of *S. enterica* in food products. The experiments were carried out using milk and lettuce stored at two temperatures, 4 °C and 25 °C. Samples contaminated with the pathogen were treated with phage lysate and the number of bacteria was determined. In the case of milk samples, application of the phage preparation allowed the number of bacteria to be reduced by 2.07 log CFU/mL at 4 °C and by 3.67 log CFU/mL at 25 °C with an MOI = 1000. When the MOI was 10,000, the activity of phages was slightly higher as the number of bacteria decreased by 2.19 log CFU/mL at 4 °C and 4.33 log CFU/mL at a temperature of 25 °C. For lettuce samples only MOI = 10,000 was tested. The number of *Salmonella* cells decreased by 2.2 log CFU/mL at 4 °C and by 2.34 log CFU/mL at 25 °C. The results confirmed that application of phages may be a promising approach in combating *Salmonella* contamination in food [45].

In similar studies of Islam et al. (2019) the effect of the phage cocktail composed of three phages, LPSTLL, LPST94 and LPST153, on *S.* Typhimurium and *S.* Enteritidis and their mixture in milk and in chicken meat was investigated. As in former studies, the experiments were performed at temperatures of 4 °C and 25 °C. The results showed that the number of *S.* Typhimurium cells in milk was reduced below the detectable limit (<1 CFU/100 µL) after 3 h and 6 h at 4 °C with MOI of 10,000 and 1000, respectively. For the mixture of *Salmonella* almost complete elimination of bacterial cells in milk was observed after 6 h and 12 h at 4 °C with MOI of 10,000 and 1000, respectively. Of note, for a single *Salmonella* strain or a mixture of *Salmonella* strains, the viable counts declined completely after 6 h (MOI = 10,000) and 24 h (MOI = 1000) at 25 °C. Similarly, for the chicken meat, there was complete elimination of bacteria after 3 h at 4 °C and 3 h and 6 h at 25 °C for MOI of 10,000 and 1000. Furthermore, the authors tested the effect of bacteriophage cocktail on biofilm structures formed by *S.* Typhimurium and the mixed culture of *S.* Typhimurium and *S.* Enteritidis. The experiments were performed using a 96-well plate and a stainless steel surface. Biofilm structures were formed and then treated with preparations with phage titer of 7 log PFU/mL and 8 log PFU/mL. After 24 h, biofilm reduction was determined for both matrices. In the case of the 96-well plate, reduction of *S.* Typhimurium biofilm structure was 48.3% and 63.25% for both tested phage titers. In analogical experiments with mixed biofilm the decrease was 44.28% and 63.25%. When using the steel surface and a single *S.* Typhimurium strain, the biofilm decreased by 5.5 log and 6.42 log for respective titers 7 log PFU/mL and 8 log PFU/mL. For a mixed biofilm, reduction of 44.28% and 51.17% compared to the control with no phage was achieved [46]. The results indicated that phage cocktails may be potentially used as biological control agents against *Salmonella*, including the removal of *Salmonella* biofilm structures from food-related equipment.

In another study, the potential of the SE07 phage specific to *S. enterica* isolated from chicken and beef meat intended for sale was tested. The effects of the phage in reducing *Salmonella* contamination was evaluated for a variety of foods, including fresh eggs, beef, poultry meat and fresh fruit juice. A significant reduction in the number of pathogens was noted after 12 h of the experiment; however, after the following hours the further decline of the bacterial count was negligible. In the example of beef, the number of bacteria decreased from the initial concentration of 4.23 log CFU/mL to 2.32 log CFU/mL after 12 h following phage application. Similarly, in the case of the chicken sample, the bacteria count dropped from 4.16 log CFU/mL to 2.34 log CFU/mL in the 12th hour of the experiment [47].

SalmoFresh™ (Intralytix, Columbia, SC, USA) is a bacteriophage preparation against *Salmonella* spp. which has been granted GRAS (Generally Recognize As Safe) status by the US Food and Drug Administration (FDA, 2013; Intralytix, 2015). According to information provided by the producer the product is specifically designed for treating foods that are at high risk for *Salmonella* contamination. Red meat and poultry in particular can be treated prior to grinding for significant reductions in pathogen count [48]. The preparation is composed of six phages targeting different serotypes of *Salmonella* spp. In the studies of Zhang et al. (2019), the surfaces of lettuce, mung bean sprouts and its seeds were covered with SalmoFresh, whereas the control group was washed with chlorinated water. In an additional experimental group, the food products were treated with a mixture of both chlorinated water and the phage preparation. The findings demonstrated the effectiveness of the cocktail in reducing *Salmonella* contamination on lettuce and sprouts as bacterial counts decreased during storage of the products by 0.76 log CFU/g and 0.83 log CFU/g, respectively. The results were inconclusive in the case of seeds as there was exponential growth of bacteria observed after their germination. Surprisingly, the most effective method turned out to be combined use of chlorinated water and a bacteriophage cocktail, which gave satisfactory results in all trials [49].

A relatively novel application of bacterial viruses is their use as components of detection systems for different bacteria. Minh et al. (2020) developed a method for the rapid detection of *Salmonella* using NanoLuc reporter phages. The gene of luciferase was introduced with homologous recombination downstream of the main capsid protein sequence. Bacteria were incubated with modified phages SEA1.NL and TSP1.NL for two hours and their presence was evaluated based on the luminescence production. A combination of the two phages provided the best results since the TSP1.NL phage gave high intensity of luminescence, while SEA1.NL showed high specificity. This method has been referred to as PhageDx for *Salmonella*. PhageDx has been proved to work flawlessly for pure cultures, and the question arose whether it could be effective for food matrices. Therefore, in subsequent studies, possible application of the method in detection of *Salmonella* contamination on ground turkey meat and a powder infant formula has been evaluated. PhageDx proved to be effective for both matrices since no false positive results were noted [50]. This suggests that application of bacteriophages in food safety may go beyond direct elimination of bacteria and phages can be potentially important tools in preventive strategies including microbiological detection systems.

1.4. Escherichia coli

Although *E. coli* is a natural inhabitant of the intestinal microflora of all mammals, at the same time it is the cause of many serious intestinal and extraintestinal diseases, including infant meningitis or sepsis [51]. There are two main groups of pathogenic strains of *E. coli*. The first is diarrheal *E. coli*, also known as enteric pathogenic *E. coli* (IPEC), which causes diarrhea or intestinal flu. This group includes well-known pathogens such as enterotoxigenic *E. coli* (ETEC), enteroinvasive *E. coli* (EIEC), enteropathogenic *E. coli* (EPEC) and shigatoxigenic or Shiga toxin-producing *E. coli* (STEC) with the subgroup of enterohemorrhagic *E. coli* (EHEC). The second group is extraintestinal pathogenic *E. coli* (ExPEC), which contributes to infections outside the intestines [52]. The bacterium is highly contagious as even a small number of bacterial cells may cause a disease. It is transmitted mainly by food and contaminated water [28].

In a recent study, Vengarai Jagannathan et al. (2021) investigated potential application of a bacteriophage cocktail in reducing the growth of pathogenic *E. coli* O157:H7 on food products. Fresh spinach leaves were rinsed for 10 min with sterile drinking water loaded with a mixture of *E. coli* and bacteriophages (MOI 2:3). Compared to the control group in which vegetables were dunk-washed with a water suspension of bacteria, a reduction in pathogen count of 99% was achieved [53].

Mangieri et al. (2020) used bacteriophages specific to STEC to reduce bacterial contamination of fresh cucumbers. A phage cocktail composed of three bacteriophages was tested on vegetables at two temperatures of 25 °C and 4 °C for 24 h. The number of pathogenic bacteria was reduced by 1.16 log CFU/g after 6 h and 2.01 log CFU/g after 24 h at a temperature of 4 °C and 1.97 CFU/g after 6 h and 2.01 log CFU/g after 24 h at 25 °C. The authors stated that phage cocktails may be potentially used in controlling bacterial contamination in fresh vegetables [54].

These observations were confirmed in another study by Dewanggana et al. (2022), who investigated the potential use of bacterial viruses to eliminate ETEC from different food products (chicken meat, fish meat, cucumber, tomato, lettuce). Food samples were rinsed with *E. coli* and then the phage lysate was used to remove the contamination. The samples were incubated at 4 °C and the number of bacteria was determined after 24 h and 6 days. For all food products a more significant effect was observed after 6 days of incubation. The highest efficiency in reducing *E. coli* was observed for the chicken meat as the number of bacteria decreased by 80.93% and 87.29% at days 1 and 6, respectively. The effect was found to be the weakest for lettuce as bacterial contamination was reduced by 46.88% and 43.38% for both tested temperatures [55].

Importantly, phages are also effective in removing the biofilm structure formed by *E. coli* on food. The influence of AZO145A bacteriophage on *E. coli* biofilms on beef was investigated by Wang et al. (2020). Meat treated with ε-polylysine was used as a control. The use of both phage and ε-polylysine showed beneficial effects in reducing biofilm transmission between pieces of meat. The elimination of biofilm structure was comparable for both disinfectants as phages reduced the bacterial count by 3.1 log CFU/coupon and ε-polylysine 2.9 log CFU/coupon. However, it was found that without increasing the doses of both preparations, biofilm reappeared in the trials. Such observations notwithstanding, AZO145A phage may be considered a promising agent in combating *E. coli* in meat products [56].

Choi et al. (2021) used a novel approach in reducing *E. coli* on raw beef meat. The authors immobilized bacteriophages on the surface of the polymer film used as a standard for food packaging. The objective of the study was to develop a material which would be safe for use and at the same time would provide antibacterial safety. The bacteriophage T4, which is probably the best known representative of phages specific to *E. coli*, was covalently attached to a polycaprolactone (PCL) film. Then, portions of raw beef contaminated with *E. coli* were packed in the packages with the phage. The bacterial growth was measured after 30 min and compared with the standard packages with no phage. The reduction of bacterial population was 2.44 log compared to the control group, demonstrating the possibility of using phages in systems of food packaging [57].

1.5. Listeria

Listeria monocytogenes is a Gram-positive, anaerobic, rod-shaped bacterium [58]. It is a foodborne zoonotic pathogen, which means that it is naturally transmissible from vertebrate animals to humans. In humans it causes listeriosis, the symptoms of which include encephalitis, meningitis, sepsis and gastrointestinal disorders. It is particularly dangerous for pregnant women since it may lead to miscarriages [59].

The recent findings confirm that bacteriophages may show high activity against *L. monocytogenes* strains found in meat and meat-derived products. Importantly, it has been observed that they do not have a detrimental effect on the natural microflora of the consumers. Moreover, phages have also been proven to be stable under refrigerated storage [60]. Undoubtedly, the aforementioned features are among those that have led to the extensive development of *Listeria* targeting commercial bacteriophage products.

Currently available phage preparations against *Listeria* include Phage LM-103a, Phage LMP-102a, Phage Ply511, ListShield (all of them Intralytix, USA), and Listex P-100 (Micreos Food Safety B.V.; The Netherlands) [61,62]. The effectiveness of these preparations in controlling *Listeria* contamination in different food products has been confirmed in several studies.

ListShield (formally known as LMP-102) was the first phage-based preparation to receive FDA approval, in 2006, for direct application on meat and poultry products that meet the ready-to-eat definition [63]. ListShield is a cocktail composed of six bacteriophages with high lytic activity against *Listeria* strains. It was shown that it significantly reduces the presence of pathogens in various types of RTE foods, with the efficiency of 82–99% [64]. Ishaq et al. (2022) reported that in the case of smoked salmon, the use of ListShield completely inhibited the growth of *L. monocytogenes* on naturally contaminated samples as well as those infected in an in vitro experimental model. Importantly, there were no organoleptic changes in the tasted samples, which is important from the perspective of the consumers. Moreover, ListShield has also been found to have bactericidal properties on surfaces, which may be of great importance in the context of protecting, for example, processing plants in the food industry against *Listeria* contamination [3].

Shortly after the ListShield approval, another *Listeria*-targeting preparation, Listex P100, was given GRAS status by the FDA and a positive opinion by the EFSA, which stated that the product is not only safe but also effective [65]. Listex P100, is composed of six *L. monocytogenes* specific phages and it is recommended for the reduction of bacterial contamination on RTE products of animal origin [66]. The product showed high specificity in trials performed at different temperatures. The largest decrease in bacterial count achieved was 4.44 log CFU/mL when the inoculum count was 3 log CFU/g [64]. Importantly, Listex P100 has also been proved to be effective in reducing the biofilm structure formed by *L. monocytogenes* on stainless steel with a reduction of 3.5–5.4 log CFU/cm^2 compared to a control group in which no antibacterial agent was applied [67]. Compared to standard antibacterials, such as nisin and sodium lactate, Listex P100 has been proved to be more effective in eliminating *L. monocytogenes* on ready-to-eat, sliced ham [68]. Notably, former observations were confirmed that application of the preparation does not affect the organoleptic properties of the product [66].

Going beyond products of animal origin, potential application of phages in combating *Listeria* contamination on fruits has also been investigated. Phages were administered by injection into the pulp of contaminated melons and apples or evenly spread on the surface of the fruits. Bacterial contamination in melons decreased by 4 log units compared to the control group in which fruits were treated with nisin. Notably, in the case of apples, the amount of bacteria declined by only 0.4 log units compared to bacteriocin treated products. Furthermore, the effectiveness of nisin should be noted, as on fruits protected with this preparation the number of *Listeria* declined by 6 log units in the case of melon and by 2 log units in the samples with apples compared to the control group in which no disinfectant was applied [69].

1.6. Campylobacter

Campylobacter jejuni, one of the best known representatives of the Campylobacteraceae family, is one of the most important foodborne pathogens. It naturally inhabits the intestines of birds and mammals. However, in particular conditions, it may cause gastroenteric infections called campylobacteriosis, and hence *C. jejuni* is among the most common causes of zoonotic illnesses worldwide [28]. Infections are frequently caused by drinking contaminated water or raw milk and eating contaminated meat, especially chicken or beef, and other animal-derived products. The most common symptom of infection is diarrhea, but infected people may sporadically develop secondary diseases, such as Guillain–Barré or Miller–Fisher syndrome [70].

Thung et al. (2020) explored the potential use of bacteriophages as microbiological tools to control *C. jejuni* contamination in chicken meat and mutton. The meat was sliced and infected with bacteria, then incubated and sprayed with the phage preparation. The samples were stored at 4 °C for 2 days and the amount of *C. jejuni* was determined. The authors observed a decrease of bacterial number by 1.68 CFU/g and 1.70 CFU/g for chicken meat and mutton, respectively. These findings suggest the possibility for use of bacterial viruses in strategies against *C. jejuni* in meat products [71].

This hypothesis was supported by the results of another study in which two phages, Φ7-izsam and Φ16-izsam, were used to counteract the development of antimicrobial-resistant *C. jejuni* in poultry. First, animals were tested in terms of natural infections with the bacterium. Then, uninfected birds were orally given a *C. jejuni* suspension. Bacteriophages were provided to the animals prior to slaughter. Compared to the control group, bacterial counts in cecal content of animals treated with phage preparations were significantly reduced. The number of *C. jejuni* decreased by 1 log CFU/g and 2 log CFU/g for two groups of animals in which the phage preparation was given at different time points [72].

In a similar study of Richard et al. (2019), bacterial viruses were used to control *C. jejuni* in broiler chickens. Four days after the birds were infected, bacteriophages CP20 and CP30A were orally administered to the animals. The chickens were sacrificed every 24 h and *Campylobacter* counts in the intestinal lumen were determined. As a result, a remarkable reduction in a bacterial number in phage-treated groups was observed. The most significant effect was noted 2 days after the bacteriophage application, as the bacterial number decreased by 2.4 log CFU/g compared to the control group of infected, untreated animals. An important objective of the study was to investigate the influence of the phage treatment on the gut microflora of the animals. The results indicated that bacteriophages did not affect birds' microbiota, which supports previous observations regarding the safety of phage treatment [73]. Of note, recent studies on bacteriophages of different specificity showed that phage application influences the microbiological balance in birds' intestines [74–77]; however, reliably answering whether this influence is harmful or harmless for the animals requires further, comprehensive research.

The efficacy of bacteriophages in eliminating *C. jejuni* in chicken meat was also confirmed in the studies of Zampara et al. (2017). The authors isolated phages capable of reducing *C. jejuni* at a chilled temperature and then created a phage cocktail composed of two phages with the highest lytic spectrum. The cocktail reduced the number of bacteria by 0.73 log units compared to the control group. Remarkably, the results confirmed that phages may be used in protecting the chicken meat against *C. jejuni* also at lower temperatures [78].

It is noteworthy that campylophages, compared to other bacterial viruses, have some features which make their application difficult. Żbikowska et al. (2020), in a recent review on potential use of phages in the poultry industry, listed in this regard: (i) the problems associated with the optimization methods for phage isolation, propagation and purification; (ii) difficulties with appropriate selection of phage candidates for application due to the significant differences between *Campylobacter* phages within groups, even they are genetically very similar; (iii) evidenced phage resistance of *Campylobacter* after phage treatment; and (iv) the cost of production [79]. Notwithstanding these limitations, phage treatment of *Campylobacter* infections remains an appealing option.

1.7. Yersinia

Yersinia enterocolitica is a Gram-negative bacillus-shaped bacterium, which belongs to the Enterobacteriaceae family [80]. To date, 28 species have been identified, 3 of which are pathogenic to humans [81]. *Yersinia* spp. are heterogeneous species represented by six biotypes (1A, 1B, 2, 3, 4, 5) and different serogroups showing specific virulence factors correlated with the geographic region in which the bacteria occur [80]. This pathogen is mainly transmitted through raw food or water sources, causing gastrointestinal disease also known as yersinosis in humans. It is the fourth most frequently reported foodborne bacterial disease and a huge threat to human life [82]. The main symptoms of yersiniosis are fever, often hemorrhagic diarrhea, lymphadenitis, abdominal pain, and nausea [83]. Combating yersinosis is a great challenge since numerous representatives of this species have developed resistance to many frontline antibiotics such as penicillin, ampicillin, cephalosporine and macrolides [84]. Clearance of *Yersinia* contamination in food and food processing equipment is also difficult due to the ability of this pathogen to form biofilm structures [85].

Compared to phages specific to other foodborne pathogens, the knowledge regarding *Yersinia* phages is rather scarce. Jun et al. (2018) isolated and characterized four virulent bacteriophages specific to *Yersinia enterocolitica*. Bacteriophage fHe-Yen3-01 was assigned to the family Podoviridae, while three other phages (fHe-Yen9-01, fHe-Yen9-02 and fHe-Yen9-03) were assigned to the Myoviridae family. In subsequent tests, isolated viruses were used to reduce contamination in selected food products. Raw pork and milk were contaminated with the *Y. enterocolitica* O:9 Ruokola/71 strain, a genetically modified strain of *Yersinia* for use in in vitro tests. The bacterial count during the experiment decreased in raw pork from 2.3×10^3 CFU/g to 2.12×10^2 CFU/g and in milk from 4.15×10^3 CFU/mL below the detection limit (<10 CFU/ml). In the case of RTE pork, the reduction of bacterial count was from 2.1×10^3 CFU/g to 3.8×10 CFU/g. Furthermore, potential application of phages against *Yersinia* contamination on kitchen utensils, such as cutting boards, wooden spoons and kitchen knives, has been evaluated. Tools were immersed in a bacterial inoculum with the concentration of 10^4 CFU/mL, then they were exposed to phages directed against *Y. enterocolitica*. The highest efficiency was demonstrated for the fHe-Yen9-01 phage, which reduced the number of bacteria by 1/3 compared to the initial bacterial count. These findings suggest potential use of phages to control the growth of *Y. enterocolitica* in food and everyday kitchen items [86].

Yersinia bacteriophages have also been used for the development of a selective tool for identification of this bacteria. Immune separation (IMS) was considered a promising approach in identifying *Yersinia* spp.; however, due to the existence of numerous serotypes of the pathogen, the method did not eventually find practical application. Therefore, many attempts have been made to modify this technique. One of them is based on use of magnetic microparticles together with RNA binding proteins (RBPs) Gp17, Gp47 and Gp37 of phages to selectively "capture" the epidemiological serotypes of *Y. enterolytica*. In the case of microparticles coated with RBP Gp17, it was possible to detect the O:3 type, which is the most virulent serotype causing yersiniosis. Moreover, use of Gp47 and Gp37 allowed for the identification of serotypes O:3, O:5, 27, O:8 and O:9. Notably, compared to the method based on antibodies, the modified IMS technique is stable to physical–chemical interactions [83]. The results confirm the potential of phages in the development of identification systems for different bacterial species, including food related pathogens.

2. Perspectives

Microbial safety is one of the priority issues in the food industry as it directly affects consumers' health. Currently applied antibacterial strategies have their limitations and thus solutions that are easy to implement, do not influence product quality, and are safe and cheap are eagerly anticipated.

Bacteriophage effectivity against foodborne pathogens has been confirmed in numerous studies, and the potential of bacterial viruses in strategies against this group of bacteria has been noted by many scientists (Table 1) [42,60,87]. It is worth emphasizing that this review presents recent studies on the effectiveness of bacterial viruses in combating only selected food-related bacteria, and research studies in this field are much more advanced. They also include such other important foodborne pathogens as e.g., *Shigella* sp. [49,88], *Staphylococcus aureus* [89,90], *Pseudomonas* [91,92] and *Vibrio* [87].

Despite the fact that food safety is the industrial area in which commercialization of bacteriophage-based products shows the fastest progress, popularization of phage application is still in its infancy (Table 2). Over the last 12 years, the number of acceptable bacteriophage preparations approved for use in the food industry has increased. In 2006, the FDA issued the first approval for a bacteriophage cocktail against *L. monocytogenes* on RTA meat and poultry products, which was the aforementioned ListShield (LMP-102) (Intralytix, Columbia, USA). Later, approvals by the FDA were issued for other preparations such as PhageGuard and Listex (Micreos Food Safety B.V., Wageningen, The Netherlands), EcoShield (Intralytix, Columbia, USA), ShigaShield (Intralytix, Columbia, USA) and SalmoFresh (Intralytix, Columbia, USA). GRAS status was given to SalmoFresh (Intralytix, Columbia, USA) and PhageGuard (Micreos Food Safety B.V., Wageningen, The Netherlands) [93]. Currently 13 phage preparations for food safety applications have been approved by different North American or European institutions, most of them dedicated to fighting *E. coli* (6) and *Salmonella* spp. (4). Other preparations are active against *L. monocytogenes* (2) and *Shigella* spp. (1) [25]. It is worth noting a phage-based preparation dedicated for applications in agriculture, AgriPhage, offered by Omnilytix (USA), is active against *Xanthomonas campestris* pv. *vesicatoria* and *Pseudomonas syringae* pv. *tomato* and is available for sale [94]. Importantly, there are also preparations which have not yet received acceptance by western institutions but have been commercialized on the Asian market. As an example, BAFASAL (Proteon, Łódź, Poland), available as a feed additive and targeting *S.* Typhimurium and *S.* Enteritidis, still awaits commercialization on the European market [95].

Despite the many advantages of bacteriophage-based antibacterial strategies, there are also some drawbacks, which have to be considered particularly in the context of widespread phage application. One of the best known is the phage ability to randomly transfer fragments of the bacterial genetic material in the process called transduction [96]. Obviously transduction may include potentially harmful genes of virulence or antibiotic resistance contributing to the environmental spread of the strains of increased risk for public health. Clearly, this potential threat has to be considered with respect to intended application of bacterial viruses.

Furthermore, the direct influence of phages on the human organism has not been comprehensively investigated. Notably, the assumption regarding safety of the use of phages results primarily from many years of clinical experience rather than scientific knowledge. Numerous research studies suggest the ability of bacterial viruses to interact with mammalian cells [97–99]; consequently, such mutual interactions should be thoroughly investigated.

As mentioned above, phage specificity is also a substantial limitation in widespread application of bacteriophages, particularly when food products are contaminated with different foodborne pathogens. However, this may be easily overcome by using a phage cocktail with a broad lytic spectrum and targeting different bacteria. Phage resistance may influence activity of a phage preparation, but including several phages with similar specificity into one cocktail should ensure activity of the product. From a purely practical point of view, isolation, propagation and application of phages for food safety are rather easy. Nevertheless, for development on a conventional product all the procedures have to be well established, standardized and repeatable. This, together with the extremely rigorous requirements concerning newly registered products, makes the pace of approval of commercial phage preparations still insufficient.

The question of whether commercialization of phage application in food safety can be expedited remains open. On one hand, there is an emerging global problem of drug resistance of many bacterial strains which demands new solutions for fighting microbial contamination. This issue also naturally concerns the food industry. On the other hand, we are at a point in history when mankind is about to face a major food crisis. Improving the safety of food and thus safeguarding it from wastage seems to be a critical challenge. Undoubtedly, multifaceted application of bacteriophages in combating foodborne pathogens is an important option.

Table 1. Exemplary studies on bacteriophage use against selected foodborne pathogens.

Bacteria	Symptoms	Transmission	Phages	Results	References
Campylobacter jejuni	Fever, muscle aches, headaches, arthralgia, abdominal pain and cramps, weakness, bloody diarrhea, gastric or intestinal pain, occurrence of Guillain-Barré syndrome	Poultry meat, milk, contaminated water, swimming in contaminated water bodies, contact with animal.	CJ01	Mutton and chicken meat were stored at 4 °C and injected with 5 mL of *C. jejuni* with a concentration of 10^4 CFU/mL. The samples prepared in this way were incubated for 4 h. Then, they were sprayed with 5 mL of bacteriophage with PFU/mL, and the samples were again incubated for 48 h. The final result was 10^2 CFU/g.	[4]
			Φ7-izsam Φ16-izsam	The influence of bacteriophages on naturally or artificially contaminated poultry was investigated. Bacteriophages were given to the animals before slaughter and resulted in a reduction of 1 log10 CFU/g and 2 log10 CFU/g for both test groups.	[72]
			CP20 and CP30A	Poultry were infected and bacteriophages were administered 4 days later. Chickens were sacrificed every 24 h and the intestinal pathogen concentration was examined. The most prominent result was obtained on the second day of incubation and caused a decrease of bacteria by 2.4 log CFU/g in relation to the control.	[73]
			12673, P22, 29C;	The contaminated skin of chickens was examined. The level of the pathogen decreased by 2 log units when using the MOI of the phage 100:1 or 1000:1.	[100]

Table 1. *Cont.*

Bacteria	Symptoms	Transmission	Phages	Results	References
Escherichia coli	Vomiting, headaches, stomach pain, low-grade fever or fever, diarrhea, weakness, bloody stools, hemolytic uremic syndrome, neonatal meningitis, pneumonia, sepsis	Pork, poultry, contaminated ruminants such as goats, deer, sheep, elk, water, milk and dairy products, direct contact with animals	FM10, DP16 and DP19	The phage cocktail was tested on fresh intubated cucumber at two temperatures of 4 °C and 25 °C for 24 h. The number of bacteria was reduced by 1.97–2.01 log CFU/g and 1.16–2.01 log CFU/g at 25 °C and 4 °C.	[54]
			DW-EC	It was tested on many matrices, such as chicken meat, lettuce meat, fish meat, and tomato. The samples were contaminated with bacteria, then they were subjected to the phage section. A significant result was obtained at 6 days of incubation. The best effect was seen in the chicken feed samples where the pathogen value decreased by 80.93% after the first day and 87.29% after the 6th day. The weakest effect was observed on lettuce leaves.	[55]
			AZO145A	The effect of phage on the biofilm was investigated. Exposure to bacteriophage at a concentration of 10^{10} PFU/cm^2 for 2 h resulted in a 4.0 log 10 PFU/mL reduction in biofilm on stainless steel. However, on the surface of beef, at 48 h incubation, the pathogen decreased by 3.1 log10 CFU/g.	[56]
			T4	The aim of the study was to design an antimicrobial package by using the immobilization of T4 phage (10^5 CFU/mL) on the surface of the PCL foil. Contaminated beef was placed in this package. The bacterial concentration applied to the meat was 10^7 CFU/mL. After 48 h of incubation, the concentration of bacteria was reduced by 3 log CFU/mL.	[84]
			FAHEc1	Contaminated raw beef as a test matrix; after using phage, the concentration of bacteria decreased by 2 or 4 log units at the appropriate storage temperatures, 24 °C and 37 °C.	[101]

Table 1. *Cont.*

Bacteria	Symptoms	Transmission	Phages	Results	References
Listeria monocytogenes	Fever, chills, muscle aches, headaches, nausea and vomiting, confusion, local infections, inflammation of the lymph nodes, inflammation of the lungs, joints, bone marrow, pericarditis and myocarditis, inflammation of the eyeball, gastrointestinal infections.	Raw vegetables and fruit, unpasteurized dairy products (milk, cheese, ices cream), raw, cooked and frozen poultry meat, raw and smoked fish, delicatessen products, semi-finished products, fast-food products, soil, sewage, water, rotting plants, silage, wild and farm animals.	FWLLm1	Bacterial levels dropped by 2 log units on the surface of the chicken that had become contaminated with *Listeria*. The samples were stored in a vacuum package at 4 °C and 30 °C. A positive result was observed only for the sample kept at 30 °C.	[7]
			A511	Bacterial levels were tested in milk chocolate, mozzarella and brie cheese. The phage were given and incubated at 6 °C. Bacteria concentration dropped by 5 log units.	[102]
Pseudomonas spp.	Pneumonia, fever, chills, severe shortness of breath, cough, confusion, chronic lower respiratory tract infection, Roth's spots, i.e., petechiae on the retina, small painless erythematous changes on the hands and feet—Janeway symptom, painful reddish lumps on the fingers—Osler's nodules, subungual petechiae	Water, soil, human and animal digestive tract.	UFJF_PfDIW6, UFJF_PfSW6	The lyophilized phage cocktail was incubated with raw milk at 4 °C for 7 days. After the incubation period, the *Pseudomonas* bacterial population decreased by 3.2 log CFU/mL.	[103]
			V523, V524, JG003	Three bacteriophages used separately and together as a cocktail were used to biocontrol bacteria in the water. The effect of the phages was tested against two bacteriophage strains: PAO1 and the environmental strain 17V1507. Of all the bacteriophages, V523 was most effective in reducing the PAO1 strain (>2.4 log10). The other strain was sensitive only to JG003, resulting in its reduction by 1.2 log10. The phage cocktail resulted in higher reductions in PAO1 (>3.4 log10) compared to using them alone. In contrast, the same reduction was observed in 17V1507 as with JG003 alone.	[104]

Table 1. *Cont.*

Bacteria	Symptoms	Transmission	Phages	Results	References
Salmonella spp.	Abdominal pain, vomiting, diarrhea, fever, headache, chills, reduced urine output, dry mucous membranes, excessive sleepiness, apathy.	Chicken, turkey, pig, duck, goose meat, eggs, soil, water, cheese, milk, fruit, vegetables, contact with contaminated animals	LPSTLL, LPST94, LPST153	The use of the bacteriophage cocktail decreased the concentration of bacteria by 3 log units. Chicken breasts were inoculated using an inoculum. The influence on the biofilm created by *Salmonella* was also examined, the administered cocktail effectively inhibited the growth after 72 h, the microplates decreased by 5.23 log units.	[46]
			LYPSET	The biocontrol was tested in milk and on lettuce leaves. In milk samples there was a decrease of bacteria by 2.19 log CFU/mL at 4 °C and 4.3 log CFU/mL at 25 °C. However, in the samples containing lettuce, there was a decrease of 2.2 log CFU/mL at 4 °C and 2.34 log CFU/mL at 25 °C at MOI = 10,000.	[45]
			SE07	The effects of phage on eggs, beef and poultry meat were tested. The best effect for beef was obtained after 48 h of incubation; the bacterial value dropped from 4.23 log CFU/mL to 2.11 log CFU/mL, while for chicken the effect was even better as there was a decrease from 4.16 log CFU/mL to 2.14 log CFU/mL also at 48 h.	[47]
			SJ2	The use of phage resulted in a significant reduction of bacteria in the soft pork and eggs. Incubation was carried out at 4 °C.	[105]
			BSPM4, BSP101, BSP22A	The phages were presented as a cocktail. The reduction of bacterial colonies was tested on lettuce leaves and fresh cucumber. There was a reduction of 4.7 log for lettuce and 5.8 log for cucumber.	[106]

Table 1. *Cont.*

Bacteria	Symptoms	Transmission	Phages	Results	References
Shigella sp.	Vomiting, anorexia, abdominal cramps, bowel urgency, severe watery diarrhea, fever, diarrhea with an admixture of mucus and blood, rapid breathing, heart rate, low blood pressure, dry mouth and skin (dehydration), pain on palpation of the abdomen	Touching skin of contaminated person, oral cavity (fecal–oral route), contaminated water and food, sexual contact, swimming in contaminated water, by insects, such as housefly.	SSE1, SGF3, SGF2,	The influence of the SFE3 phage and its combination as a cocktail with other phages on the reduction of biofilm on polystyrene surfaces was investigated. It was found that the single SGF2 phage (isolated from wastewater) had the greatest impact on the development of biofilm; it caused growth inhibition by 26.6%. The lowest results were obtained for SGF3, while adding it to a phage cocktail increased its effectiveness by 25%. The phage is active against strains such as *S. dysenteriae*, *S. baumannii*, and *S. flexneri*.	[107]
			SD-11, SF-A2, SS-92	The number of pathogens was decreased by 4 log in chicken meat, when they applied phage cocktail. It was stored at 4 °C.	[108]
Staphylococcus sp.	Infections of the skin and subcutaneous tissue, which are characterized by the presence of purulent discharge, impetigo, folliculitis, boils, furunculosis (multiple boils), abscesses, inflammation of the sweat glands and inflammation of the mammary gland, high fever, drop in blood pressure, organ dysfunction	Transmission mainly by direct contact. Patients after surgery are most at risk	MDR, ME18, ME126	Reducing biofilm in UHT milk at 25 °C using ME18 (MOI = 10) and MDR. They reduce biofilm in milk. However ME126 (MOI = 10) at 37 °C reduces CFU/mL by 87.2% compared to control sample.	[109]
Vibrio parahaemolyticus	Watery diarrhea, abdominal cramps, nausea, vomiting, fever or chills, abscess formation, otitis media, otitis media and conjunctivitis	Contact with contaminated water, fruit, seafood	PVP1 and PVP2	They treated sea cucumber contaminated with pathogen. MOI = 10 or MOI = 100. Test were performed at in 20 °C and it increased survival of sea cucumber to 80% compared with control sample without phage cocktail treatment, which was only 30%.	[110]

Table 1. *Cont.*

Bacteria	Symptoms	Transmission	Phages	Results	References
Yersinia	Mild or high fever, cramping abdominal pain, loose stools often with mucus or blood, vomiting, right-hand stomach pain, tenderness when examining the abdominal cavity, fast heartbeat, joint pains, mainly in the knee, ankle and wrist, rapid breathing	Pork and pork offal, milk, water, raw vegetables and fruits.	fHe-Yen3-01 fHe-Yen9-01, fHe-Yen9-02 and fHe-Yen9-03	Infected raw pork and cooking tools with the Rukola/71 strain. The kitchen tools were immersed in an inoculum at a concentration of 10^4 CFU/mL. The best effect was obtained for the phage fHe-Yen9-01, which reduced the number of bacteria by 1/3.	[86]
			PY100	The given phage reduced the amount of bacteria in the meat MOI = 10^2 by 3 log10 units after 24 h incubation and at a MOI = 10^4 by 5 log10 units after 1.5 h incubation at 37 °C. However, when incubated at 4 °C, the bacteria count decreased by 2 log units after 24 h.	[111]

Table 2. Selected commercial bacteriophage preparations.

Company	Product	Target	Reference	Regulatory & Certifications
Micreos Food Safety (The Netherlands)	PhageGuard Listex	*Listeria* sp.	[65,112,113]	Halal, OMRI, Kosher, Skal, FSSC 2200; FDA, GRN 198/21, EFSA; Swiss BAG; Israel Ministry of Health; Health Canada
	PhageGuard S	*Salmonella enterica*	[114]	Halal, FSSC 22000; FDA, GRN 468; USDA, FSIS Directive 7120.1, Swiss BAG
	PhageGuard E	*Escherichia coli* O157:H7	-	FSSC 22000
Intralytix (USA)	ListShield	*Listeria monocytogenes*	[3,66,115–117]	Kosher; Halal; OMRI; FDA, 21 CFR 172.785; FDA, GRN 528;
	SalmoFresh	*Salmonella enterica*	[5,49,118–120]	Kosher; Halal; OMRI FDA, GRN 435; USDA, FSIS Directive 7120.1
	ShigaShield	*Shigella* sp.	[88,121,122]	FDA, GRN 672
	EcoShield PX	*Escherichia coli*	[9,123–125]	FDA, GRN 834; USDA, FSIS Directive 7120.1
	CampyloShield	*Campylobacter* spp.	[126]	GRAS
Proteon Pharmaceuticals SA (Poland)	Bafasal	*Salmonella enterica*	[96,127]	-
	Bafador	*Pseudomonas* sp., *Aeromonas* sp.	-	-
Passport Food Safety Solutions	Finalyse	*E. coli* O157:H7	-	USDA, FSIS Directive 7120.1
Phagelux	SalmoPro	*Salmonella* spp.	-	FDA, GRN 603; USDA
FINK TEC GmbH (Hamm, Germany)	Secure Shield E1	*E. coli*	[93]	FDA,GRN 724
Arm and Hammer Animal & Food Production (USA)	Finalyse SAL	*Salmonella*	[128]	-

3. Conclusions

Food products are one of the main routes of transmission of infectious diseases throughout the human population. Therefore, effective, safe and easily implemented methods ensuring protection of consumers are highly desired.

Bacteriophages have properties that make them excellent candidates in antibacterial approaches, especially in the food safety sector. Numerous studies confirm the high efficiency of phages in eliminating various foodborne pathogens. This includes drug-resistant bacteria and highly stable biofilm structures, both representing a particular concern for food producers. At the same time, application of phages does not influence the quality of food products and is easy and safe. Although commercialization of phage-based products in food safety has been steadily progressing, much still needs to be done to achieve widespread use of phage preparations. Nevertheless, considering the limitations and nearly exhausted potential of currently applied antibacterial methods, new strategies are highly desirable. Undoubtedly, application of phage products may be considered an attractive choice.

Funding: This research received no external funding.

Institutional Review Board Statement: Not applicable.

Informed Consent Statement: Not applicable.

Data Availability Statement: Not applicable.

Conflicts of Interest: The authors declare no conflict of interest.

References

1. García, P.; Martínez, B.; Obeso, J.M.; Rodríguez, A. Bacteriophages and their application in food safety. *Lett. Appl. Microbiol.* **2008**, *47*, 479–485. [CrossRef]
2. Sillankorva, S.M.; Oliveira, H.; Azeredo, J. Bacteriophages and Their Role in Food Safety. *Int. J. Microbiol.* **2012**, *2012*, 863945. [CrossRef]
3. Ishaq, A.; Ebner, P.D.; Syed, Q.A.; Rahman, H.U. Employing list-shield bacteriophage as a bio-control intervention for *Listeria monocytogenes* from raw beef surface and maintain meat quality during refrigeration storage. *LWT* **2020**, *132*, 109784. [CrossRef]
4. Thung, T.Y.; Lee, E.; Mahyudin, N.A.; Anuradha, K.; Mazlan, N.; Kuan, C.H.; Pui, C.F.; Ghazali, F.M.; Rashid, N.-K.M.A.; Rollon, W.D.; et al. Evaluation of a lytic bacteriophage for bio-control of *Salmonella* Typhimurium in different food matrices. *LWT* **2019**, *105*, 211–214. [CrossRef]
5. Li, C.; Shi, T.; Sun, Y.; Zhang, Y. A Novel Method to Create Efficient Phage Cocktails via Use of Phage-Resistant Bacteria. *Appl. Environ. Microbiol.* **2022**, *88*, e02323-21. [CrossRef]
6. Ramos-Vivas, J.; Elexpuru-Zabaleta, M.; Samano, M.L.; Barrera, A.P.; Forbes-Hernández, T.Y.; Giampieri, F.; Battino, M. Phages and Enzybiotics in Food Biopreservation. *Molecules* **2021**, *26*, 5138. [CrossRef]
7. Bigot, B.; Lee, W.-J.; McIntyre, L.; Wilson, T.; Hudson, J.A.; Billington, C.; Heinemann, J.A. Control of *Listeria monocytogenes* growth in a ready-to-eat poultry product using a bacteriophage. *Food Microbiol.* **2011**, *28*, 1448–1452. [CrossRef]
8. Bandara, N.; Jo, J.; Ryu, S.; Kim, K.P. Bacteriophages BCP1-1 and BCP8-2 require divalent cations for efficient control of *Bacillus cereus* in fermented foods. *Food Microbiol.* **2012**, *31*, 9–16. [CrossRef]
9. Ferguson, S.; Roberts, C.; Handy, E.; Sharma, M. Lytic bacteriophages reduce *Escherichia coli* O157: H7 on fresh cut lettuce introduced through cross-contamination. *Bacteriophage* **2013**, *3*, e24323. [CrossRef]
10. Kang, H.W.; Kim, J.W.; Jung, T.S.; Woo, G.J. wksl3, a New Biocontrol Agent for *Salmonella enterica* Serovars Enteritidis and Typhimurium in foods: Characterization, Application, Sequence Analysis, and Oral Acute Toxicity Study. *Appl. Environ. Microbiol.* **2013**, *79*, 1956–1968. [CrossRef]
11. Strange, J.E.; Leekitcharoenphon, P.; Møller, F.D.; Aarestrup, F.M. Metagenomics analysis of bacteriophages and antimicrobial resistance from global urban sewage. *Sci. Rep.* **2021**, *11*, 1600. [CrossRef]
12. Chanishvili, N. Phage therapy—history from Twort and d'Herelle through Soviet experience to current approaches. *Adv. Vir. Res.* **2012**, *83*, 3–40. [CrossRef]
13. Sharma, S.; Chatterjee, S.; Datta, S.; Prasad, R.; Dubey, D.; Prasad, R.K.; Vairale, M.G. Bacteriophages and its applications: An overview. *Folia Microbiol.* **2017**, *62*, 17–55. [CrossRef]
14. Kasman, L.M.; Porter, L.D. *Bacteriophages*; StatPearls Publishing: Treasure Island, FL, USA, 2022.
15. Kochhar, R. The virus in the rivers: Histories and antibiotic afterlives of the bacteriophage at the sangam in Allahabad. *Notes Rec. R Soc. Lond.* **2020**, *74*, 625–651. [CrossRef]
16. Dowah, A.S.; Clokie, M.R. Review of the nature, diversity and structure of bacteriophage receptor binding proteins that target Gram-positive bacteria. *Biophys. Rev.* **2018**, *10*, 535–542. [CrossRef]

17. Abdelsattar, A.S.; Dawooud, A.; Rezk, N.; Makky, S.; Safwat, A.; Richards, P.J.; El-Shibiny, A. How to Train Your Phage: The Recent Efforts in Phage Training. *Biologics* **2021**, *1*, 70–88. [CrossRef]

18. Abedon, S.T.; García, P.; Mullany, P.; Aminov, R. Phage Therapy: Past, Present and Future. *Front. Microbiol.* **2017**, *8*, 981. [CrossRef]

19. Principi, N.; Silvestri, E.; Esposito, S. Advantages and limitations of bacteriophages for the treatment of bacterial infections. *Front. Pharmacol.* **2019**, *10*, 513. [CrossRef]

20. Domingo-Calap, P.; Delgado-Martínez, J. Bacteriophages: Protagonists of a Post-Antibiotic Era. *Antibiotics* **2018**, *7*, 66. [CrossRef]

21. Galtier, M.; De Sordi, L.; Maura, D.; Arachchi, H.; Volant, S.; Dillies, M.A.; Debarbieux, L. Bacteriophages to reduce gut carriage of antibiotic resistant uropathogens with low impact on microbiota composition. *Environ. Microbiol.* **2016**, *18*, 2237–2245. [CrossRef]

22. Arumugam, S.N.; Manohar, P.; Sukumaran, S.; Sadagopan, S.; Loh, B.; Leptihn, S.; Nachimuthu, R. Antibacterial efficacy of lytic phages against multidrug-resistant *Pseudomonas aeruginosa* infections in bacteraemia mice models. *BMC Microbiol.* **2022**, *22*, 187. [CrossRef] [PubMed]

23. Ryan, E.M.; Gorman, S.P.; Donnelly, R.F.; Gilmore, B.F. Recent advances in bacteriophage therapy: How delivery routes, formulation, concentration and timing influence the success of phage therapy. *J. Pharm. Pharmacol.* **2011**, *63*, 1253–1264. [CrossRef] [PubMed]

24. Wang, Z.; Zhao, X. The application and research progress of bacteriophages in food safety. *J. Appl. Microbiol.* **2022**, *133*, 2137–2147. [CrossRef]

25. Vikram, A.; Woolston, J.; Sulakvelidze, A. Phage Biocontrol Applications in Food Production and Processing. *Curr. Issues Mol. Biol.* **2021**, *40*, 267–302. [CrossRef]

26. Örmälä, A.M.; Jalasvuori, M. Phage therapy: Should bacterial resistance to phages be a concern, even in the long run? *Bacteriophage* **2013**, *3*, e24219. [CrossRef]

27. Molina, F.; Menor-Flores, M.; Fernández, L.; Vega-Rodríguez, M.A.; García, P. Systematic analysis of putative phage-phage interactions on minimum-sized phage cocktails. *Sci. Rep.* **2022**, *12*, 2458. [CrossRef]

28. Bintsis, T. Foodborne pathogens. *AIMS Microbiol.* **2017**, *I*, 529–563. [CrossRef]

29. Anvar, A.A.; Ahari, H.; Ataee, M. Antimicrobial properties of food nanopackaging: A new focus on foodborne pathogens. *Front. Microbiol.* **2021**, *12*, 690706. [CrossRef]

30. Khare, S.; Tonk, A.; Rawat, A. Foodborne diseases outbreak in India: A Review. *Int. J. Food Sci. Nutr.* **2018**, *3*, 9–10.

31. Schirone, M.; Visciano, P.; Tofalo, R.; Suzzi, G. Foodborne Pathogens: Hygiene and Safety. *Front. Microbiol.* **2019**, *10*, 1974. [CrossRef]

32. Karygianni, L.; Ren, Z.; Koo, H.; Thurnheer, T. Biofilm Matrixome: Extracellular Components in Structured Microbial Communities. *Trends Microbiol.* **2020**, *28*, 668–681. [CrossRef] [PubMed]

33. Bai, X.; Nakatsu, C.H.; Bhunia, A.K. Bacterial Biofilms and Their Implications in Pathogenesis and Food Safety. *Foods* **2021**, *10*, 2117. [CrossRef] [PubMed]

34. Mizan, M.F.R.; Cho, H.R.; Ashrafudoulla, M.; Cho, J.; Hossain, M.I.; Lee, D.U.; Ha, S.D. The effect of physico-chemical treatment in reducing *Listeria monocytogenes* biofilms on lettuce leaf surfaces. *Biofouling* **2020**, *36*, 1243–1255. [CrossRef] [PubMed]

35. European Food Safety Authority; European Centre for Disease Prevention and Control. The European Union summary report on trends and sources of zoonoses, zoonotic agents and food-borne outbreaks in 2015. *EFSA J.* **2016**, *14*, e04634. [CrossRef]

36. Oliver, S.P. Foodborne Pathogens and Disease Special Issue on the National and International PulseNet Network. *Foodborne Pathog. Dis.* **2019**, *16*, 439–440. [CrossRef]

37. Pandiselvam, R.; Subhashini, S.; Banuu Priya, E.P.; Kothakota, A.; Ramesh, S.V.; Shahir, S. Ozone based food preservation: A promising green technology for enhanced food safety. *Ozone Sci. Eng.* **2019**, *41*, 17–34. [CrossRef]

38. Silva, A.; Silva, S.A.; Lourenço-Lopes, C.; Jimenez-Lopez, C.; Carpena, M.; Gullón, P.; Fraga-Corral, M.; Domingues, V.F.; Fátima Barroso, M.; Simal-Gandara, J.; et al. Antibacterial Use of Macroalgae Compounds against Foodborne Pathogens. *Antibiotics* **2020**, *9*, 712. [CrossRef]

39. Cho, I.H.; Ku, S. Current Technical Approaches for the Early Detection of Foodborne Pathogens: Challenges and Opportunities. *Int. J. Mol. Sci.* **2017**, *18*, 2078. [CrossRef] [PubMed]

40. Koch, B.J.; Hungate, B.A.; Price, L.B. Food-animal production and the spread of antibiotic resistance: The role of ecology. *Front. Ecol. Environ.* **2017**, *15*, 309–318. [CrossRef]

41. Koutsoumanis, K.; Allende, A.; Alvarez-Ordóñez, A.; Bolton, D.; Bover-Cid, S.; Chemaly, M.; Davies, R.; De Cesare, A.; Herman, L.; Lindqvist, R.; et al. Evaluation of public and animal health risks in case of a delayed post-mortem inspection in ungulates. *EFSA J.* **2020**, *18*, 12. [CrossRef]

42. Wernicki, A.; Nowaczek, A.; Urban-Chmiel, R. Bacteriophage therapy to combat bacterial infections in poultry. *Virol. J.* **2017**, *14*, 179. [CrossRef] [PubMed]

43. Khan, M.A.S.; Rahman, S.R. Use of Phages to Treat Antimicrobial-Resistant *Salmonella* Infections in Poultry. *Vet. Sci.* **2022**, *9*, 438. [CrossRef] [PubMed]

44. Pelyuntha, W.; Vongkamjan, K. Combined effects of *Salmonella* phage cocktail and organic acid for controlling *Salmonella* Enteritidis in chicken meat. *Food Control* **2022**, *133*, 108653. [CrossRef]

45. Yan, T.; Liang, L.; Yin, P.; Zhou, Y.; Mahdy Sharoba, A.; Lu, Q.; Dong, X.; Liu, K.; Connerton, I.; Li, J. Application of a Novel Phage LPSEYT for Biological Control of *Salmonella* in Foods. *Microorganisms* **2020**, *8*, 400. [CrossRef]

46. Islam, M.S.; Zhou, Y.; Liang, L.; Nime, I.; Yan, T.; Wang, X.; Li, J. Application of a Phage Cocktail for Control of *Salmonella* in Foods and Reducing Biofilms. *Viruses* **2019**, *11*, 841. [CrossRef]
47. Thung, T.Y.; Premarathne, J.M.K.J.K.; San Chang, W.; Loo, Y.Y.; Chin, Y.Z.; Kuan, C.H.; Tan, C.W.; Basri, D.F.; Radzi, C.W.J.W.M.; Radu, S. Use of a lytic bacteriophage to control *Salmonella* Enteritidis in retail food. *LWT* **2017**, *78*, 222–225. [CrossRef]
48. Available online: http://www.intralytix.com/files/prod/02SP/02SP-Desc.pdf (accessed on 1 October 2022).
49. Zhang, X.; Niu, Y.D.; Nan, Y.; Stanford, K.; Holley, R.; McAllister, T.; Narváez-Bravo, C. SalmoFresh™ effectiveness in controlling *Salmonella* on romaine lettuce, mung bean sprouts and seeds. *Int. J. Food Microbiol.* **2019**, *305*, 108250. [CrossRef]
50. Nguyen, M.M.; Gil, J.; Brown, M.; Cesar Tondo, E.; Soraya Martins de Aquino, N.; Eisenberg, M.; Erickson, S. Accurate and sensitive detection of *Salmonella* in foods by engineered bacteriophages. *Sci. Rep.* **2020**, *10*, 17463. [CrossRef]
51. Alonso, C.A.; González-Barrio, D.; Tenorio, C.; Ruiz-Fons, F.; Torres, C. Antimicrobial resistance in faecal *Escherichia coli* isolates from farmed red deer and wild small mammals. Detection of a multiresistant *E. coli* producing extended-spectrum beta-lactamase. *Comp. Immunol. Microbiol. Infect. Dis.* **2016**, *45*, 34–39. [CrossRef]
52. Riley, L.W. Extraintestinal Foodborne Pathogens. *Annu. Rev. Food Sci. Technol.* **2020**, *11*, 275–294. [CrossRef]
53. Vengarai Jagannathan, B.; Kitchens, S.; Priyesh Vijayakumar, P.; Price, S.; Morgan, M. Efficacy of Bacteriophage Cocktail to Control *E. coli* O157: H7 Contamination on Baby Spinach Leaves in the Presence or Absence of Organic Load. *Microorganisms* **2021**, *9*, 544. [CrossRef] [PubMed]
54. Mangieri, N.; Picozzi, C.; Cocuzzi, R.; Foschino, R. Evaluation of a Potential Bacteriophage Cocktail for the Control of Shiga-Toxin Producing *Escherichia coli* in Food. *Front. Microbiol.* **2020**, *11*, 1801. [CrossRef] [PubMed]
55. Dewanggana, M.N.; Evangeline, C.; Ketty, M.D.; Waturangi, D.E.; Magdalena, S. Isolation, characterization, molecular analysis and application of bacteriophage DW-EC to control Enterotoxigenic *Escherichia coli* on various foods. *Sci. Rep.* **2022**, *12*, 495. [CrossRef]
56. Wang, C.; Hang, H.; Zhou, S.; Niu, Y.D.; Du, H.; Stanford, K.; McAllister, T.A. Bacteriophage biocontrol of Shiga toxigenic *Escherichia coli* (STEC) O145 biofilms on stainless steel reduces the contamination of beef. *Food Microbiol.* **2020**, *92*, 103572. [CrossRef] [PubMed]
57. Choi, I.; Chang, Y.; Kim, S.Y.; Han, J. Polycaprolactone film functionalized with bacteriophage T4 promotes antibacterial activity of food packaging toward *Escherichia coli*. *Food Chem.* **2021**, *346*, 128883. [CrossRef]
58. Kaptchouang Tchatchouang, C.D.; Fri, J.; De Santi, M.; Brandi, G.; Schiavano, G.F.; Amagliani, G.; Ateba, C.N. Listeriosis Outbreak in South Africa: A Comparative Analysis with Previously Reported Cases Worldwide. *Microorganisms* **2020**, *8*, 135. [CrossRef]
59. Swaminathan, B.; Gerner-Smidt, P. The epidemiology of human listeriosis. *Microbes Infect.* **2007**, *9*, 1236–1243. [CrossRef] [PubMed]
60. Matle, I.; Mbatha, K.R.; Madoroba, E. A review of *Listeria* monocytogenes from meat and meat products: Epidemiology, virulence factors, antimicrobial resistance and diagnosis. *Onderstep. J. Vet. Res.* **2020**, *87*, 1–20. [CrossRef]
61. Strydom, A. Control of Listeria Monocytogenes in an Avocado Processing Facility. Doctoral dissertation, University of the Free State, Bleomfontei, South Africa, 2015.
62. Dhama, K.; Karthik, K.; Tiwari, R.; Shabbir, Z.; Barbuddhe, S.; Malik, S.V.S.; Singh, R.K. Listeriosis in animals, its public health significance (food-borne zoonosis) and advances in diagnosis and control: A comprehensive review. *Vet. Q.* **2015**, *35*, 211–235. [CrossRef]
63. Jakobsen, R.R.; Trinh, J.T.; Bomholtz, L.; Brok-Lauridsen, S.K.; Sulakvelidze, A.; Nielsen, D.S. A Bacteriophage Cocktail Significantly Reduces *Listeria monocytogenes* without Deleterious Impact on the Commensal Gut Microbiota under Simulated Gastrointestinal Conditions. *Viruses* **2022**, *14*, 190. [CrossRef]
64. Osek, J.; Lachtara, B.; Wieczorek, K. *Listeria monocytogenes*-How This Pathogen Survives in Food-Production Environments? *Front. Microbiol.* **2022**, *1*, 66462. [CrossRef] [PubMed]
65. Kawacka, I.; Olejnik-Schmidt, A.; Schmidt, M.; Sip, A. Effectives of Phage- Based Inhibiotion of *Listria monocytogenes* in Food Products and Food Processing Environments. *Microorganism* **2020**, *8*, 1764. [CrossRef] [PubMed]
66. Perera, M.N.; Abuladze, T.; Li, M.; Woolston, J.; Sulakvelidze, A. Bacteriophage cocktail significantly reduces or eliminates Listeria monocytogenes contamination on lettuce, apples, cheese, smoked salmon and frozen foods. *Food Microbiol.* **2015**, *52*, 42–48. [CrossRef] [PubMed]
67. Gray, J.A.; Chandry, P.S.; Kaur, M.; Kocharunchitt, C.; Bowman, J.P.; Fox, E.M. Novel Biocontrol Methods for Listeria monocytogenes Biofilms in Food Production Facilities. *Front. Microbiol.* **2018**, *9*, 605. [CrossRef] [PubMed]
68. Rogovski, P.; Cadamuro, R.D.; da Silva, R.; de Souza, E.B.; Bonatto, C.; Viancelli, A.; Michelon, W.; Elmahdy, E.M.; Treichel, H.; Rodríguez-Lázaro, D.; et al. Uses of Bacteriophages as Bacterial Control Tools and Environmental Safety Indicators. *Front. Microbiol.* **2021**, *12*, 3793135. [CrossRef]
69. Wójcicki, M.; Błażejak, S.; Gientka, I.; Brzezicka, K. The concept of using bacteriophages to improve the microbiological quality of minimally processed foods. *Acta Sci. Pol. Technol. Aliment.* **2019**, *18*, 373–383. [CrossRef]
70. Harrer, A.; Bücker, R.; Boehm, M.; Zarzecka, U.; Tegtmeyer, N.; Sticht, H.; Schulzke, J.D.; Backert, S. *Campylobacter jejuni* enters gut epithelial cells and impairs intestinal barrier function through cleavage of occludin by serine protease HtrA. *Gut Pathog.* **2019**, *11*, 4. [CrossRef]

71. Thung, T.Y.; Lee, E.; Mahyudin, N.A.; Wan Mohamed Radzi, C.W.J.; Mazlan, N.; Tan, C.W.; Radu, S. Partial characterization and in vitro evaluation of a lytic bacteriophage for biocontrol of *Campylobacter jejuni* in mutton and chicken meat. *J. Food Saf.* **2020**, *40*, e12770. [CrossRef]

72. D'angelantonio, D.; Scattolini, S.; Boni, A.; Neri, D.; Di Serafino, G.; Connerton, P.; Connerton, I.; Pomilio, F.; Di Giannatale, G.; Migliorati, G.; et al. Bacteriophage Therapy to Reduce Colonization of *Campylobacter jejuni* in Broiler Chickens before Slaughter. *Viruses* **2021**, *13*, 1428. [CrossRef]

73. Richards, P.J.; Connerton, P.L.; Connerton, I.F. Phage Biocontrol of *Campylobacter jejuni* in Chickens Does Not produce Collateral Effects on the Gut Microbiota. *Front. Microbiol.* **2019**, *10*, 476. [CrossRef]

74. Zhao, H.; Li, Y.; Lv, P.; Huang, J.; Tai, R.; Jin, X.; Wang, J.; Wang, X. Salmonella Phages Affect the Intestinal Barrier in Chicks by Altering the Composition of Early Intestinal Flora: Association With Time of Phage Use. *Front. Microbiol.* **2022**, *13*, 947640. [CrossRef] [PubMed]

75. Sarrami, Z.; Sedghi, M.; Mohammadi, I.; Kim, W.K.; Mahdavi, A.H. Effects of bacteriophage supplement on the growth performance, microbial population, and *PGC-1α* and *TLR4* gene expressions of broiler chickens. *Sci. Rep.* **2022**, *12*, 14391. [CrossRef] [PubMed]

76. Ahmadi, M.; Karimi Torshizi, M.A.; Rahimi, S.; Dennehy, J.J. Prophylactic Bacteriophage Administration More Effective than Post-infection Administration in Reducing *Salmonella enterica* serovar Enteritidis Shedding in Quail. *Front. Microbiol.* **2016**, *7*, 1253. [CrossRef] [PubMed]

77. Upadhaya, S.D.; Ahn, J.M.; Cho, J.H.; Kim, J.Y.; Kang, D.K.; Kim, S.W.; Kim, H.B.; Kim, I.H. Bacteriophage cocktail supplementation improves growth performance, gut microbiome and production traits in broiler chickens. *J. Anim. Sci. Biotechnol.* **2021**, *12*, 49. [CrossRef]

78. Zampara, A.; Sørensen, M.C.H.; Elsser-Gravesen, A.; Brøndsted, L. Significance of phage-host interactions for biocontrol of *Campylobacter jejuni* in food. *Food Control* **2017**, *73*, 1169–1175. [CrossRef]

79. Żbikowska, K.; Michalczuk, M.; Dolka, B. The Use of Bacteriophages in the Poultry Industry. *Animals* **2020**, *10*, 872. [CrossRef]

80. Shoaib, M.; Shehzad, A.; Raza, H.; Niazi, S.; Khan, I.M.; Akhtar, W.; Safdar, W.; Wang, Z. A comprehensive review on the prevalence pathogenesis and detection of *Yersinia enterocolitica*. *RSC* **2019**, *9*, 41010–41021. [CrossRef]

81. Terentjeva, M.; Ķibilds, J.; Mesitere, I.; Gradovska, S.; Alksne, L.; Streikiša, M.; Ošmjana, J.; Valciņa, O. Virulence Determinants and Genetic Diversity of *Yersinia* Species Isolated from Retail Meat. *Pathogens* **2022**, *11*, 37. [CrossRef]

82. Available online: https://www.efsa.europa.eu/en/news/campylobacter-and-salmonella-cases-stable-eu (accessed on 25 February 2021).

83. Leon-Velarde, C.G.; Jun, J.W.; Skurnik, M. *Yersinia* Phages and Food Safety. *Viruses* **2019**, *11*, 1105. [CrossRef]

84. Gwak, K.M.; Choi, I.Y.; Lee, J.; Oh, J.H.; Park, M.K. Isolation and Characterization of a Lytic and Highly Specific Phage against *Yersinia enterocolitica* as a Novel Biocontrol Agent. *J. Microbiol. Biotechnol.* **2018**, *28*, 1946–1954. [CrossRef]

85. Pyra, A.; Filik, K.; Szemer-Olearnik, B.; Czarny, A.; Brzozowska, E. New Insights on the Feature and Function of Tail Tubular Protein B and Tail Fiber Protein of the Lytic Bacteriophage φYeO3-12 Specific for *Yersinia enterocolitica* Serotype O:3. *Molecules* **2020**, *25*, 4392. [CrossRef] [PubMed]

86. Jun, J.W.; Park, S.C.; Wicklund, A.; Skurnik, M. Bacteriophages reduce *Yersinia enterocolitica* contamination of food and kitchenware. *Int. J. Food Microbiol.* **2018**, *271*, 33–47. [CrossRef] [PubMed]

87. Alomari, M.M.M.; Dec, M.; Urban-Chmiel, R. Bacteriophage as an Alternative Method for Control of Zoonotic and Foodborne Pathogens. *Viruses* **2021**, *13*, 2348. [CrossRef] [PubMed]

88. Soffer, N.; Woolston, J.; Li, M.; Das, C.; Sulakvelidze, A. Bacteriophage preparation lytic for *Shigella* significantly reduces *Shigella sonnei* contamination in various foods. *PLoS ONE* **2017**, *12*, e0175256. [CrossRef]

89. Kolenda, C.; Josse, J.; Medina, M.; Fevre, C.; Lustig, S.; Ferry, T.; Laurent, F. Evaluation of the activity of a Combination of Three Bacteriophages Alone or in Association with Antibiotics on *Saphylococcus aureus* Embedded in Biofilm or Internalized in Osteoblasts. *Antimicrob. Agents Chemother.* **2020**, *64*, e02231-19. [CrossRef]

90. Bueno, E.; Garciía, P.; Martiínez, B.; Rodriíguez, A. Phage inactivation of *Staphylococcus aureus* in fresh and hard-type cheeses. *Int J. Food Microbiol.* **2012**, *158*, 23–27. [CrossRef]

91. Heir, E.; Moen, B.; Åsli, A.W.; Sunde, M.; Langsrud, S. Antibiotic Resistance and Phylogeny of *Pseudomonas* spp. Isolated over Three Decades from Chicken Meat in the Norwegian Food Chain. *Microorganism* **2021**, *9*, 207. [CrossRef]

92. Hungaro, H.M.; Vidigal, P.M.P.; do Nascimento, E.C.; da Costa Oliveira, G.F.; Gontijo, M.T.P.; Lopez, M.E.S. Genomic Characterisation of UFJF_PfDIW6: A Novel Lytic *Pseudomonas fluorescens*-Phage with Potential for Biocontrol in the Dairy Industry. *Viruses* **2022**, *14*, 629. [CrossRef]

93. Moye, Z.D.; Woolston, J.; Sulakvelidze, A. Bacteriophage Applications for Food Production and Processing. *Viruses* **2018**, *10*, 205. [CrossRef]

94. Svircev, A.; Roach, D.; Castle, A. Framing the Future with Bacteriophages in Agriculture. *Viruses* **2018**, *10*, 218. [CrossRef]

95. FEEDAP; Bampidis, V.; Azimonti, G.; Bastos, M.D.L.; Christensen, H.; Dusemund, B.; Kouba, M.; Durjava, M.F.; Lopez-Alonso, M.; Puente, S.L.; et al. Safety and efficacy of a feed additive consisting on the bacteriophages PCM F/00069, PCM F/00070, PCM F/00071 and PCM F/00097 infecting *Salmonella* Gallinarum B/00111 (Bafasal®) for all avian species (Proteon Pharmaceuticals S.A.). *EFSA J.* **2021**, *19*, 6534. [CrossRef]

96. Colavecchio, A.; Cadieux, B.; Lo, A.; Goodridge, L.D. Bacteriophages Contribute to the Spread of Antibiotic Resistance Genes among Foodborne Pathogens of the *Enterobacteriaceae* Family–A Review. *Front. Microbiol.* **2017**, *8*, 1108. [CrossRef]

97. Putra, R.D.; Lyrawati, D. Interactions between Bacteriophages and Eukaryotic Cells. *Scientifica* **2020**, *2020*, 3589316. [CrossRef]

98. Żaczek, M.; Górski, A.; Skaradzińska, A.; Łusiak-Szelachowska, M.; Weber-Dąbrowska, B. Phage penetration of eukaryotic cells: Practical implications. *Future Med.* **2020**, *14*, 745–760. [CrossRef]

99. Podlacha, M.; Grabkowski, Ł.; Kosznik-Kawśnicka, K.; Zdrojewska, K.; Stasiłojć, M.; Węgrzyn, G.; Węgrzyn, A. Interactions of Bacteriophages with Animal and Human Organisms-Safety Issues in the Ligth of Phage Therapy. *Int. J. Mol. Sci.* **2021**, *22*, 8937. [CrossRef] [PubMed]

100. Goode, D.; Allen, V.M.; Barrow, P.A. Reduction of Experimental *Salmonella* and *Campylobacter* Contamination of Chicken Skin by Application of Lytic Bacteriophages. *Appl. Env. Microbiol.* **2003**, *69*, 5032–5036. [CrossRef] [PubMed]

101. Hudson, J.A.; Billington, C.; Wilson, T.; On, S.L. Effect of phage and host concentration on the inactivation of Escherichia coli O157:H7 on cooked and raw beef. *Food Sci. Technol. Int.* **2013**, *21*, 104–109. [CrossRef] [PubMed]

102. Guenther, S.; Huwyler, D.; Richard, S.; Loessner, M.J. Virulent bacteriophage for efficient biocontrol of *Listeria monocytogenes* in ready-to-eat foods. *Appl. Env. Microbiol.* **2009**, *75*, 93–100. [CrossRef]

103. Do Nascimento, E.C.; Sabino, M.C.; da Roza Corguinha, L.; Targino, B.N.; Lange, C.C.; de Oliveira Pinto, C.L.; de Faria Pinto, P.; Vidigal, P.M.P.; Sant'Ana, A.S.; Hungaro, H.M. Lytic bacteriophages UFJF_PfDIW6 and UFJF_PfSW6 prevent *Pseudomonas fluorescens* growth in vitro and the proteolytic-caused spoilage of raw milk during chilled storage. *Food Microbiol.* **2022**, *101*, 103892. [CrossRef]

104. Kauppinen, A.; Siponen, S.; Pitkänen, T.; Holmfeldt, K.; Pursiainen, A.; Torvinen, E.; Miettinen, I.T. Phage biocontrol of *Pseudomonas aeruginosa* in water. *Viruses* **2021**, *13*, 928. [CrossRef]

105. Hong, Y.; Schmidt, K.; Marks, D.; Hatter, S.; Marshall, A.; Albino, L.; Ebner, P. Treatment of *Salmonella*-contaminated Eggs and Pork with a Broad-Spectrum, Single Bacteriophage: Assessment of Efficacy and Resistance Development. *Foodborne Pathog. Dis.* **2016**, *13*, 679–688. [CrossRef] [PubMed]

106. Bai, J.; Jeon, B.; Ryu, S. Effective inhibition of *Salmonella* Typhimurium in fresh produce by a phage cocktail targeting multiple host receptors. *Food Microbiol.* **2019**, *77*, 52–60. [CrossRef]

107. Lu, H.; Liu, H.; Lu, M.; Wang, J.; Liu, X. Isolation and Characterization of a Novel *myovirus* Infecting *Shigella dysenteriae* from Aeration Tank Water. *Appl. Biochem. Biotechnol.* **2020**, *192*, 120–131. [CrossRef]

108. Zhang, H.; Wang, R.; Bao, H. Phage inactivation of foodborne *Shigella* on ready-to-eat spiced chicken. *Poult. Sci.* **2013**, *92*, 211–217. [CrossRef] [PubMed]

109. Gharieb, R.M.A.; Saad, M.F.; Mohamed, A.S.; Tartor, Y.H. Characterization of two novel lytic bacteriophages for reducing biofilms of zoonotic multidrug-resistant *Staphylococcus aureus* and controlling their growth in milk. *LWT* **2020**, *124*, 109145. [CrossRef]

110. Ren, H.; Li, Z.; Xu, Y.; Wang, L.; Li, X. Protective effectiveness of feeding phage cocktails in controlling *Vibrio parahamolyticus* infection of sea cucumber *Apostichopus japonicus*. *Aquaculture* **2019**, *503*, 322–329. [CrossRef]

111. Orquera, S.; Hertwig, S.; Alter, T.; Hammerl, J.A.; Jirova, A.; Golz, G. Development of transient phage resistance in *Campylobacter coli* against the group II phage CP84. *Berl. Munch Tierarztl. Wochenschr.* **2015**, *128*, 141–147. [CrossRef]

112. Sommer, J.; Trautner, C.; Witte, A.K.; Fister, S.; Schoder, D.; Rossmanith, P.; Mester, P.J. Don't Shut the Stable Door after the Phage Has Botled–The Imporatace of bacteriophage Inactivaction in Food Environments. *Viruses* **2019**, *11*, 468. [CrossRef]

113. Silva, E.N.; Figueiredo, A.C.; Miranda, F.A.; de Castro Almeida, R.C. Control of *Listeria monocytogenes* growth in soft cheeses by bacteriophage P100. *Braz. J. Microbiol.* **2014**, *45*, 11–16. [CrossRef] [PubMed]

114. Yeh, Y.; de Moura, F.H.; Van Den Broek, K.; de Mello, A.S. Effect of ultraviolet light, organic acids and bacteriophage on *Salmonella* populations in ground beef. *Meat Sci.* **2018**, *139*, 44–48. [CrossRef] [PubMed]

115. Yang, S.; Sadekuzzaman, M.; Ha, S.D. Reduction of *Listeria monocytogenes* on chicken breast by combined treatment with UV-C light and bacteriophage List Shield. *LWT* **2017**, *86*, 193–200. [CrossRef]

116. Gutiérrez, D.; Rodríguez-Rubio, L.; Fernández, L.; Martínez, B.; Rodríguez, A.; García, P. Applicability of commercial phage-based products against *Listeria monocytogenes* for improvement of food safety in Spanish dry-cured ham and food contact surfaces. *Food Cont.* **2017**, *73*, 1474–1482. [CrossRef]

117. Sadekuzzaman, M.; Yang, S.; Kim, H.S.; Mizan, M.F.R.; Ha, S.D. Evaluation of a novel atimicrobiological (lauric arginate ester) susbstance against biofilm of *Escherichia coli* O157:H17, *Listeria monocytogenes* and *Salmonella* spp. *Int. J. Food Sci. Technol.* **2017**, *52*, 2058–2067. [CrossRef]

118. Wottlin, L.R.; Edrington, T.S.; Anderson, R.C. *Salmonella* Carriage in Peripheral Lymph Nodes and Feces of Cattle at Slaughter is Affected by Cattle Type, Region, and Season. *Front. Anim. Sci.* **2022**, *3*, 859800. [CrossRef]

119. Moye, Z.D.; Das, C.R.; Tokman, J.I.; Fanelli, B.; Karathia, H.; Hasan, N.A.; Marek, P.J.; Senecal, A.G.; Sulakvelidze, A. Treatment of fresh produce with a *Salmonella*- targeted bacteriophage cocktail is compatible with chlorine or peracetic acid and more consistently preserves the microbial community on produce. *J. Food Saf.* **2020**, *40*, e12763. [CrossRef]

120. Sharma, M.; Dashiell, G.; Handy, E.T.; East, C.; Reynnells, R.; White, C.; Nyarko, E.; Micallef, S.; Hashem, F.; Millner, P.D. Survival of *Salmonella* Newport on whole and fresh-cut cucumbers treated with lytic bacteriophages. *J. Food Prot.* **2017**, *80*, 668–673. [CrossRef] [PubMed]

121. Mai, V.; Ukhanova, M.; Reinhard, M.K.; Li, M.; Sulakvelidze, A. Bacteriophage administration significantly reduces *Shigella* colonization and shedding by *Shigella*-challenged mice without deleterious side effects and distortions in the gut microbiota. *Bacteriophage* **2015**, *5*, e1088124. [CrossRef]

122. Magnone, P.J.; Marek, J.P.; Sulakvelidze, A.; Senecal, A. Additive Approach for Inactivation of *Escherichia coli* O157:H7, *Salmonella*, and *Shigella* spp. on Contaminated Fresh Fruits and Vegetables Using Bacteriophage Cocktail and Produce Wash. *JFP* **2013**, *76*, 1336–1341. [CrossRef]
123. Vikram, A.; Tokman, J.; Woolston, J.; Sulakvelidze, A. Phage biocontrol improves food safety by significantly reducing both the concentration and occurrence of *Escherichia coli* O157:H7 in various foods. *J. Food Prot. Press* **2020**, *83*, 668–676. [CrossRef]
124. Boyacioglu, O.; Sharma, M.; Sulakvelidze, A.; Goktepe, I. Biocontrol of *Escherichia coli* O157:H7 on fresh-cut leafy greens. *Bacteriophage* **2013**, *3*, e24620. [CrossRef]
125. Carter, C.D.; Parks, A.; Abuladze, T.; Li, M.; Woolston, J.; Magnone, J.; Senecal, A.; Kropinski, A.M.; Sulakvelidze, A. Bacteriophage cocktail significantly reduces *Escherichia coli* O157:H7 contamination of lettuce and beef, but does not protect against recontamination. *Bacteriophage* **2012**, *2*, 178–185. [CrossRef] [PubMed]
126. Available online: http://www.intralytix.com/index.php?page=prod&id=12 (accessed on 20 September 2022).
127. Mayne, J.; Zhang, X.; Butcher, J.; Walker, K.; Ning, Z.; Wójcik, K.; Bastych, J.; Stintzi, A.; Figeys, D. Examining the Effects of an Anti-Salmonella Bacteriophage Preparation BASFAL, on Ex-Vivo Human Gut Microbiome Composton an Fuction Using a Multi-Omics Approoach. *Viruses* **2021**, *13*, 1734. [CrossRef] [PubMed]
128. Available online: https://ahfoodchain.com/en/about/news/2020/08/finalyse-sal-release (accessed on 6 September 2022).

Article

Microbial Profiling of Potato-Associated Rhizosphere Bacteria under Bacteriophage Therapy

Samar Mousa [1,2,3], Mahmoud Magdy [4], Dongyan Xiong [1], Raphael Nyaruabaa [1,2], Samah Mohamed Rizk [4], Junping Yu [1] and Hongping Wei [1,*]

1   CAS Key Laboratory of Special Pathogens and Biosafety, Center for Biosafety Mega-Science, Wuhan Institute of Virology, Chinese Academy of Sciences, Wuhan 430071, China
2   International College, University of Chinese Academy of Sciences, Beijing 101408, China
3   Agricultural Botany Department, Faculty of Agriculture, Suez Canal University, Ismailia 41522, Egypt
4   Genetics Department, Faculty of Agriculture, Ain Shams University, Cairo 11241, Egypt
*   Correspondence: hpwei@wh.iov.cn

Abstract: Potato soft rot and wilt are economically problematic diseases due to the lack of effective bactericides. Bacteriophages have been studied as a novel and environment-friendly alternative to control plant diseases. However, few experiments have been conducted to study the changes in plants and soil microbiomes after bacteriophage therapy. In this study, rhizosphere microbiomes were examined after potatoes were separately infected with three bacteria (*Ralstonia solanacearum*, *Pectobacterium carotovorum*, *Pectobacterium atrosepticum*) and subsequently treated with a single phage or a phage cocktail consisting of three phages each. Results showed that using the phage cocktails had better efficacy in reducing the disease incidence and disease symptoms' levels when compared to the application of a single phage under greenhouse conditions. At the same time, the rhizosphere microbiota in the soil was affected by the changes in micro-organisms' richness and counts. In conclusion, the explicit phage mixers have the potential to control plant pathogenic bacteria and cause changes in the rhizosphere bacteria, but not affect the beneficial rhizosphere microbes.

Keywords: bacteriophage treatments; rhizosphere microbiota; *Solanum tuberosum*; single phage therapy; phage cocktail therapy

Citation: Mousa, S.; Magdy, M.; Xiong, D.; Nyaruabaa, R.; Rizk, S.M.; Yu, J.; Wei, H. Microbial Profiling of Potato-Associated Rhizosphere Bacteria under Bacteriophage Therapy. *Antibiotics* **2022**, *11*, 1117. https://doi.org/10.3390/antibiotics11081117

Academic Editors: Aneta Skaradzińska and Anna Nowaczek

Received: 6 July 2022
Accepted: 10 August 2022
Published: 18 August 2022

Publisher's Note: MDPI stays neutral with regard to jurisdictional claims in published maps and institutional affiliations.

1. Introduction

Potatoes (*Solanum tuberosum* L.) are considered to be the third most consumed crop globally and the main food for more than one billion people in the world [1–3]. This means that potatoes contribute significantly to the global food security and economy when used as cash crops [4]. Among these threats, potato infection by bacterial diseases is serious as it may lead to a tremendous crop loss of up to 80% [5,6]. Two major forms of potato bacterial disease exist including potato soft rot and wilt [7]. The potato soft rot is caused by a range of bacteria, including *Pectobacterium carotovorum*, *Pectobacterium atrosepticum*, and *Dickeya* spp., while potato wilt is caused by *Ralstonia solanacearum* [8–10]. These bacterial pathogens are soilborne and can infect plants during growth, causing severe damage [11]. Therefore, effective and environmentally friendly control agents can be used to combat these diseases and their associated bacterial pathogens [12–14].

Several strategies, including the use of bactericides [15–18], antimicrobials [16,19], and bacterial inoculants, have been adopted to control potato soft rot and wilt [17,18]. Despite their efficacy, each of these methods has its own demerits [19]. For example, bactericides such as copper compounds, 5-nitro-8-hydroxyquinoline, chlorine dioxide, and mercuric chloride can cause environmental pollution, increase resistant bacterial strains and heighten the price of agricultural production [5,17,19]. Additionally, the use of antimicrobials such as oxolinic acid, streptomycin, and validamycin A for controlling bacteria that can cause soft rot and wilt can lead to resistant strains that ultimately contribute to the already alarming list of antimicrobial-resistant strains [6,20]. Biocontrol using bacterial inoculants to modify the composition of plant rhizosphere microbiota has been proposed as an alternative to pesticides for pathogen elimination [21–23]. However, bacterial inoculants are often ineffective owing to their poor establishment in the rhizosphere, competition with native microbiota for resources, and interference with native microbiota [20,24]. As a result, new approaches, including the use of bacteriophages as potential biocontrol agents, are being explored [25].

Bacteriophages (phages) are viruses that infect and propagate within bacterial cells [13,26]. The growing interest in applying phages in the biocontrol of plant pathogens stems from their advantages, including host specificity, environmental friendliness, self-replication, non-toxicity, ability to overcome antimicrobial resistance, cost-effectiveness, ease of production, and the ability to be used as cocktails to improve their efficacy [12,14,25]. Owing to these advantages and more, studies have shown that phages can be used to control soft-rot Enterobacteriaceae (SRE) and potato wilt with satisfactory accomplishment in field trials [10,11,20]. Despite this, experimental evidence on the effects of phages on the native rhizosphere, as well as on the properties of the soil such as pH and organic contents, is still scarce. Additionally, phages can be used as single variants or as cocktails to improve their efficacy. The use of cocktails may further have an additional effect on soil properties and native rhizosphere microbiota [27–29].

Recent advancements in molecular diagnostic tools such as sequencing, metagenomics, and bioinformatics can be used to answer these questions [30–32]. Using these tools, studies can be conducted to determine how evolutionary trade-offs or phage-mediated pathogen density reduction may affect the composition and functions of the native rhizosphere microbiome [27,29]. For example, a decrease in pathogen density of one bacterium mediated by phages may result in an increased competition of niche space and nutrient uptake by other native bacteria, consequently leading to changes in native rhizosphere and microorganism diversity [29,30]. These changes may have beneficial secondary effects on the plant owing to a reduction in bacterial loads associated with plant diseases [27].

Therefore, in this study, we determined the effects of phage therapy on potato bacterial diseases using three pathogenic bacteria, *R. solanacearum*, *P. carotovorum*, and *P. atrosepticum*. Using greenhouse experiments and metagenomic analysis, we assessed the effects of single and cocktail phages against potato bacterial soft rot and wilt in complex microbial communities and tested whether these effects extend to other microbes within the rhizosphere area.

2. Results

2.1. Efficacy of Bacteriophage Therapy on Potato Bacterial Diseases In Vivo

The phages used in this study were previously isolated and used to assess their biocontrol efficacy on potato infecting phytobacteria in vitro [1,5]. In this study, we designed a greenhouse experiment using the same phages as single and cocktails (Supplementary Figure S1) to determine if the phages can also control potato bacterial disease in vivo (Supplementary Figure S2).

Three bacteria, *R. solanacearum* (Rs), *P. carotovorum* (Pc), and *P. atrosepticum* (Pa), were inoculated to cause potato wilt and soft rot diseases, respectively. As shown in Figure 1, all of the five plants in the positive control group (inoculated with bacteria Rs, Pc, or Pa, without phage treatment) showed signs of bacterial infection (the ratios were 0:5 in terms of healthy to infected plants for Rs, Pc, and Pa). The single phage treatments showed differences in disease incidence, with ratios of 4:1, 5:0, and 3:2 in terms of healthy to infected plants for SRs, SPc, and SPa, respectively. On the other hand, the phage cocktail treatments were more effective for the reduction of the diseases' incidence, with ratios of 5:0, 5:0, and 4:1 in terms of healthy to infected plants for RsPck, PcPck, and PaPck, respectively (Figure 1A).

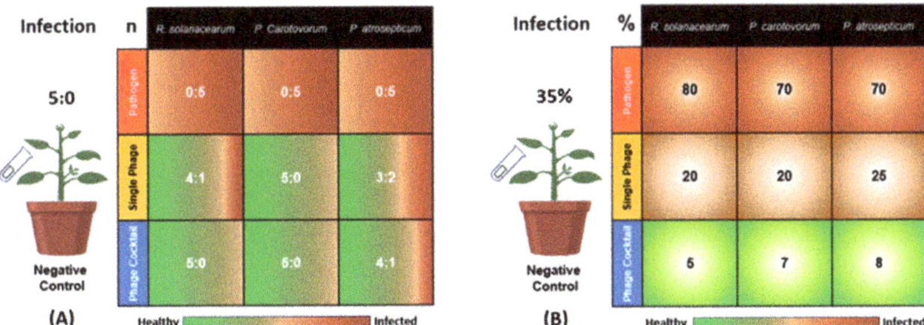

Figure 1. Effects of phage therapy on the incidence of potato bacterial disease in terms of healthy: infected plants for all treatments (**A**) and Percentage of disease symptoms after different treatments (**B**). Positive control or pathogen groups: inoculated with *R. solanacearum* (Rs), *P. carotovorum* (Pc), or *P. atrosepticum* (Pa), respectively; Single phage groups (SRs, SPc, and SPa): inoculated with *R. solanacearum* (Rs), *P. carotovorum* (Pc) or *P. atrosepticum* (Pa) and then treated with a single phage, respectively; Phage cocktail groups (RsPck, PcPck, and PaPck): inoculated with *R. solanacearum* (Rs), *P. carotovorum* (Pc) or *P. atrosepticum* (Pa) and then treated with phage cocktails, respectively. "n" is representing the number of healthy: infected plants and "%" is indicating the percentage of the disease symptoms revealed on the infected plants.

Notably, the phage cocktails treatments showed a remarkable plant growth than all groups indicating that it may have killed the three bacteria causing potato wilt and soft rot diseases. In detail, the percentage of disease symptoms revealed in Rs-treated plants was 80%, the percentage ranged between 10–20% when the single-phage (SRs) was applied, and decreased down to 0–5% when the phage cocktail (RsPck) was used. For the Pc-treated plants, the percentage of disease symptoms was 70% and the percentage of disease symptoms revealed after inoculation with SPc ranged between 10–25% and decreased down to 0–7% when the PcPck treatment was applied. The percentages of disease symptoms caused by Pa was about 70%, decreased down to 25% with the application of SPa and reduced down to 8% when the PaPck treatment was applied. The negative control remained asymptomatic during the experiment period. Reductions were significant (*p*-value < 0.001) in all the applied phage treatments. Data indicated that phage cocktails were more effective than single phage treatments (Figure 1B).

2.2. Microbial Communities: Pathogens and Phage Therapy

2.2.1. Rhizosphere Microbiome Profiling

The microbial profiling of the soils yielded average total OTU (Operational Taxonomic Unit) counts of 1294, 1272, and 1280 for the Rs, Pc, and Pa groups, respectively. After phage treatments, the total OTUSs were 1286, 1272, and 1280 for SRs, SPc, and SPa, respectively, and 1285, 1207, and 1291 for RsPck, PcPck, and PaPck, respectively, versus the average total OTU of the native soils which was 794. Among all samples, the common OTU count was 1075, separated into 695 OTUs shared with the native soil sample and 380 OTUs exclusively shared among the treated samples (Figure 2).

The average Shannon index (i.e., an index to measure the diversity of species in a community) for replicates per treatment was applied to estimate the detected diversity within each sample (i.e., alpha diversity). Among all phage therapy treatments, the phage cocktail (PaPck) and the single phage (SPa) had the highest diversity, followed by the single phage (SRs) and phage cocktail (RsPck). The difference was significant with *p* < 0.01.

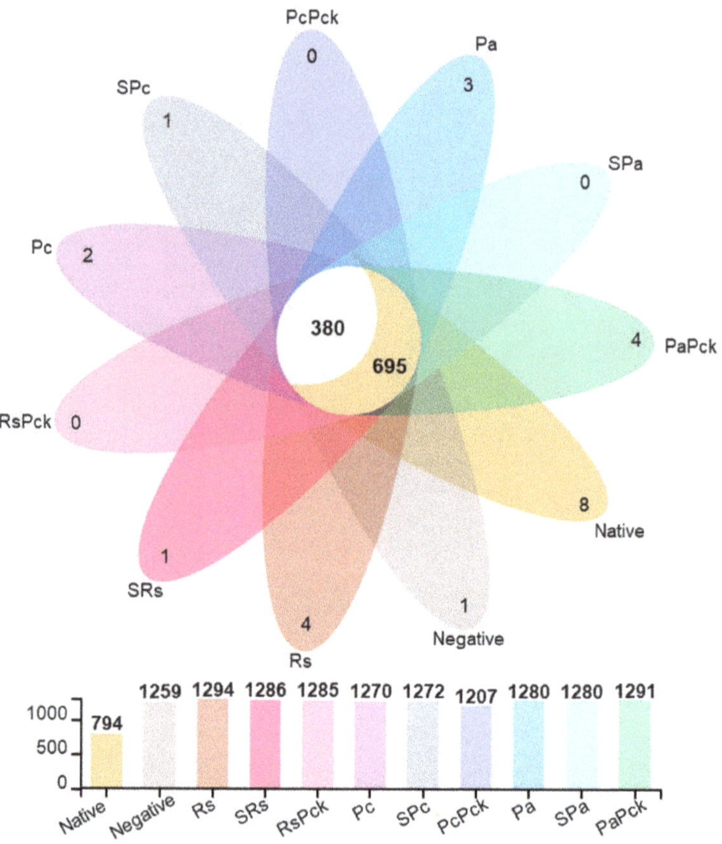

Figure 2. The OTU diversity of three phage therapy groups, positive control, or pathogens group (*R. solanacearum* (Rs), *P. carotovorum* (Pc), and *P. atrosepticum* (Pa)), single phage group (SRs, SPc, and SPa) and phage cocktail groups (RsPck, PcPck, and PaPck) compared to the negative control (no treatments added) and native soil samples, represented by different colors.

The single phage (SPc) and phage cocktail (PcPck) had the lowest diversity compared to native soil sample. The Shannon index ranged from 4.06 to 5.21. In detail, the Shannon diversity index values of Rs, SRs, and RsPck were 5.21, 5.04, and 4.86, respectively, and 4.99, 4.68, and 4.32 for Pc, SPc, and PcPck, respectively, while the values for Pa, Spa, and PaPck were 4.51, 4.25, and 4.28, respectively, when compared to the native soil (4.06).

2.2.2. Rhizosphere Microbial Communities

Proteobacteria were highly abundant among all phage treatments (percentages of 61, 59, 57, 55%, for SRs, RsPck, PcPck, and SPc, respectively) when compared to the negative control (57%). Firmicutes was highly abundant in the phage therapy treatments (31, 25%), for Spa and PaPck, respectively, compared to the negative control (2%). Additionally, Bacteriodota was highly abundant in the phage therapy treatments (19, 17, 10%), for PcPck, SPc and RsPck, respectively, compared to the negative control (7%). In contrast, Actinobacteriota had relatively low abundance in the phage therapy treatments (10, 8, 7%) for RsPck, PcPck, and PaPck, respectively, compared to the negative control (13%; Figure 3A).

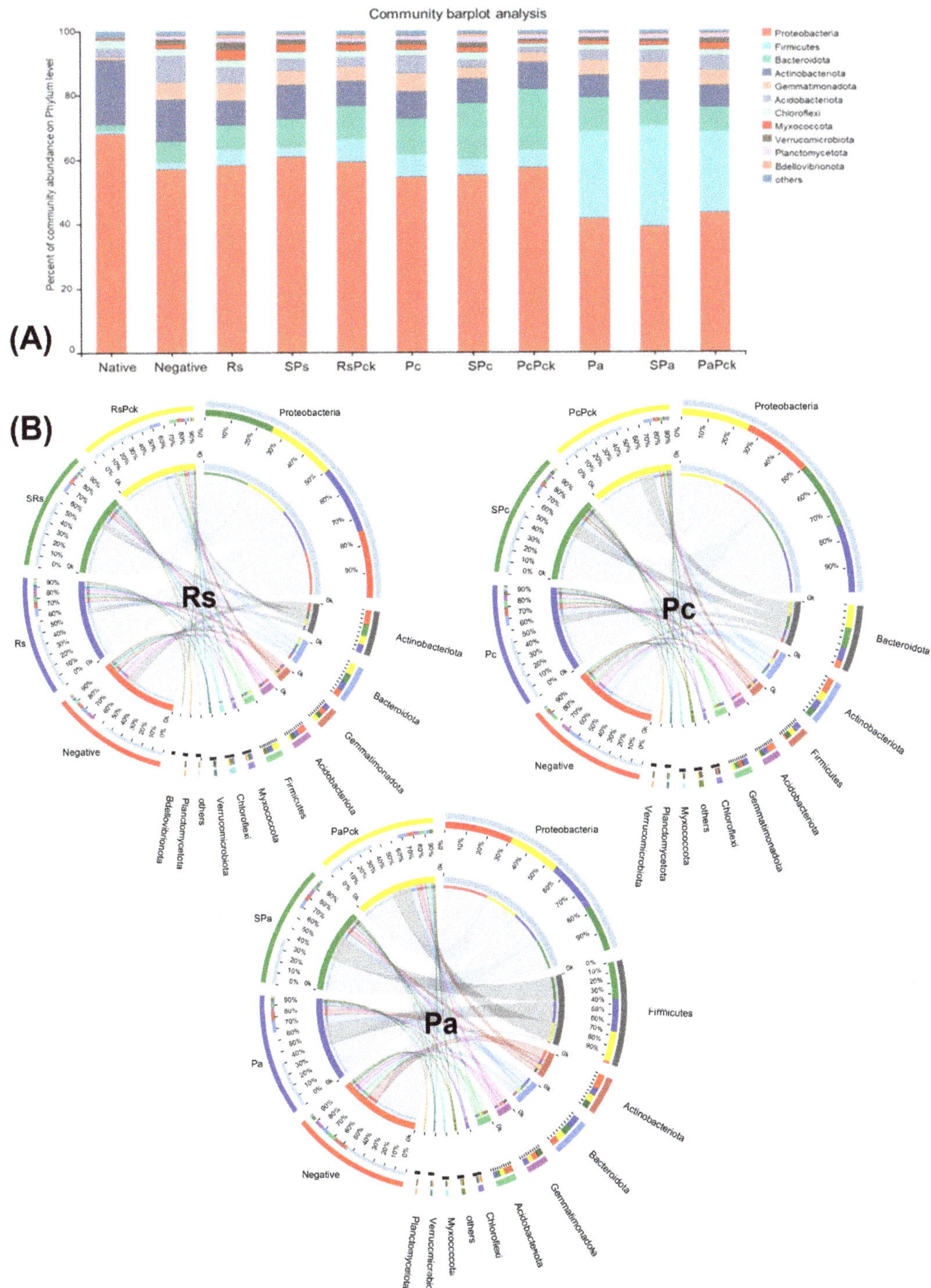

Figure 3. The microbial composition of the surveyed microbiota at phyla level, presented as (**A**) barplot and (**B**) comparative circus plot for each of the three phage therapy treatments (Rs, Pc, and Pa).

The commonly shared OTUs among the phage therapy treatments revealed significant differences in six microbial phyla which included Proteobacteria, Firmicutes, Acidobacteriota, Actinobacteriota, Bacteriodota, and Gemmatimonadota. Among all samples of the three phage therapy groups, the phage therapy group of *R. solanacearum* (Rs) revealed the abundance of highly bacterial phyla, generally being Proteobacteria, followed by Actinobacteriota, Bacteriodota, Gemmatimonadota, Acidobacteriota, Firmicutes, Acidobacteriota, Myxococcota, and Chloroflexi, while the most identified bacterial phyla of the phage therapy group of *P. carotovorum* (Pc) generally was Proteobacteria, followed by Bacteriodota, Actinobacteriota, Firmicutes, Acidobacteriota and Gemmatimonadota. In contrast with the phage therapy group of *P. atrosepticum* (Pa), the most identified bacterial phylum was Proteobacteria, followed by Firmicutes, Actinobacteriota, Bacteriodota, Gemmatimonadota and Acidobacteriota.

Regardless of the phage therapy type, the abundance of Firmicutes was significant in the phage therapy treatments compared to the negative control among all groups. The Actinobacteriota, Bacteriodota and Firmicutes phyla were the most presented among all with almost an equal distribution among different treatments (Figure 3B).

2.2.3. Phage Therapy-Related Microbial Communities

After initial screening, all detected genera (nodes) were retained in two clusters and compared to uncultivated soil. On average, phage cocktail (RsPck, PcPck, and PaPck) networks were more connected and had shorter path lengths. Instead, most of the taxa associations were completely different between the phage cocktail and the three pathogen (Rs, Pc and Pa) communities, and the number of significant associations increased with the number of phages when compared to uncultivated soil (potato) (Figure 4A).

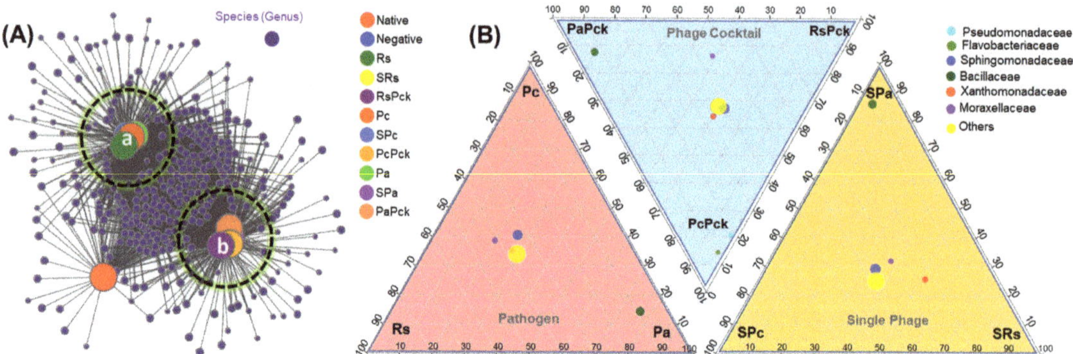

Figure 4. (**A**) Bacterial co-occurrence networks between single, phage cocktail and pathogens for the three phage therapy experiment (a: cluster of the negative, and pathogen treated samples, and b: cluster of phage treated samples) compared to an uncultivated soil sample (native). (**B**) Three triangular comparisons at the family level among the different pathogens (Rs, Pc and Pa), phage cocktails (RsPck, PcPck and PaPck), and single phages (SRs, SPc and SPa) are shown.

The microbiome composition and diversity at the family level was investigated among the three phage-therapy treatments at the pathogen, single-phage, and phage cocktail treatments, independently (Figure 4B). The bacterial species belonging to Bacillaceae family were common among Pa groups. In comparison, Pseudomonadaceae and Flavobacteriaceae were common among PcPck. In contrast, Sphingomonadaceae, Xanthomonadaceae and Moraxellaceae were shared between all three groups.

2.2.4. Phenotypic Prediction of Phage Treated Groups

Based on the recorded metadata for microbial species in databases, phenotypic categories were defined. The phenotypic profiles of the rhizosphere of phage therapy treatments and negative control were compared and controlled by the phage therapy groups. The rhizosphere microbial community showed a significant difference among the phage treatments. For the first group of the experiment (Rs), the phage cocktail (RsPck) presented a significant difference to the negative control, being highly effective with the pathogen (Rs) among facultatively anaerobic microbes. Moreover, significant differences between RsPck and the negative control with potentially pathogenic microbes (Figure 5A) were found.

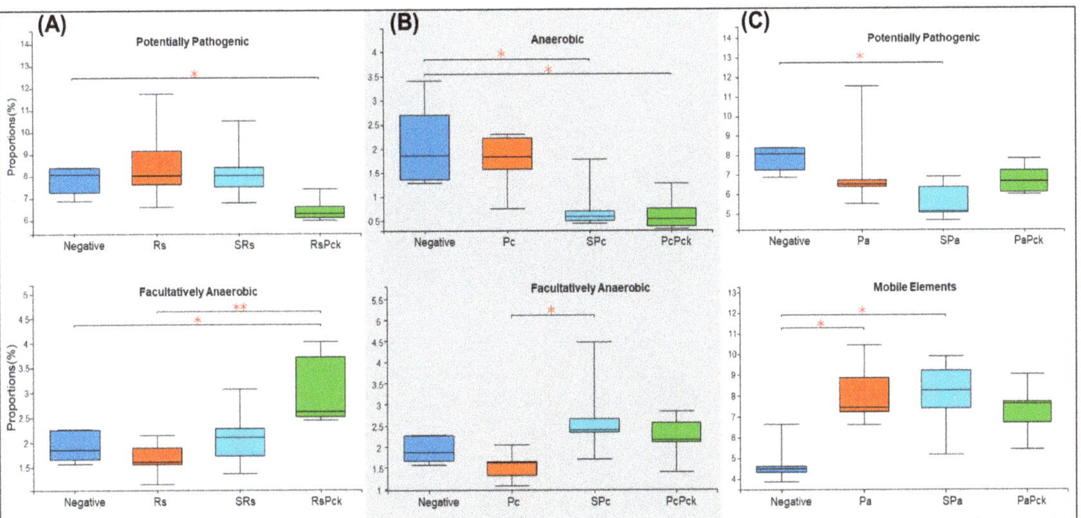

Figure 5. Boxplot diversity explained by phenotypic prediction among the phage therapy treatments. (**A**) Proportions of the potentially pathogenic and facultatively anaerobic phenotype of *R. solanacearum* phage therapy (Rs) treatments. (**B**) Proportions of anaerobic and facultatively anaerobic phenotypes of *P. carotovorum* phage therapy. (**C**) Proportions of the potentially pathogenic and facultatively anaerobic phenotype of *P. atrosepticum* phage therapy (* *p*-value < 0.05, ** *p*-value < 0.001).

The second group (Pc) showed that both single phage (SPc) and phage cocktail (PcPck) had significant difference with the negative control samples among the anaerobic group of bacteria. Furthermore, significant differences between single phages (SPc) and pathogens (Pc) among facultatively anaerobic microbes were found (Figure 5B).

In contrast, the third group (Pa) showed significant differences among the single phage (SPa) treatment and the pathogen (Pa) with the negative control samples at mobile elements, and significant difference between single phage (SPa) and the negative control, being potentially pathogenic (Figure 5C).

2.2.5. Functional Prediction of Phage Treated Groups

The functional properties of the detected bacterial taxa were investigated in relation to the different phage therapy treatments based on KEGG pathways. Based on the enriched pathways values of the negative samples (*x*-axis) in contrast to all the other samples (*y*-axis) the most represented functional pathways were detected (Figure 6). The most represented pathways were related to the organism's metabolism for all samples, followed by the biosynthesis of secondary metabolites, microbial metabolism in diverse environment, biosynthesis of amino acids and carbon metabolism. In the case of the environmental information processes group, the two component system and ABC transporters were distinguished, while ribosomes' formation was highly presented as the genetic information processes group. It was observed that the PaPck was highly represented when compared to the other treatments (Figure 6A). Additionally, the purine metabolism, oxidative phosphorylation, pyruvate metabolism and glyoxylate dicarboxylate metabolism of the metabolism group were presented at lower levels, as well as the quorum sensing of the cellular processes group. Equally, the PaPck was more highly represented than the other treatments, followed by SPa and Pa. The RsPck and PaPck were the only treatments that showed higher enrichment levels of the previous groups when compared to their pathogen, or single phage-treated samples, in contrast to the PcPck, which showed the lowest enrichment pathways among all (Figure 6B).

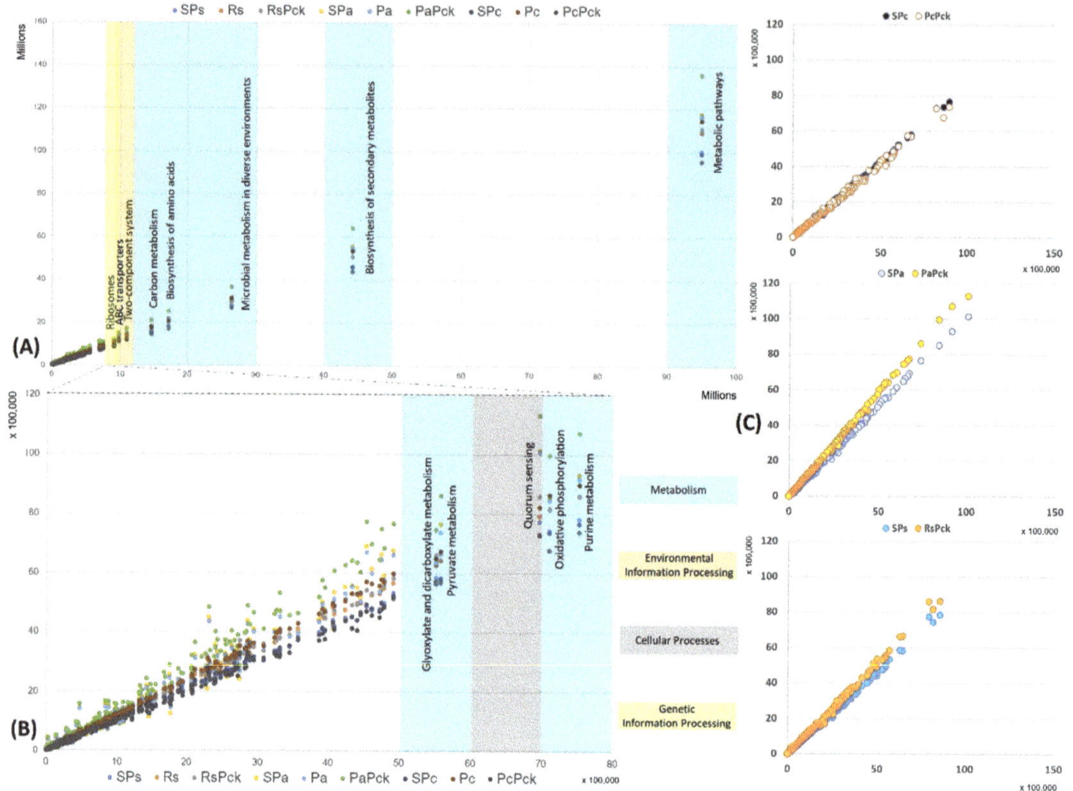

Figure 6. The functional prediction KEGG pathways for all samples at three levels, (**A**) top pathway above 10 million enriched genes, (**B**) between 5500K to 8000K enriched genes and (**C**) the gene enrichment plot of the phage therapy treatments versus the untreated soil.

3. Discussion

The impact of microbe–microbe interactions on the host–microbial pathogen interaction outcomes is a relevant subject in microbiology and plant pathology. Studies have shown that the microbiome structure, assemblage, and compositions are directly influenced by soil biotic and abiotic factors [31]. Phage therapy is a common practice that has been previously reported in agriculture and plant protection fields [12–14]. However, the commercial use of phages in agriculture is still limited [12–14]. This study has shown that phages can be used effectively as biocontrol agents while improving overall plant health. This effectiveness was observed after an in vivo evaluation of different phage treatments following a specific sampling design. It was apparent that all the tested phages were able to control potato wilt bacteria *R. solanacearum* and the soft rot bacteria *P. carotovorum* and *P. atrosepticum* as previously reported [1,5]. Although a single phage decreased the occurrence of bacterial wilt and soft rot diseases in contrast to the control, the occurrence of diseases was reduced more by the phage cocktail that contained three different phages under greenhouse conditions. The decline in occurrence of infections could be clarified by a decline in pathogens densities and this impact became stronger with the use of phage cocktails.

Bacteriophages are known for their specificity to bacteria, thus, the phage therapy should affect its specific host (i.e., pathogen) and show an insignificant effect on the natural rhizosphere microbiota [26]. The clear divergence in both species' richness and counts in the pathogen-treated samples confirm the association of the pathogen with different microbial groups. Thus, the reduction or elimination of this pathogen by phages would eventually cause differences in the existing rhizosphere microbial community represented in the anaerobic microbes that will contribute to facilitate phosphate solubilization and promote the precipitation of soluble Cd in the soil, as well as the facultative anaerobes capable of reducing Fe (III). Effectively, in the current study, the observed changes in the rhizosphere microbiota confirmed the vital role of phages in shaping the potato-related rhizosphere microbiome. The enhancement of plant health after the application of different types of phages may not be only limited to the elimination of the pathogen but also due to the new shifts in the microbial composition.

For example, the NGS metabarcoding-based microbiome profiling revealed the predominance of the species belonging to the phylum Proteobacteria regardless of the treatment group. The Proteobacteria were previously found to be associated with bioremediation of environmental contaminants and the production of highly beneficial phytohormones such as the indole-3-pyruvate pathway for synthesis of the auxinic phytohormone indole acetic acid (IAA) in Azospirillum and Enterobacter genera which belong to the phylum Proteobacteria [32–35]. Moreover, Azospirillum, Burkholderia, and other genera have the ability for nitrogen fixation by nitrogenase-encoding genes nifHDK [32,36]. However, Pseudomonas belonging to Proteobacteria can synthesize the pyrroloquinoline quinone-encoding genes pqqBCDEFG that can contribute to mineral phosphate solubilization [37], production of the 1-aminocyclopropane-1-carboxylate (ACC) deaminase gene acdS that enables the degradation of the plant's ethylene precursor [38,39], and synthesis of antimicrobial compounds by the genes hcnABC (hydrogen cyanide) and phlACBD (2,4-diacetylphloroglucinol) [40].

However, Firmicutes are capable of producing ACC deaminase and suppress pathogens which leads to enhanced plant growth and pathogen suppression [41]. Members of the genus Bacillus, which belongs to the phylum Firmicutes, secrete exopolysaccharides and siderophores that inhibit or stop the movement of toxic ions and help maintain an ionic balance [42]. As well as this, they are the direct synthesis of antimicrobial compounds, phytohormones, and siderophores that inhibit or stop the movement of toxic ions and help maintain an ionic balance [43]. An additional feature of the Bacillus genus is its ability to form biofilms, as the biofilm provides a matrix on which the community can develop [42]. This bacterial genus belongs to the phylum Actinobacteriota which contributes to the rotation of soil components into organic components through the decomposition of a complex combination of organic matter in lifeless plants, and animals, in addition to fungal material [44]. The most abundant genus belonging to Actinobacteriota are Streptomyces, which are a prolific source of antimicrobial, and extracellular enzymes. They have the ability to produce secondary metabolites of biotechnological and clinical importance that can play a major role in nutrient cycling. The Streptomyces importance is revealed as biocontrol agents, plant growth-promoting bacteria, and efficient biofertilizers [45].

Therefore, the phylum Bacteriodota contributes to mineral phosphate solubilization as well as the family Cyclobacteriaceae [46]. The bacterial species of the phylum Acidobacteriota have genes that probably help in survival and competitive colonization of the rhizosphere, leading to the establishment of beneficial relationships with plants, regulation of biogeochemical cycles, decomposition of biopolymers, exopolysaccharide secretion, and plant growth promotion [47]. The species belonging to the phylum Chloroflexi are known as anaerobic microbes that can co-exist with methane-metabolizing microbes and are crucial organic matter degraders under anoxic conditions [48]. Methane metabolism is used for the bioremediation of Cd contamination and promotes the precipitation of soluble Cd in soil [48]. Gemmatimonadota is known as the eighth-most abundant bacterial phylum in soils, representing about 1–2% of the soil bacteria worldwide. They are capable of anoxygenic photosynthesis and are associated with the plants and the rhizosphere, treatment plants, and biofilms [49]. The phylum Myxococcota, is broadly distributed in soil with the ability to produce diverse secondary metabolites acting as antimicrobials, antiparasitic, antivirals, cytotoxins, and anti-blood coagulants [50].

Notably, we found that most of the phyla which are presented correlated positively with the functional prediction. The Proteobacteria and Firmicutes both have the ability to produce ACC deaminase, antimicrobial compounds, and phytohormones, while the phylum Bacteriodota facilitates phosphate solubilization in the soil, as well as the phylum Proteobacteria. Therefore, Actinobacteriota and Myxococcota can produce secondary metabolites of biotechnological and antimicrobials. As well as this, Gemmatimonadota, Firmicutes and Acidobacteriota phyla are known microbes for association with the plants and the rhizosphere, treatment plants and plant growth promotion. These function predictions of the detected bacterial taxa support the hypothesis that phage mixers have the potential to control plant pathogenic bacteria and cause changes in the rhizosphere bacteria but not affect the beneficial rhizosphere microbes.

The limitations of these types of treatments are the application of the phage therapy to the field without studying its effect over many plant generations, which will require more time and effort and to be tested over different climatic conditions and soil types. Our continuous plan to overcome this limitation includes: testing it in the field over different seasons from different cultivation spots; observing the phage biocontrol effect on *R. solanacearum*, *P. carotovorum*, and *P. atrosepticum* for longer time periods, such as 1 or 2 years; studying the histopathology of plants at the cellular level. Additionally, we are planning to apply a whole genome metagenomic analysis to study the bacteriophage therapy effect on the wider microbiome community, including protists and fungi.

4. Materials and Methods

4.1. Bacterial Isolates and Culture Conditions

The bacterial phytopathogens used in this study included *Ralstonia solanacearum* GIM1.74 (Rs), *Pectobacterium carotovorum* subsp *carotovorum* KPM17 (Pc), and *Pectobacterium atrosepticum* WHG10001 (Pa). The *R. solanacerum* GIM1.74 (Rs) strain was cultured on CTG agar plates and in broth (1% Casamino acid hydroxylate, 1% tryptone and 1.5% *w/v*, agar) at 28 °C with shaking (170 r.p.m.). The bacterial species of the genus *Pectobacterium* were cultured on Luria Bertani (LB) agar plates (1.5% *w/v*, agar) and in broth at 28 °C [51]. After the incubation, the bacterial culture count in the suspensions ranged between 10^7 to 10^8 CFU/mL.

4.2. Phage Isolates, Amplification, and Tittering Conditions

Three phages, PSG11, WC4, and CX5, that were previously reported as specific bacteriophages for *R. solanacearum*, *P. carotovorum*, and *P. atrosepticum*, respectively, were used as single phages [1,5]. Three bacteriophage cocktails that each included three different types of bacteriophages were prepared individually: the PSG2/PSG3/PSG11, WC1/WC2/WC4 and CX2/CX3/CX5 phage cocktails. A list of the used bacteria and phages is provided in Table 1.

All the used bacteriophages were prepared in Tris-HCl phage buffer at pH 7.5 (50 mM Tris-base, 150 mM NaCl, 10 mM MgCl$_2$.6H$_2$O and 2 mM CaCl$_2$). Purified phages were amplified by mixing 500 μL of the host bacteria with 10 μL of their respective phage. The mixture was vortexed at 160 rpm and incubated at 28 °C for 15 min. Thereafter, 4 mL of soft agar were added to the phage–bacteria mixture, poured on LB and CTG agar plates and incubated overnight at 28 °C. The overlay agar was scrapped off from the double agar plate into a 15 mL centrifuge tube containing 2 mL phage buffer, followed by vertexing for 2 min and centrifugation at 5500 rpm for 15 min at 4 °C. The phage lysate was then filtered through a syringe-driven filter (0.22 µm). The titer of the phages was determined through 10-fold serial dilutions and placing a spot of 10 µL of the lysate on a double agar layer containing the host bacteria.

Table 1. Bacterial isolates and sources and well as phages are shown.

Bacteria		Phage	
Strain	**Source**	**Single**	**Cocktail**
Ralstonia solanacearum strain GIM1.74 (Rs)	Purchased from Guangdong Microbiology Culture Center, China	P-PSG-11 (SRs)	Rs 2, 3,11 cocktail (RsPck)
Pectobacterium carotovorum subsp *carotovorum* strain KPM17 (Pc)	Isolated from Molo, Kenya	Wc4 (SPc)	Wc1, 2, 4 cocktail (PcPck)
Pectobacterium atrosepticum strain WGH10001 (Pa)	Isolated from Mongolia, China	CX5 (SPa)	CX 2, 3, 5 cocktail (PaPck)

4.3. Greenhouse Experiment Design and Treatments

The efficacy of single and cocktail phages for controlling potato bacterial wilt and soft rot were tested in pots. All experiments were carried out in the greenhouse of the Wuhan Institute of Virology (Wuhan, China) in the period between August to October 2019. The temperature, relative humidity, and light density fluctuated between 28–37 °C, 58–85%, and 15–20 Klux, respectively.

Soil materials that were used in the present study were collected from a field located in the Wuhan Institute of Virology (Zhendian street, Jianxia district, Wuhan, China) at 0–30 cm depth. Then, the soil was air-dried, grinded and sieved through a 2 mm sieve. Some properties of the soil including pH, particle size distribution, soluble cations and anions were determined according to the Olsen method [52] (Table S1). The nonsterile soil was uniformly packed in plastic pots of 18 cm height and 26.5 cm mean diameter at a rate of 5 kg soil pot^{-1} (with a 1 cm drainage hole). The soil in each pot was mixed with 50 g cattle manure (CM) (1% *w/w*) as an organic fertilizer. Potato seeds were surface sterilized with 3% NaClO for 5 min and followed by 70% ethyl alcohol for 1 min prior to cultivation. The seeds were then germinated on water–agar plates for two days and further transplanted to each pot. A suspension of pathogens (10^7 cells/gram of soil) were inoculated onto the plants after 4 days from transplanting, while the phage treatments (10^6 particles/gram of soil) were inoculated 2 days after the pathogen inoculation. All pots received the same P fertilization at a rate of 1.0 g superphosphate (15.5% P_2O_5) per pot, an equivalent to 31.0 kg P_2O_5 per feed mixed with the soil before cultivation. Thereafter, three potato seeds were sown in each pot and irrigated to about soil field capacity using tap water. After two weeks, the seedlings were thinned to 1 plant per pot. Then, ammonium sulphate and potassium sulphate at rates of 0.60 g N and 0.25 g K_2O pot^{-1} (equivalent to 120 kg N and 50 kg K_2O fed^{-1}, respectively) were applied to all pots twice with 20 and 50% of the total amounts after 25 and 50 days from sowing date, respectively.

The experiment was designed as a randomized complete block where ten treatments for the three pathogens were categorized as follows: negative control (i.e., potato seeds cultivated in untreated soil), pathogen-treated samples (i.e., potato seeds cultivated in soil treated with specific bacterial pathogen), single phage treatment (i.e., potato seeds cultivated in soil treated with the specific bacterial pathogen and treated with a single bacteriophage species), and phage-cocktail treatment (i.e., potato seeds cultivated in soil treated with the specific bacterial pathogen and treated with three mixed bacteriophage species), compared to native soil. Treatments were replicated five times and rearranged randomly every four days. Each replicate contained one potato plant per pot. The pathogen treatments were named by pathogen initials, as Rs, Pc and Pa. The single phage treatments were prefixed by the letter "S" (i.e., SRs, SPc and SPa), while the phage cocktail treatments were suffixed by "Pck" (i.e., RsPck, PcPck and PaPck; Figure 7).

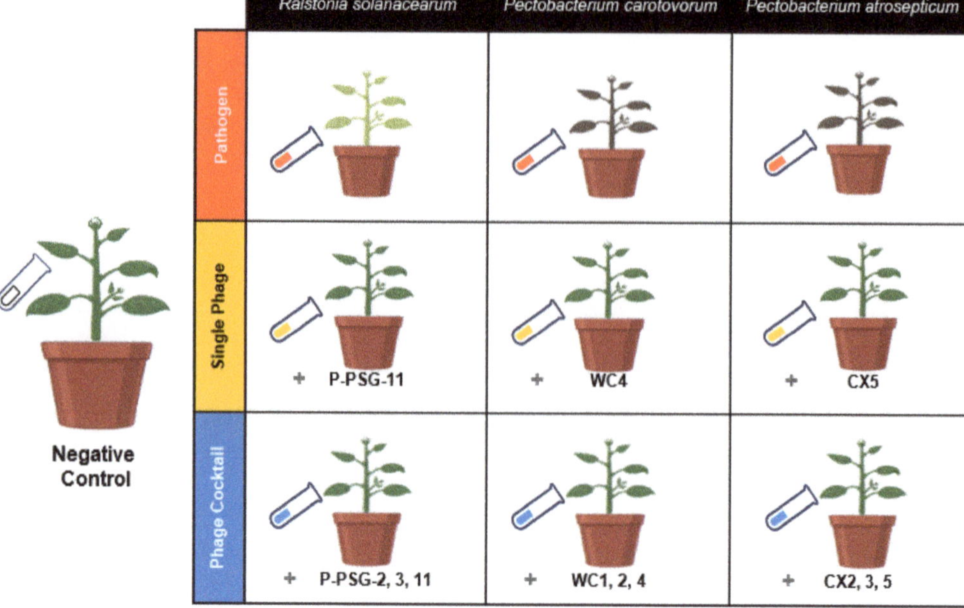

Figure 7. Illustration model for the biocontrol of the three groups, pathogens group *Ralstonia solanacearum* (Rs), *Pectobacterium carotovorum* (Pc) and *Pectobacterium atrosepticum* (Pa), single phage group SRs (PSG11), SPc (WC4) and SPa (CX5) and phage cocktail groups RsPck (P-PSG-2, 3, 11), PcPck (WC1, 2, 4) and PaPck (CX2, 3, 5), compared with negative control (no bacteria or phage added) in a greenhouse experiment.

4.4. Metabarcoding Analysis

For every pot, soil samples were collected randomly before the beginning of the experiment and at the end of the greenhouse experiment from the plant–rhizosphere area and kept in plastic bags for determining the changes in the rhizosphere microbiome composition. Samples were then sent to the company Sangon Biotech (Shanghai, China) for DNA extraction and metabarcoding analysis. A total of 50 samples were collected along with one sample from uncultivated soil (i.e., the source of all the soil used to conduct the experiment). DNA extraction was performed using the Power Soil MoBio DNA Isolation Kit (MO BIO Laboratories Inc., Carlsbad, CA, USA) at a final elution volume of 150 mL.

The bacterial communities in the soil were assessed by sequencing amplicons of the V3–V4 variable region of the 16S rRNA gene, with the primer pair 338F (5′-ACT CCT ACG GG AGG CAG CAG-3′) and 806R (5′- 489 GGA CTA CHV GGG TWT CTA AT-3′). The PCR reaction was performed using the TransStart FastPfu DNA Polymerase mixture. The reaction mixture (20 µL) was composed of 4 µL of 5x FastPfu Buffer, 2 µL of 2.5 mM (each) dNTPs, 0.8 µL of 5 µM Bar-PCR primer F, 0.8 µL of 5 µM primer R, 0.4 µL of FastPfu polymerase, 0.2 µL of BSA and 10 ng of template DNA. Amplification conditions for PCR were as follows: 3 min at 98 °C to denature the DNA, followed by 27 cycles of denaturation at 98 °C for 10 s, primer annealing at 60 °C for 30 s, and strand extension at 72 °C 45 s, followed by 7 min at 72 °C on an ABI Gene Amp 9700 thermocycler (IET, London, UK). Electrophoresis on 2% agarose gel was used to check the quality of the PCR products and purified using Agencourt AMPure XP beads (Beckman, Brea, CA, USA). The pooled DNA product was used to construct an Illumina pair-end library followed by Illumina-adapter ligation and sequencing by Illumina (MiSeq, PE 2 × 300 bp mode), following the manufacturer's instructions.

Paired-end data were demultiplexed into each sample based on the index sequences downloaded from the Illumina MiSeq platform. Hence, the paired-end sequences of each sample were trimmed based on their quality and length using Trimmomatic [53] and FLASH [54] software. The metabarcoding analysis was performed using the online Majorbio Cloud Platform (http://en.majorbio.com/ (accessed on 1 June 2022)). Uparse V7.1 (http://drive5.com/uparse/ (accessed on 1 June 2022)) was used to detect and remove chimera sequences. Mothur v.1.9.0 software [55] was used to infer richness and to perform library comparisons. The Operational Taxonomic Unit OTU (is the basic unit in numerical taxonomy and can be used to classify groups of closely related species, individuals, or genes) was clustered at a sequence similarity of 0.99, while the taxonomy was identified by the RDP classifier algorithm (http://rdp.cme.msu.edu/ (accessed on 1 June 2022)) versus the Silva 16S rRNA database (version 138) at a 70% confidence threshold. The PICRUSt (http://huttenhower.sph.harvard.edu/galaxy/ (accessed on 1 June 2022)) was employed to predict the functional characters of the detected microbial communities and functions. The co-occurrence network was analyzed using Orange V3.24.1 (https://orange.biolab.si/ (accessed on 1 June 2022)). The counts were analyzed and visualized using Venn diagrams (vegan R-package) and Circos plots (Circos -0.6; http://circos.ca/ (accessed on 1 June 2022)).

Supplementary Materials: The following supporting information can be downloaded at: https://www.mdpi.com/article/10.3390/antibiotics11081117/s1, Table S1: Soil properties including pH, particle size distribution, soluble cations, and anions, Figure S1: Phage plaque assay, Figure S2: Phage biocontrol experiment.

Author Contributions: Conceptualization, S.M. and H.W.; methodology, S.M.; software, S.M. and M.M.; validation, S.M., D.X., R.N. and H.W.; formal analysis, M.M.; investigation, S.M.; resources, S.M. and J.Y.; data curation, S.M. and M.M.; writing—original draft preparation, S.M.; writing—review and editing, S.M., M.M., D.X., R.N., S.M.R. and J.Y.; visualization, S.M. and H.W.; supervision, H.W.; project administration, S.M. and H.W.; funding acquisition, H.W. All authors have read and agreed to the published version of the manuscript.

Funding: This research was financially supported by the National Biosafety Laboratory, Wuhan Institute of Virology, Chines Academy of Sciences, China (ZDRW-ZS-2016-4). The authors are grateful to the CAS-TWAS program for funding S.M.A.M. during her PhD studies in the University of Chinese academy of Science (Beijing, China) and Wuhan Institute of Virology (Wuhan, China).

Institutional Review Board Statement: Not applicable.

Informed Consent Statement: Not applicable.

Data Availability Statement: The raw data is available at NCBI BioProject: PRJNA867554.

Acknowledgments: This research was financially supported by Wuhan Institute of Virology, Chines Academy of Sciences, China (ZDRW-ZS-2016-4). The authors are grateful to Jun Hua, Kelvin K. Kering, Mysara Griesh, and Liu Huan for their support and assistance in the lab during the conduction of the study.

Conflicts of Interest: The authors declare no conflict of interest.

References

1. Wei, C.; Liu, J.; Maina, A.N.; Mwaura, F.B.; Yu, J.; Yan, C.; Zhang, R.; Wei, H. Developing a bacteriophage cocktail for biocontrol of potato bacterial wilt. *Virol. Sin.* **2017**, *32*, 476–484. [CrossRef]
2. Vleeshouwers, V.G.; Raffaele, S.; Vossen, J.H.; Champouret, N.; Oliva, R.; Segretin, M.E.; Rietman, H.; Cano, L.M.; Lokossou, A.; Kessel, G.; et al. Understanding and exploiting late blight resistance in the age of effectors. *Annu. Rev. Phytopathol.* **2011**, *49*, 507–531. [CrossRef]
3. Spooner, D.M.; Ghislain, M.; Simon, R.; Jansky, S.H.; Gavrilenko, T. Systematics, Diversity, Genetics, and Evolution of Wild and Cultivated Potatoes. *Bot. Rev.* **2014**, *80*, 283–383. [CrossRef]
4. Devaux, A.; Goffart, J.P.; Petsakos, A.; Kromann, P.; Gatto, M.; Okello, J.; Suarez, V.; Hareau, G. Global Food Security, Contributions from Sustainable Potato Agri Food Systems. In *The Potato Crop*; Springer: Cham, Switzerland, 2020; pp 3–35
5. Muturi, P.; Yu, J.; Maina, A.N.; Kariuki, S.; Mwaura, F.B.; Wei, H. Bacteriophages Isolated in China for the Control of *Pectobacterium carotovorum* Causing Potato Soft Rot in Kenya. *Virol. Sin.* **2019**, *34*, 287–294. [CrossRef] [PubMed]

6. La Torre, A.; Iovino, V.; Caradonia, F. Copper in plant protection: Current situation and prospects. *Phytopathol. Mediterr.* **2018**, *57*, 201–236. [CrossRef]

7. Charkowski, A.; Sharma, K.; Parker, M.; Secor, G.; Elphinstone, J. Bacterial Diseases of Potato. In *The Potato Crop*; Springer: Cham, Switzerland, 2020; pp. 351–388.

8. Pérombelon, M. Potato diseases caused by soft rot erwinias: An overview of pathogenesis. *Plant Pathol.* **2002**, *51*, 1–12.

9. Wei, Z.; Hu, J.; Gu, Y.A.; Yin, S.; Xu, Y.; Jousset, A.; Shen, Q.; Friman, V.-P. Ralstonia solanacearum pathogen disrupts bacterial rhizosphere microbiome during an invasion. *Soil Biol. Biochem.* **2018**, *118*, 8–17. [CrossRef]

10. Buttimer, C.; Hendrix, H.; Lucid, A.; Neve, H.; Noben, J.P.; Franz, C.; O'Mahony, J.; Lavigne, R.; Coffey, A. Novel N4-Like Bacteriophages of Pectobacterium atrosepticum. *Pharmaceuticals* **2018**, *11*, 45. [CrossRef] [PubMed]

11. Xue, H.; Lozano-Duran, R.; Macho, A. Insights into the Root Invasion by the Plant Pathogenic Bacterium *Ralstonia solanacearum*. *Plants* **2020**, *9*, 516. [CrossRef] [PubMed]

12. Alvarez, B.; Biosca, E.G. Bacteriophage-Based Bacterial Wilt Biocontrol for an Environmentally Sustainable Agriculture. *Front. Plant Sci.* **2017**, *8*, 1218. [CrossRef] [PubMed]

13. Buttimer, C.; McAuliffe, O.; Ross, R.P.; Hill, C.; O'Mahony, J.; Coffey, A. Bacteriophages and Bacterial Plant Diseases. *Front. Microbiol.* **2017**, *8*, 34. [CrossRef] [PubMed]

14. Balogh, B.; Jones, J.; Iriarte, F.; Momol, M. Phage Therapy for Plant Disease Control. *Curr. Pharm. Biotechnol.* **2009**, *11*, 48–57. [CrossRef]

15. Lamichhane, J.R.; Osdaghi, E.; Behlau, F.; Köhl, J.; Jones, J.B.; Aubertot, J.-N. Thirteen decades of antimicrobial copper compounds applied in agriculture. A review. *Agron. Sustain. Dev.* **2018**, *38*, 28. [CrossRef]

16. Sundin, G.W.; Wang, N. Antibiotic Resistance in Plant-Pathogenic Bacteria. *Annu. Rev. Phytopathol.* **2018**, *56*, 161–180. [CrossRef] [PubMed]

17. Cooksey, D.A. Genetics of Bactericide Resistance in Plant Pathogenic Bacteria. *Annu. Rev. Phytopathol.* **2003**, *28*, 201–219. [CrossRef]

18. Behlau, F.; Canteros, B.I.; Minsavage, G.V.; Jones, J.B.; Graham, J.H. Molecular Characterization of Copper Resistance Genes from *Xanthomonas citri* subsp. citri and *Xanthomonas alfalfae* subsp. citrumelonis. *Appl. Environ. Microbiol.* **2011**, *77*, 4089–4096. [CrossRef] [PubMed]

19. Ritchie, D. Copper- and Streptomycin-Resistant Strains and Host Differentiated Races of *Xanthomonas campestris* pv. vesicatoria in North Carolina. *Plant Dis.* **1991**, *75*, 733. [CrossRef]

20. Berendsen, R.L.; Pieterse, C.M.; Bakker, P.A. The rhizosphere microbiome and plant health. *Trends Plant Sci.* **2012**, *17*, 478–486. [CrossRef] [PubMed]

21. Wei, Z.; Yang, T.; Friman, V.P.; Xu, Y.; Shen, Q.; Jousset, A. Trophic network architecture of root-associated bacterial communities determines pathogen invasion and plant health. *Nat. Commun.* **2015**, *6*, 8413. [CrossRef]

22. Hu, J.; Wei, Z.; Friman, V.P.; Gu, S.H.; Wang, X.F.; Eisenhauer, N.; Yang, T.J.; Ma, J.; Shen, Q.R.; Xu, Y.C.; et al. Probiotic Diversity Enhances Rhizosphere Microbiome Function and Plant Disease Suppression. *mBio* **2016**, *7*, e01790-16. [CrossRef] [PubMed]

23. Brodeur, J. Host specificity in biological control: Insights from opportunistic pathogens. *Evol. Appl.* **2012**, *5*, 470–480. [CrossRef] [PubMed]

24. Kyselková, M.; Moënne-Loccoz, Y. Pseudomonas and other Microbes in Disease-Suppressive Soils. In *Organic Fertilisation, Soil Quality and Human Health*; Sustainable Agriculture Reviews; Springer: Berlin/Heidelberg, Germany, 2012; pp. 93–140.

25. Kering, K.K.; Kibii, B.J.; Wei, H. Biocontrol of phytobacteria with bacteriophage cocktails. *Pest Manag. Sci.* **2019**, *75*, 1775–1781. [CrossRef]

26. Hill, C. Bacteriophages: Viruses That Infect Bacteria. *Front. Young Minds* **2019**, *7*, 146. [CrossRef]

27. Wang, X.; Wei, Z.; Yang, K.; Wang, J.; Jousset, A.; Xu, Y.; Shen, Q.; Friman, V.P. Phage combination therapies for bacterial wilt disease in tomato. *Nat. Biotechnol.* **2019**, *37*, 1513–1520. [CrossRef]

28. Yu, L.; Wang, S.; Guo, Z.; Liu, H.; Sun, D.; Yan, G.; Hu, D.; Du, C.; Feng, X.; Han, W.; et al. A guard-killer phage cocktail effectively lyses the host and inhibits the development of phage-resistant strains of *Escherichia coli*. *Appl. Microbiol. Biotechnol.* **2018**, *102*, 971–983. [CrossRef] [PubMed]

29. Gu, Y.; Wei, Z.; Wang, X.; Friman, V.-P.; Huang, J.; Wang, X.; Mei, X.; Xu, Y.; Shen, Q.; Jousset, A. Pathogen invasion indirectly changes the composition of soil microbiome via shifts in root exudation profile. *Biol. Fertil. Soils* **2016**, *52*, 997–1005. [CrossRef]

30. Li, M.; Wei, Z.; Wang, J.; Jousset, A.; Friman, V.P.; Xu, Y.; Shen, Q.; Pommier, T. Facilitation promotes invasions in plant-associated microbial communities. *Ecol. Lett.* **2019**, *22*, 149–158. [CrossRef] [PubMed]

31. Jacobs, J.; Carroll, T.L.; Sundin, G.W. The Role of Pigmentation, Ultraviolet Radiation Tolerance, and Leaf Colonization Strategies in the Epiphytic Survival of Phyllosphere Bacteria. *Microb. Ecol.* **2005**, *49*, 104–113. [CrossRef] [PubMed]

32. Bashan, Y.; de-Bashan, L. Chapter Two—How the Plant Growth-Promoting Bacterium Azospirillum Promotes Plant Growth—A Critical Assessment. *Adv. Agron.* **2010**, *108*, 77–136.

33. Spaepen, S.; Versées, W.; Rother, D.; Pohl, M.; Steyaert, J.; Vanderleyden, J. Characterization of Phenylpyruvate Decarboxylase, Involved in Auxin Production of *Azospirillum brasilense*. *J. Bacteriol.* **2007**, *189*, 7626–7633. [CrossRef] [PubMed]

34. Ryu, R.; Patten, C. Aromatic Amino Acid-Dependent Expression of Indole-3-Pyruvate Decarboxylase Is Regulated by TyrR in Enterobacter cloacae UW5. *J. Bacteriol.* **2008**, *190*, 7200–7208. [CrossRef] [PubMed]

35. Zimmer, W.; Hundeshagen, B.; Niederau, E. Demonstration of the indolepyruvate decarboxylase gene in different auxin-producing species of the Enterobacteriaceae. *Can. J. Microbiol.* **1995**, *40*, 1072–1076. [CrossRef]
36. Suarez-Moreno, Z.; Caballero-Mellado, J.; Gonçalves Coutinho, B.; Mendonça-Previato, L.; James, E.; Venturi, V. Common Features of Environmental and Potentially Beneficial Plant-Associated Burkholderia. *Microb. Ecol.* **2011**, *63*, 249–266. [CrossRef] [PubMed]
37. Miller, S.; Browne, P.; Prigent-Combaret, C.; Combes-Meynet, E.; Morrissey, J.; O'Gara, F. Biochemical and genomic comparison of inorganic phosphate solubilization in Pseudomonas species. *Environ. Microbiol. Rep.* **2009**, *2*, 403–411. [CrossRef]
38. Blaha, D.; Prigent-Combaret, C.; Mirza, M.; Moenne-Loccoz, Y. Phylogeny of the 1-aminocyclopropane-1-carboxylic acid deaminase-encoding gene acdS in phytobeneficial and pathogenic Proteobacteria and relation with strain biogeography: AcdS phylogeny in Proteobacteria. *FEMS Microbiol. Ecol.* **2006**, *56*, 455–470. [CrossRef] [PubMed]
39. Glick, B.; Jacobson, C.; Schwarze, M.; Pasternak, J. 1-Aminocyclopropane-1-carboxylic acid deaminase mutants of the plant growth promoting rhizobacterium Pseudomonas putida GR12-2 do not stimulate canola root elongation. *Can. J. Microbiol.* **1994**, *40*, 911–915. [CrossRef]
40. Haas, D.; Keel, C. Regulation of Antibiotic Production in Root-Colonizing *Pseudomonas* spp. and Relevance for Biological Control of Plant Disease. *Annu. Rev. Phytopathol.* **2003**, *41*, 117–153. [CrossRef] [PubMed]
41. Compant, S.; Samad, A.; Faist, H.; Sessitsch, A. A review on the plant microbiome: Ecology, functions and emerging trends in microbial application. *J. Adv. Res.* **2019**, *19*, 29–37. [CrossRef] [PubMed]
42. Hashem, A.; Tabassum, B.; Abd Allah, E.F. Bacillus subtilis: A plant-growth promoting rhizobacterium that also impacts biotic stress. *Saudi J. Biol. Sci.* **2019**, *26*, 1291–1297. [CrossRef] [PubMed]
43. Barea, J.-M.; Pozo, M.; Azcón, R.; Azcon-Aguilar, C. Microbial co-operation in the rhizosphere. *J. Exp. Bot.* **2005**, *56*, 1761–1778. [CrossRef] [PubMed]
44. Hazarika, S. *Beneficial Microbes in Agro-Ecology*; Academic Press: Cambridge, MA, USA, 2020.
45. Olanrewaju, O.; Babalola, O. Streptomyces: Implications and interactions in plant growth promotion. *Appl. Microbiol. Biotechnol.* **2019**, *103*, 1179–1188. [CrossRef]
46. Wu, X.; Rensing, C.; Han, D.; Xiao, K.-Q.; Dai, Y.; Tang, Z.; Liesack, W.; Peng, J.; Cui, Z.; Zhang, F. Genome-Resolved Metagenomics Reveals Distinct Phosphorus Acquisition Strategies between Soil Microbiomes. *mSystems* **2022**, *7*, e0110721. [CrossRef] [PubMed]
47. Chen, X.; Wang, J.; You, Y.; Wang, R.; Chu, S.; Chi, Y.; Hayat, K.; Hui, N.; Liu, X.; Zhang, D.; et al. When nanoparticle and microbes meet: The effect of multi-walled carbon nanotubes on microbial community and nutrient cycling in hyperaccumulator system. *J. Hazard. Mater.* **2021**, *423*, 126947. [CrossRef] [PubMed]
48. Meng, D.; Li, J.; Liu, T.; Liu, Y.; Yan, M.; Hu, J.; Li, X.; Liu, X.; Liang, Y.; Liu, H.; et al. Effects of redox potential on soil Cadmium solubility: Insight into microbial community. *J. Environ. Sci.* **2018**, *75*, 224–232. [CrossRef] [PubMed]
49. Mujakic, I.; Piwosz, K.; Koblizek, M. Phylum Gemmatimonadota and Its Role in the Environment. *Microorganisms* **2022**, *10*, 151. [CrossRef]
50. Garcia, R.; Müller, R. *Simulacricoccus ruber* gen. nov., sp. nov., a microaerotolerant, non-fruiting, myxospore-forming soil myxobacterium and emended description of the family Myxococcaceae. *Int. J. Syst. Evol. Microbiol.* **2018**, *68*, 3101–3110. [CrossRef] [PubMed]
51. Kering, K.; Zhang, X.; Nyaruaba, R.; Yu, J.; Wei, H. Application of Adaptive Evolution to Improve the Stability of Bacteriophages during Storage. *Viruses* **2020**, *12*, 423. [CrossRef]
52. Sparks, D.L.; Page, A.; Helmke, P.A.; Loeppert, R.; Soltanpour, P.N.; Tabatabai, M.; Johnston, C.; Sumner, M. *Methods of Soil Analysis. Part III. Chemical Methods*; John Wiley & Sons: New York, NY, USA, 1996; Volume 5.
53. Bolger, A.; Lohse, M.; Usadel, B. Trimmomatic: A Flexible Trimmer for Illumina Sequence Data. *Bioinformatics* **2014**, *30*, 2114–2120. [CrossRef]
54. Reyon, D.; Tsai, S.; Khayter, C.; Foden, J.; Sander, J.; Joung, J. FLASH assembly of TALENs for high-throughput genome editing. *Nat. Biotechnol.* **2012**, *30*, 460–465. [CrossRef] [PubMed]
55. Schloss, P.; Westcott, S.; Ryabin, T.; Hall, J.; Hartmann, M.; Hollister, E.; Lesniewski, R.; Oakley, B.; Parks, D.; Robinson, C.; et al. Introducing mothur: Open-Source, Platform-Independent, Community-Supported Software for Describing and Comparing Microbial Communities. *Appl. Environ. Microbiol.* **2009**, *75*, 7537–7541. [CrossRef]

antibiotics

Article

Adaptive Phage Therapy for the Prevention of Recurrent Nosocomial Pneumonia: Novel Protocol Description and Case Series

Fedor Zurabov [1,*], Marina Petrova [2], Alexander Zurabov [1], Marina Gurkova [1], Petr Polyakov [2], Dmitriy Cheboksarov [2], Ekaterina Chernevskaya [2], Mikhail Yuryev [2], Valentina Popova [1], Artem Kuzovlev [2], Alexey Yakovlev [2] and Andrey Grechko [2]

[1] Research and Production Center "MicroMir", 107031 Moscow, Russia; azurabov@micromir.bio (A.Z.); mgurkova@micromir.bio (M.G.); val.popova@micromir.bio (V.P.)

[2] Federal Research and Clinical Center of Intensive Care Medicine and Rehabilitology, 10703 Moscow, Russia; mpetrova@fnkcrr.ru (M.P.); p.polyakov@fnkcrr.ru (P.P.); dcheboksarov@fnkcrr.ru (D.C.); echernevskaya@fnkcrr.ru (E.C.); myurev@fnkcrr.ru (M.Y.); artem_kuzovlev@fnkcrr.ru (A.K.); ayakovlev@fnkcrr.ru (A.Y.); avgrechko@fnkcrr.ru (A.G.)

* Correspondence: fmzurabov@micromir.bio

Abstract: Nowadays there is a growing interest worldwide in using bacteriophages for therapeutic purposes to combat antibiotic-resistant bacterial strains, driven by the increasing ineffectiveness of drugs against bacterial infections. Despite this fact, no novel commercially available therapeutic phage products have been developed in the last two decades, as it is extremely difficult to register them under the current legal regulations. This paper presents a description of the interaction between a bacteriophage manufacturer and a clinical institution, the specificity of which is the selection of bacteriophages not for an individual patient, but for the entire spectrum of bacteria circulating in the intensive care unit with continuous clinical and microbiological monitoring of efficacy. The study presents the description of three clinical cases of patients who received bacteriophage complex via inhalation for 28 days according to the protocol without antibiotic use throughout the period. No adverse effects were observed and the elimination of multidrug-resistant microorganisms from the bronchoalveolar lavage contents was detected in all patients. A decrease in such inflammatory markers as C-reactive protein (CRP) and procalcitonin was also noted. The obtained results demonstrate the potential of an adaptive phage therapy protocol in intensive care units for reducing the amount of antibiotics used and preserving their efficacy.

Keywords: phage therapy; bacteriophages; antimicrobial resistance; regulatory framework; personalized medicine; clinical case; intensive care

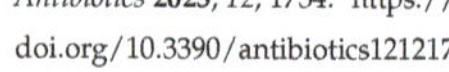

Citation: Zurabov, F.; Petrova, M.; Zurabov, A.; Gurkova, M.; Polyakov, P.; Cheboksarov, D.; Chernevskaya, E.; Yuryev, M.; Popova, V.; Kuzovlev, A.; et al. Adaptive Phage Therapy for the Prevention of Recurrent Nosocomial Pneumonia: Novel Protocol Description and Case Series. *Antibiotics* **2023**, *12*, 1734. https://doi.org/10.3390/antibiotics12121734

Academic Editors: Aneta Skaradzińska and Anna Nowaczek

Received: 18 October 2023
Revised: 5 December 2023
Accepted: 11 December 2023
Published: 14 December 2023

1. Introduction

One of the challenges of the 21st century is the ineffectiveness of drugs against bacterial infections [1]. Hospital-acquired infections currently cause millions of deaths per year, and the prognosis for the future is even worse, as evidenced by current research. In 2009, over 20,000 people in the US died due to the lack of effective antibiotics. The same statistics have been observed in Europe. Worldwide, more than 100,000 people have died from infections caused by antibiotic-resistant bacteria [2]. In 2019, the annual number of deaths associated with antibiotic-resistant bacteria surged to 4.95 million [3].

Phage therapy is based on the therapeutic use of bacteriophages to treat bacterial infections. Since Felix D'Hérelle's introduction of the term "bacteriophage" in 1917, phages have been used extensively in combatting bacterial pathogens. However, in Western Europe and the USA, phage therapy was soon abandoned due to questionable treatment outcomes, lack of standardization, and the discovery of antibiotics. Bacteriophages continued to be used for therapy in Eastern European nations like the USSR, supported by the establishment of the Bacteriophage Institute in Tbilisi in 1934 by George Eliava together with Felix D'Hérelle [4]. A reduction in treatment options for patients due to the spread of antibiotic resistance among bacterial pathogens has led researchers to explore new methods of managing bacterial populations and restoring the natural balance of the microbiota. Currently, there is a growing interest worldwide in using bacteriophages for therapeutic purposes to combat antibiotic-resistant bacterial strains, and they are increasingly being used by scientists and clinicians [5]. Published clinical cases and individual clinical trials of phage therapy suggest a high level of safety when using different routes of administration in humans, although some studies do not prove efficacy [6]. Humans co-exist with phages throughout their lives, and they are an inherited element of the human microbiome [7], so the occurrence of unwanted side effects, including toxicity or allergy, is highly unlikely. Furthermore, phage therapy has proven to have practically no side effects [8].

In spite of their importance, no new commercially available therapeutic phage products have been developed in the last two decades. This is largely due to the regulatory restrictions that exist for bacteriophages, which make it impossible to register these products under current legal frameworks [9]. In both Europe and the USA, there is no specific regulation of bacteriophages. Since 2011, phages have been classified as a drug in the USA or as a medicinal product in the European Union [10]. The qualification of phages as a medicinal product for human use was approved in 2015 at a workshop of the European Medicines Agency (EMA), despite warnings from phage researchers about the inadequacy of such an assessment. Following the event, researchers expressed their disagreement in a letter highlighting the need for a novel regulatory structure [11].

Clinical trials of phage products have revealed a range of problems associated with double-blind studies, which are costly and time-consuming for the testing of at least part of the drug formulation [2]. Additionally, a significant downside of commercializing phage-based products is the requirement to create phage cocktails that target the largest number of strains of the given bacterium. Furthermore, these cocktails require regular updates to ensure they are effective against currently circulating clinical strains. Unfortunately, this strategy cannot be implemented due to current drug approval regulations [9]. Nonetheless, studies indicate that authorities responsible for licensing phage therapy treatments in Europe and the United States are attempting to streamline the licensing process [2].

It is important to be able to use newly discovered phages quickly. One way to do this would be through the reduction of licensing requirements to the definition of a full production cycle. There is ongoing discussion regarding possible licensing pathways for phage products. A licensing pathway model should be created with consideration given to the fact that bacteriophages are natural, biological agents that can specifically combat pathogenic bacteria, while self-regulating [2]. Researchers suggest that the licensing process for phage therapy should be adapted to its unique characteristics, rather than the other way around [12].

Because of the aforementioned difficulties in registering bacteriophage products, they are often used within the confines of Article 37 of the Declaration of Helsinki [13]. Article 37 of the Declaration of Helsinki states that physicians may use an unproven intervention for an individual patient only when proven or known methods are ineffective, the physician has sought expert advice, has obtained informed consent from the patient or the patient's legal representative, and believes that the unproven method offers hope of saving the life, restoring the health, or alleviating the suffering of the patient. According to Pirnay et al. [14], only 28 patients were treated with personalized phage therapy between 2018 and 2022 (based on published case reports and case series). These sparse patient numbers suggest that the personalized approach to phage therapy, although effective in many cases, is difficult to scale up and does not provide access to phage therapy for a wide range of patients.

The most significant advancement in regulating phage therapy took place in Belgium, where in 2017, the government, under the guidance of a team of researchers from the Queen Astrid Military Hospital in Brussels, decided to classify phages not as industrial drugs, but as active components in magistral preparations. A "magistral preparation" is defined as "any medicinal product prepared in a pharmacy according to a physician's prescription for an individual patient" [15]. This process enabled a Belgian hospital to produce phages for the treatment of bacterial infections in humans. However, there are still questions that need to be clarified, as there is no clear consensus on the requirements and standards for the production of such preparations.

Scientists and clinicians in the Russian Federation are developing a comparable approach to legalize the administration of bacteriophages. The difference is that a "cocktail" of bacteriophages is prepared not for an individual patient but for a specific medical institution/department based on the results of bacterial composition monitoring, which is called "adaptive phage therapy" [16]. In order to expand and legitimize the approach, pharmaceutical substances must be registered and manufactured into final forms by compounding pharmacies. The first pharmaceutical substance, "Bacteriophages specific to *Klebsiella pneumoniae*", was registered in July 2023 by the Research and Production Centre (RPC) "Micromir".

The present article focuses on the description of the adaptive phage therapy protocol for patients in intensive care units, the specificity of which is the selection of bacteriophages not for an individual patient, but for the entire spectrum of bacteria circulating in the intensive care unit with continuous clinical and microbiological monitoring of efficacy, and the description of case series of patients who received bacteriophage complex via inhalation for 28 days according to the protocol without antibiotic use throughout the period.

2. Results

2.1. Adaptive Phage Therapy Protocol

The protocol was created to describe the interaction between the Research and Production Centre (RPC) "Micromir" acting as a phage center and a production site certified under the rules of GMP, and the intensive care units (ICU) of the Federal Research and Clinical Center of Intensive Care Medicine and Rehabilitology (FRCC ICMR). The main goal is to start using an adapted bacteriophage cocktail in all patients admitted to the ICU to prevent and treat nosocomial pneumonia. The interaction cycle is continuous with time constraints on the execution of every stage. A schematic description of the protocol is presented in Figure 1.

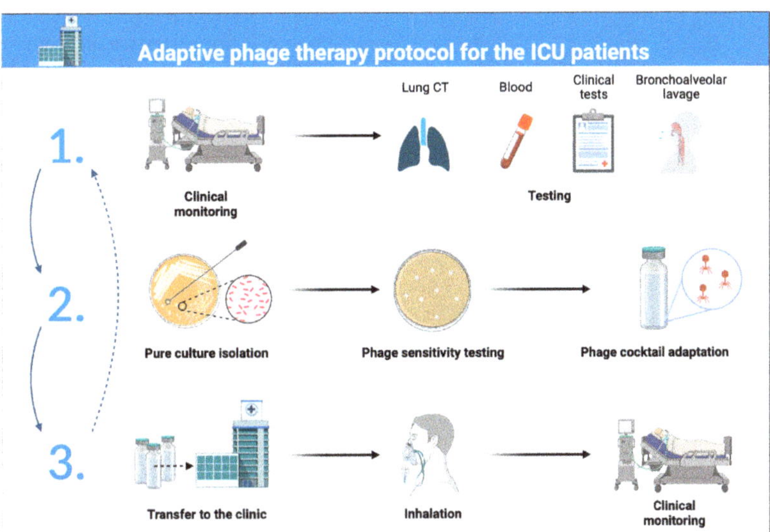

Figure 1. A schematic description of the adaptive phage therapy protocol for the ICU patients; step 1: the medical facility conducts clinical monitoring and collects samples for microbiological monitoring; step 2: the medical facility isolates pure cultures with a minimum of 50 isolates of each target bacterial species, and phage production site conducts sensitivity testing and produces an adapted phage cocktail; step 3: an adapted phage cocktail is transferred to the medical facility for administration to patients under continuous clinical monitoring. CT—computed tomography.

FRCC ICMR collects clinical material samples from all patients and isolates pure cultures with a minimum of 50 isolates of each target bacterial species. The period of sample collection is 30 days. The research and production center (RPC) "Micromir" conducts sensitivity testing of pure cultures to bacteriophages from its phage bank with an execution period of up to 21 days. After that the RPC "Micromir" produces an adapted phage cocktail with an efficacy of at least 70% to each species of target bacteria and transfers the adapted cocktail to the FRCC ICMR, where it is applied in intensive care units 2–3 times a day in a dose of 5 mL with a nebulizer.

Clinical evaluation of the efficacy of the adapted bacteriophage complex application in patients is carried out by FRCC ICMR, which analyses lung CT scans, blood tests, clinical data, and microbiological and PCR tests of bronchoalveolar fluid and summarizes the results once a month. To verify the sensitivity of cultures to the adapted complex, FRCC ICMR weekly transfers up to five samples of pure cultures to the laboratory of RPC "Micromir" for sensitivity tests of bacterial isolates to the adapted bacteriophage cocktail. Based on the results of monitoring in one time interval (1 month), if clinical efficacy is absent in more than 50% of patients and/or less than 60% of each bacterial species isolates are sensitive to the current adapted phage cocktail, a decision to start a new cycle of adaptation is made.

2.2. Adapted Phage Cocktail Application

This section presents the first clinical cases of intensive care unit patients who received adaptive phage therapy without the need for antibiotics.

1. Patient 92/23, 42 years old, was admitted to the intensive care unit of FRCC ICMR from another medical center. Based on the patient's history, antimicrobials (AMPs) were administered to treat hospital-acquired pneumonia, urological infection, and consequent colitis. According to the data of computed tomography of the chest organs on the day of admission, the presence of focal masses and infiltrative changes in the lungs were not

detected. A CT scan shows hypostatic and post-inflammatory changes in the right lung. These changes are indicative of a previous case of pneumonia. Clinical parameters of the patient are presented in Table 1.

The results of PCR-diagnostics on the day of admission revealed the presence of multidrug-resistant *A. baumannii* (10^5 CFU/mL) in bronchoalveolar lavage and multidrug-resistant (except for Trimethoprim and Fosfomycin) *K. pneumoniae* (10^5 CFU/mL). No other pathogenic bacteria were detected in the investigated samples. The results of sensitivity testing of the isolated bacteria to antibiotics and the detection of resistance markers are presented in Table S1.

The effectiveness of complex phage therapy is confirmed by the data of microbiological studies and laboratory-instrumental data. No growth of pathological bacteria was detected on the 28th day. According to computed tomography data, no evidence of pneumonia is detected.

During the whole period of treatment and rehabilitation measures in the intensive care unit, no cases of infectious and septic complications were observed in the patient, which fully allowed to implement the range of rehabilitation measures and qualitatively improve the patient's condition.

Precisely 28 days after admission, the course of therapeutic and rehabilitative measures in the ICU was completed, and the patient was transferred to the neurorehabilitation department with improvement for further therapy. No antibiotics were administered to the patient throughout the stay in the ICU.

2. Patient 62/22, 59 years old, was admitted to the intensive care unit of FRCC ICMR from another medical center. Based on history, the patient was prescribed antimicrobials (AMPs) to treat hospital-acquired pneumonia and colitis. Computed tomography of the chest organs at the time of admission indicates inflammatory-atelectatic changes in the lower lobe of the right lung and post-inflammatory changes in both lungs. Clinical parameters of the patient are presented in Table 2.

Table 1. Clinical parameters of the patient 92/23. "-" means no data is available.

Parameters	Reference Values	Results				
		1 Day	7 Days	14 Days	21 Days	28 Days
Bronchoalveolar lavage, PCR	-	1. *K. pneumoniae*—10^5 CFU/mL; 2. *A. baumannii*—10^5 CFU/mL	-	1. *K. pneumoniae*—10^6 CFU/mL; 2. *A. baumannii*—10^6 CFU/mL	-	No detection
Body temperature, °C	36.6	36.6	36.8	36.7	36.8	36.8
Nature of bronchial secretion, Clinical Pulmonary Infection Score (CPIS)	0—minimal mucous; 1—moderate mucopurulent; 2—purulent	1	1	1	1	1
White blood cells, 10^9/L	4.5–11	9	7.1	7.7	7.3	7.3
Neutrophils, 10^9/L	1.9–8.5	5.7	-	-	-	4.4
Platelets, 10^9/L	152–420	595	-	-	-	461
Lymphocytes, 10^9/L	1.3–3.1	1.8	-	-	-	1.8
Procalcitonin, ng/mL	<0.1	0.1	0.1	0.1	<0.05	0.1
C-reactive protein (CRP), mg/L	<5	28.65	7.41	10.5	10.7	25.73
Bilirubin, μmol/L	1.7–20	7.3	6.8	6.8	6.9	5
Ventilation mode	-	no	no	no	no	No
Sequential Organ Failure Assessment (SOFA) score	-	2	2	2	2	2

CFU—colony forming units.

Table 2. Clinical parameters of the patient 62/22. "-" means no data is available.

Parameters	Reference Values	Results				
		1 Day	7 Days	14 Days	21 Days	28 Days
Bronchoalveolar lavage, PCR	-	1. *A. baumannii*—10^4 CFU/mL; 2. *K. pneumoniae*—10^5 CFU/mL; 3. *P. aeruginosa*—10^5 CFU/mL; 4. *S. aureus*—10^5 CFU/mL	-	1. *P. putida*—10^6 CFU/mL	-	No detection
Body temperature, °C	36.6	36.6	36.5	36.6	36.7	36.4
Nature of bronchial secretion, Clinical Pulmonary Infection Score (CPIS)	0—minimal mucous; 1—moderate mucopurulent; 2—purulent)	1	1	1	1	0

Table 2. *Cont.*

Parameters	Reference Values	Results				
		1 Day	**7 Days**	**14 Days**	**21 Days**	**28 Days**
White blood cells, 10^9/L	4.5–11	6.9	6.7	6.9	8.9	8.2
Neutrophils, 10^9/L	1.9–8.5	2.9	-	-	-	3.7
Platelets, 10^9/L	152–420	156	-	-	-	211
Lymphocytes, 10^9/L	1.3–3.1	2.7	-	-	-	2.8
Procalcitonin, ng/mL	<0.1	<0.05	<0.05	<0.05	<0.05	<0.05
C-reactive protein (CRP), mg/L	<5	29.25	20.13	23.75	19.87	11.5
Bilirubin, μmol/L	1.7–20	8.8	7.7	6.4	5.7	6.5
Ventilation mode	-	no	no	no	no	no
Sequential Organ Failure Assessment (SOFA) score	-	3	5	4	4	1

The results of PCR-diagnostics on the day of admission revealed the presence of multidrug-resistant *A. baumannii* (10^4 CFU/mL), *K. pneumoniae* (10^5 CFU/mL), *P. aeruginosa* (10^5 CFU/mL), and *S. aureus* (10^5 CFU/mL) in bronchoalveolar lavage. No other pathogenic bacteria were detected in the investigated samples. The results of sensitivity testing of the isolated bacteria to antibiotics and the detection of resistance markers are presented in Tables S2 and S3.

The effectiveness of complex phage therapy is confirmed by the data of microbiological studies and laboratory-instrumental data. No growth of pathological bacteria was detected on the 28th day. According to computed tomography data, improvement of bronchial patency on the right side is noted.

During the whole period of treatment and rehabilitation measures in the intensive care unit, no cases of infectious and septic complications were observed in the patient, which fully allowed the implementation of the range of rehabilitation measures to qualitatively improve the patient's condition.

Precisely 28 days after admission, the course of therapeutic and rehabilitative measures in the ICU was completed, and the patient was transferred to the neurorehabilitation department with improvement for further therapy. No antibiotics were administered to the patient throughout the stay in the ICU.

3. Patient 847/21, 73 years old, was admitted to the intensive care unit of FRCC ICMR from another medical center. Based on history, the patient was prescribed antimicrobials (AMPs) to treat hospital-acquired pneumonia. Computed tomography of the chest organs at the time of admission indicates bilateral lower lobe pneumonia. Clinical parameters of the patient are presented in Table 3.

The results of PCR-diagnostics on the day of admission revealed the presence of multidrug-resistant *A. baumannii* (10^5 CFU/mL) and *K. pneumoniae* (10^4 CFU/mL) in bronchoalveolar lavage. No other pathogenic bacteria were detected in the investigated samples. The results of sensitivity testing of the isolated bacteria to antibiotics and the detection of resistance markers are presented in Table S4.

The effectiveness of complex phage therapy is confirmed by the data of microbiological studies and laboratory-instrumental data. No growth of pathological bacteria was detected on the 28th day. According to computed tomography data, in the posterior sections of the lower lobes of both lungs, the areas of "ground glass" in combination with reticular changes are preserved; positive dynamics in the form of reduction in size and severity is noted. A decrease in the extent and density of "ground glass" type areas and consolidation in the upper lobe of the right lung is noted; in the upper lobe of the left lung a single focal area of ground glass type is preserved, positive dynamics is noted.

Table 3. Clinical parameters of the patient 847/21. "-" means no data is available.

Parameters	Reference Values	Results				
		1 Day	7 Days	14 Days	21 Days	28 Days
Bronchoalveolar lavage, PCR	-	1. *A. baumannii*—10^5 CFU/mL; 2. *K. pneumoniae*—10^4 CFU/mL	-	-	-	No detection
Body temperature, °C	36.6	37.8	37.3	36.7	36.6	36.8
Nature of bronchial secretion, Clinical Pulmonary Infection Score (CPIS)	0—minimal mucous; 1—moderate mucopurulent; 2—purulent)	1	1	1	1	0
White blood cells, 10^9/L	4.5–11	10.5	9.4	10.7	13.6	10.9
Neutrophils, 10^9/L	1.9–8.5	8.0	-	-	-	8.4
Platelets, 10^9/L	152–420	311	-	-	-	416
Lymphocytes, 10^9/L	1.3–3.1	1.5	-	-	-	1.9
Procalcitonin, ng/mL	<0.1	0.5	-	-	-	0.1
C-reactive protein (CRP), mg/L	<5	150	90.5	77.9	99.6	28.2
Bilirubin, µmol/L	1.7–20	27.4	36.3	29.9	26.4	19
Ventilation mode	-	BIPAP	BIPAP	BIPAP	CPAP	CPAP
Sequential Organ Failure Assessment (SOFA) score	-	4	5	4	4	4

BIPAP—biphasic positive airway pressure; CPAP—continuous positive airway pressure.

Precisely 28 days after admission, the course of therapeutic and rehabilitation measures in the ICU was completed. For further therapy the patient was transferred with improvement to the palliative care unit, where no further infectious and septic complications were observed. No antibiotics were administered to the patient throughout the stay in the ICU.

3. Discussion

Currently, the use of phages in healthcare facilities is mainly limited to individualized selection of bacteriophages for each patient suffering from an antibiotic-resistant infection [2]. This process is very challenging to scale up, leading to a loss of time that can be life-threatening for some patients. Magistral phage production in Belgium is a more convenient process as it allows phages to be selected from stock banks of pre-purified bacteriophage lines, then transferred to suitable GMP manufacturing sites and then to physicians for use according to prescription in individual patients [15].

The proposed adaptive phage therapy technology implies strict compliance of a set of bacteriophages to the needs of a particular intensive care unit rather than a particular patient. This avoids the necessity of individual bacteriophage selection, as the cocktail is prepared based on sensitivity testing of bacteria obtained from many patients of the same ICU. This reduces the decision time required for immediate initiation of therapy, which will help to improve the effectiveness of treatment for critically ill patients. We aim for this technology to decrease the usage of antibiotics in ICUs and enhance their efficacy when required. Studies indicate that phage application can, in certain cases, restore the susceptibility of bacterial strains to antibiotics, as antibiotic resistance mechanisms can be lost in the process of bacterial population adaptation to phage infection [17]. Moreover, in the conducted in vivo study, the combination of phage and antibiotic has demonstrated a higher bactericidal effect against severe *A. baumannii* infection, compared to each agent individually. Phage øFG02 has been shown to consistently stimulate the in vivo evolution of *A. baumannii* towards a capsule-deficient, phage-resistant phenotype that became sensitive to ceftazidime [18]. This mechanism highlights the clinical potential of phage therapy in combination with antimicrobial therapy for restoring antimicrobial activity and reducing the amount of antibiotics used.

In the present study, the cocktail of bacteriophages for inhalation including 3–4 virulent bacteriophages to each bacterial species was administered to patients, as it is recommended to use multiple phages in therapy to prevent the development of bacterial resistance to phages as well as to expand the phage-host range and increase the number of target pathogens [19]. In some clinical cases, researchers have noted that the use of several phages simultaneously may negatively affect the efficacy of individual phages, but detailed data on antagonism were not presented, and the primary factor contributing to the negative clinical outcome was multi-organ failure [20].

All patients included in this study received a cocktail of bacteriophages according to one regimen: 5.0 mL of the solution per inhalation two times a day for 28 days. It is noteworthy that patients did not receive any antibiotics during the whole course of phage cocktail administration. During the therapy, no adverse events and side effects were reported. Instead, all patients showed improvement on day 28 of the therapy. The primary outcome was the elimination of multidrug-resistant microorganisms (*K. pneumoniae*, *S. aureus*, *A. baumannii*) from the BAL contents in all patients. Following bacterial elimination, all patients demonstrated positive dynamics according to lung computed tomography data; 2/3 of patients also showed improvement in nature of bronchial secretion according to Clinical Pulmonary Infection Score (CPIS). Moreover, a decrease in such inflammatory markers as CRP was observed in all patients. In the case of initially increased level of procalcitonin, there was also a decrease in this indicator. A retrospective analysis of bacteriophage administration in 37 patients also showed a significant decrease in mean CRP values measured between days 9 and 32 [21]. This may be due to a reduction in the intensity of the inflammatory response due to a decrease in bacterial load. In conducted studies, a decrease in CRP levels during bacteriophage administration was also noted [16,22].

The major limitation of this study is the small sample size to ascertain the statistical validity of the obtained results. Case reports are generally not the basis for testing statistical hypotheses but are used to create hypotheses for future research. The main purpose of this manuscript is to describe the protocol of adaptive phage therapy and individual clinical experience. The findings suggest the potency of adaptive phage therapy and the feasibility of extending the study to larger groups. Despite the demonstrated efficacy of adaptive phage therapy, we do not exclude cases where in some individuals this approach will result in a lack of clinical efficacy, as bacteriophages are not selected for the individual patient. Therefore, this approach involves close cooperation between the clinical institution and the bacteriophage manufacturer, regular clinical monitoring, sensitivity testing of bacterial strains to the action of bacteriophages, and adaptation of the phage cocktail. The main advantage of the adaptive phage therapy approach is the possibility to start phage cocktail administration from the first days of the patient's admission to the ICU.

An additional limiting factor was the exclusive use of EUKAST standards in the evaluation of sensitivity to protected β-lactam antibiotics (Amoxicillin/Clavulanate). The use of CLSI standards will enable the evaluation of different antimicrobial to beta-lactamase inhibitor ratios in future studies. It has been demonstrated that the CLSI and EUCAST methodologies showed poor concordance in determining the MIC of amoxicillin/clavulanate [23]. MIC values obtained using the EUCAST methodology were more predictive of failure than those obtained using the CLSI methodology. EUCAST-derived MIC values >16/2 mg/L were independently associated with therapeutic failure. The described method may be a promising way to reduce the amount of administered antibiotics and maintain their efficacy. It appears to be more convenient and faster than classical individual phage therapy. However, scaling up this approach may cause some difficulties, as it requires the shipment of pure cultures from the health care facility to the phage center. Regrettably, not all medical facilities have sufficient medical personnel and equipment to qualitatively isolate and characterize pure bacterial cultures. Moreover, sending materials over long distances also involves logistical difficulties and safety risks. The solution may be the establishment of a network of phage centers and authorized laboratories that will work in cooperation with medical institutions. For large-scale application of adaptive phage therapy without the need for approval of the ethical committee of each individual hospital, registration of a sufficient number of phage pharmaceutical substances is essential.

The conducted study demonstrates the potential of an adaptive phage therapy protocol in intensive care units. We will continue to investigate this method and its impact on the quality of patient care, as well as on the amount of antibiotics used in the ICU and their efficacy.

4. Materials and Methods

4.1. Participants

All patients at the time of inclusion in the study had no clinical, laboratory, or instrumental signs of systemic inflammatory complications requiring the prescription of antimicrobial drugs. Patients did not receive antibiotic therapy during the 28-day stay. Treatment and rehabilitation measures were performed by specialists who had no information about the inclusion of patients in this study. Patients were selected randomly according to the following criteria.

Inclusion criteria:

1.   Patient age >18 years;
2.   Chronic critical condition;
3.   Absence of acute systemic infection requiring the use of antimicrobial therapy (AMT) at the time of hospitalization in Federal Research and Clinical Center of Intensive Care Medicine and Rehabilitology (FRCC ICMR);
4.   Antimicrobial therapy at the previous stages of hospitalization;
5.   Informed consent from the patient or next of kin for inclusion in the study.

Exclusion criteria:

1.   Low chance of survival, Simplified Acute Physiology Score (SAPS) II score of more than 65;
2.   Treatment with immunosuppressants or corticosteroids;
3.   Oncological diseases;
4.   Evidence of systemic severe infection (Sepsis-3 criteria);
5.   Candidemia.

Three patients were included in the present study: Patient 92/23, female, 42 years old; Patient 62/22, male, 59 years old; Patient 847/21, male, 73 years old. Prior to admission to FRCC ICMR, Patient 92/23 underwent treatment with the diagnosis: consequences of subarachnoid haemorrhage from a saccular aneurysm of the supraclinoid aneurysm of the right internal carotid artery. She underwent surgery: pterional craniotomy on the right side, clipping of the supraclinoid aneurysm of the right internal carotid artery. Patient 62/22 was treated for intracerebral haemorrhage in the left thalamus region prior to admission to FRCC ICMR. The course of the disease was complicated by the development of occlusive deep vein thrombosis on the right side, hospital-acquired lower lobe pneumonia, and multi-organ failure. A puncture-dilatation tracheostomy was performed. Prior to admission to FRCC ICMR, patient 847/21 was treated for subcortical haemorrhage in the left cerebral hemisphere with blood breakthrough into the liquor system. The course of the disease was complicated by the development of hospital-acquired lower lobe pneumonia; puncture-dilatation tracheostomy was performed.

Patients underwent a set of therapeutic and rehabilitation measures: maintenance of functions of vital organs and systems, pharmacological correction of the level of consciousness, nutritional and metabolic therapy, symptomatic treatment. Acid-base status analyses, including PaO2 analysis, were taken every 7 days to assess the severity of the patients' condition. The analyses were performed using a GemPremier 3500 analyzer (Version 7.2.5, Instrumentation Laboratory, Bedford, MA, USA).

To prevent recurrence of nosocomial pneumonia, patients received an adapted complex of bacteriophages. Phage therapy was carried out from the first day of the patient's admission to the intensive care unit (ICU) of FRCC ICMR by aerosol therapy using a nebulizer with 5.0 mL of the solution per inhalation 2 times a day.

The cocktail for inhalation included 3–4 virulent phages to each bacterial species active against clinical strains of *Acinetobacter baumannii, Stenotrophomonas maltophilia, K. pneumoniae, K. pneumoniae subsp. ozanae, Pseudomonas aeruginosa, Staphylococcus aureus, Staphylococcus epidermidis, Staphylococcus warneri, Staphylococcus haemolyticus, Staphylococcus capitis, Staphylococcus caprae, Staphylococcus succinus, Streptococcus pyogenes, Streptococcus agalactiae.* Each bacteriophage was produced in a separate production and purification series in accordance with good manufacturing practice (GMP) standards. The preparation consisted of a sterile suspension of phage particles in a physiological solution. The titer of each bacteriophage was 10^5–10^6 PFU/mL.

4.2. Clinical Monitoring

Patients were under constant clinical and laboratory monitoring with evaluation of indicators of the cardiovascular system, neurological status, respiratory, liver, kidney function, and the level of organ dysfunction. The levels of inflammatory biomarkers in serum (C-reactive protein, procalcitonin) were measured in dynamics. The determination of transferrin and C-reactive protein (CRP) levels was performed on an automatic biochemical analyzer AU 480 (Beckman Coulter, Brea, CA, USA) using original reagents. Procalcitonin level was determined on immunological analyzer VIDAS (bioMerieux SA, Marcy-l'Étoile, Lyon, France). General clinical blood parameters (leukocytes, neutrophils, platelets, lymphocytes) were determined on an automatic hematological analyzer Sysmex XN550 (Sysmex, Kobe, Hyogo, Japan).

To analyze the results of computed tomography of the chest organs, the method of automatic calculation of the volume of the damaged lung tissue according to the ground glass type using the software "Ground glass" (InfoRad 3.0 DICOM Viewer, Moscow, Russia) was used. Segmentation of the right and left lungs and trachea with a threshold of −250 Hounsfield units (HU) was performed. Within the lungs, lesion regions were highlighted with densities in a custom range (default −785 HU to 150 HU). Small vessels that were assumed to be lesions were excluded using a morphological "closure" operation.

For microbiological examination, samples of bronchoalveolar fluid were collected into sterile tubes following aseptic rules. The morning portion of bronchoalveolar fluid was examined. Identification of microorganisms and determination of antibiotic sensitivity were performed on the automated system BD Phoenix-100 (BDBiosciences, San Jose, CA, USA). To assess the taxonomic composition of BAL, a reagent set for DNA isolation from clinical material "RIBO-prep" and reagent sets for detection and quantification of DNA of *Enterobacteriaceae* family, *Pseudomonas aeruginosa, Staphylococcus* spp. and *Streptococcus* spp. were used. Qualitative assessment of antibiotic resistance genes was performed using reagent kits for detection of genes of acquired carbapenemases of KPC and OXA-48-like groups (types OXA-48 and OXA-162), genes of acquired carbapenemases of MBL class of VIM, IMP, and NDM groups (Amplisens, Moscow, Russia) by PCR with hybridization-fluorescence detection of amplification products in "real time" mode. The measurements were performed on a CFX 96 Real-Time PCR Detection System (BioRad, Hercules, CA, USA).

5. Conclusions

The introduction of adaptive phage therapy in the intensive care units of FRCC ICMR allowed clinicians to apply phage cocktail from the first day of patient admission to the ICU without the use of antibiotics. The elimination of multidrug-resistant microorganisms from the BAL contents and the improvement of lung condition according to CT data, as well as general condition, was achieved. Moreover, a decrease in such inflammatory markers as CRP and procalcitonin was noted. The implementation of the described protocol demonstrates potential as an approach to reduce the number of antibiotics used in intensive care units and maintain their efficacy. Extensive research and large-scale trials are essential to confirm the findings. Moreover, to advance adaptive phage therapy and scale up the approach, registration of a sufficient number of bacteriophage pharmaceutical substances to a wide range of bacteria is required.

6. Patents

RU 2 794 585 C2.

Supplementary Materials: The following supporting information can be downloaded at: https://www.mdpi.com/article/10.3390/antibiotics12121734/s1; Table S1. Antibiotic sensitivity and resistance markers of *K. pneumoniae* and *A. baumannii* isolated from patient 92/23's bronchoalveolar lavage sample. Determination of antibiotic sensitivity was performed on a BD Phoenix-100 automated bacteriological analyzer (BDBiosciences, San Jose, CA, USA) and data interpretation was conducted according to the installed EUCAST protocols; Table S2. Antibiotic sensitivity and resistance markers of *K. pneumoniae*, *A. baumannii* and *P. aeruginosa* isolated from patient 62/22's bronchoalveolar lavage sample. Determination of antibiotic sensitivity was performed on a BD Phoenix-100 automated bacteriological analyzer (BDBiosciences, USA) and data interpretation was conducted according to the installed EUCAST protocols; Table S3. Antibiotic sensitivity and resistance markers of *S. aureus* isolated from patient 62/22's bronchoalveolar lavage sample. Determination of antibiotic sensitivity was performed on a BD Phoenix-100 automated bacteriological analyzer (BDBiosciences, USA) and data interpretation was conducted according to the installed EUCAST protocols; Table S4. Antibiotic sensitivity and resistance markers of *K. pneumoniae* and *A. baumannii* isolated from patient 847/21's bronchoalveolar lavage sample. Determination of antibiotic sensitivity was performed on a BD Phoenix-100 automated bacteriological analyzer (BDBiosciences, USA) and data interpretation was conducted according to the installed EUCAST protocols.

Author Contributions: Conceptualization, F.Z., M.P., A.Z., E.C., M.G., V.P., A.K. and A.G.; investigation, M.P., M.Y., D.C. and A.Y.; validation, F.Z., A.Z., P.P., M.Y. and A.G.; visualization, F.Z.; writing—original draft, F.Z.; writing—review and editing, E.C., V.P. and A.K.; supervision, A.Z., A.Y., M.P., M.G., V.P. and A.G. All authors have read and agreed to the published version of the manuscript.

Funding: This research received no external funding.

Institutional Review Board Statement: The study was conducted in accordance with the Declaration of Helsinki and approved by the Ethics Committee of Federal Research and Clinical Center of Intensive Care Medicine and Rehabilitology (protocol code PP #4/20 from 22 September 2020).

Informed Consent Statement: Informed consent was obtained from all subjects involved in the study.

Data Availability Statement: Data are contained within the article and Supplementary Materials.

Conflicts of Interest: The authors declare no conflict of interest.

References

1. World Health Organization. *Antimicrobial Resistance: Global Report on Surveillance*; WHO: Geneva, Switzerland, 2014.
2. Rohde, C.; Wittmann, J.; Kutter, E. Bacteriophages: A Therapy Concept against Multi-Drug-Resistant Bacteria. *Surg. Infect.* **2018**, *19*, 737–744. [CrossRef] [PubMed]
3. Antimicrobial Resistance Collaborators. Global burden of bacterial antimicrobial resistance in 2019: A systematic analysis. *Lancet* **2022**, *399*, 629–655. [CrossRef] [PubMed]
4. Summers, W.C. The strange history of phage therapy. *Bacteriophage* **2012**, *2*, 130–133. [CrossRef] [PubMed]
5. Brives, C.; Pourraz, J. Phage therapy as a potential solution in the fight against AMR: Obstacles and possible futures. *Palgrave Commun.* **2020**, *6*, 100. [CrossRef]

6. Luong, T.; Salabarria, A.C.; Roach, D.R. Phage Therapy in the Resistance Era: Where Do We Stand and Where Are We Going? *Clin. Ther.* **2020**, *42*, 1659–1680. [CrossRef] [PubMed]

7. Reyes, A.; Semenkovich, N.P.; Whiteson, K.; Rohwer, F.; Gordon, J.I. Going viral: Next-generation sequencing applied to phage populations in the human gut. *Nat. Rev. Microbiol.* **2012**, *10*, 607–617. [CrossRef]

8. Miedzybrodzki, R.; Borysowski, J.; Weber-Dąbrowska, B.; Fortuna, W.; Letkiewicz, S.; Szufnarowski, K.; Pawelczyk, Z.; Rogoz, P.; Klak, M.; Wojtasik, E.; et al. Clinical aspects of phage therapy. *Adv. Virus Res.* **2012**, *83*, 73–121. [CrossRef]

9. Fernández, L.; Gutiérrez, D.; Rodríguez, A.; García, P. Application of Bacteriophages in the Agro-Food Sector: A Long Way Toward Approval. *Front. Cell Infect. Microbiol.* **2018**, *8*, 296. [CrossRef]

10. Fauconnier, A. Phage therapy regulation: From night to dawn. *Viruses* **2019**, *11*, 352. [CrossRef]

11. Debarbieux, L.; Pirnay, J.P.; Verbeken, G.; De Vos, D.; Merabishvili, M.; Huys, I.; Patey, O.; Schoonjans, D.; Vaneechoutte, M.; Zizi, M.; et al. Bacteriophage journey at the European Medicines Agency. *FEMS Microbiol. Lett.* **2015**, *363*, 225. [CrossRef]

12. Fauconnier, A. Regulating phage therapy: The biological master file concept could help to overcome regulatory challenge of personalized medicines. *EMBO Rep.* **2017**, *18*, 198–200. [CrossRef] [PubMed]

13. Pirnay, J.P.; De Vos, D.; Verbeken, G. Clinical applications of bacteriophages in Europe. *Microbiol. Aust.* **2019**, *40*, 8–15. [CrossRef]

14. Pirnay, J.P.; Ferry, T.; Resch, G. Recent progress toward the implementation of phage therapy in Western medicine. *FEMS Microbiol. Rev.* **2022**, *46*, fuab040. [CrossRef] [PubMed]

15. Pirnay, J.P.; Verbeken, G.; Ceyssens, P.J.; Huys, I.; De Vos, D.; Ameloot, C.; Fauconnier, A. The Magistral Phage. *Viruses* **2018**, *10*, 64. [CrossRef]

16. Beloborodova, N.V.; Grechko, A.V.; Gurkova, M.M.; Zurabov, A.Y.; Zurabov, F.M.; Kuzovlev, A.N.; Megley, A.Y.; Petrova, M.V.; Popova, V.M.; Redkin, I.V.; et al. Adaptive Phage Therapy in the Treatment of Patients with Recurrent Pneumonia (Pilot Study). *Gen. Reanimatol.* **2021**, *17*, 4–14. [CrossRef]

17. Chan, B.K.; Sistrom, M.; Wertz, J.E.; Kortright, K.E.; Narayan, D.; Turner, P.E. Phage selection restores antibiotic sensitivity in MDR Pseudomonas aeruginosa. *Sci. Rep.* **2016**, *6*, 26717. [CrossRef]

18. Gordillo Altamirano, F.L.; Kostoulias, X.; Subedi, D.; Korneev, D.; Peleg, A.Y.; Barr, J.J. Phage-antibiotic combination is a superior treatment against Acinetobacter baumannii in a preclinical study. *EBioMedicine* **2022**, *80*, 104045. [CrossRef]

19. Oechslin, F. Resistance Development to Bacteriophages Occurring during Bacteriophage Therapy. *Viruses* **2018**, *10*, 351. [CrossRef]

20. Onallah, H.; Hazan, R.; Nir-Paz, R.; Brownstein, M.J.; Fackler, J.R.; Horne, B.; Hopkins, R.; Basu, S.; Yerushalmy, O.; Alkalay-Oren, S.; et al. Refractory Pseudomonas aeruginosa infections treated with phage PASA16: A compassionate use case series. *Med* **2023**, *4*, 600–611. [CrossRef]

21. Miedzybrodzki, R.; Fortuna, W.; Weber-Dabrowska, B.; Górski, A.A. Retrospective analysis of changes in inflammatory markers in patients treated with bacterial viruses. *Clin. Exp. Med.* **2009**, *9*, 303–312. [CrossRef]

22. Zurabov, F.M.; Chernevskaya, E.A.; Beloborodova, N.V.; Zurabov, A.Y.; Petrova, M.V.; Yadgarov, M.Y.; Popova, V.M.; Fatuev, O.E.; Zakharchenko, V.E.; Gurkova, M.M.; et al. Bacteriophage Cocktails in the Post-COVID Rehabilitation. *Viruses* **2022**, *14*, 2614. [CrossRef] [PubMed]

23. Delgado-Valverde, M.; Valiente-Mendez, A.; Torres, E.; Almirante, B.; Gómez-Zorrilla, S.; Borrell, N.; Aller-García, A.I.; Gurgui, M.; Almela, M.; Sanz, M.; et al. MIC of amoxicillin/clavulanate according to CLSI and EUCAST: Discrepancies and clinical impact in patients with bloodstream infections due to Enterobacteriaceae. *J. Antimicrob. Chemother.* **2017**, *72*, 1478–1487. [CrossRef] [PubMed]

MDPI
St. Alban-Anlage 66
4052 Basel
Switzerland
www.mdpi.com

Antibiotics Editorial Office
E-mail: antibiotics@mdpi.com
www.mdpi.com/journal/antibiotics